Let's Review Regents:

Algebra I 2020

Gary M. Rubinstein
B.A. Mathematics
Tufts University
M.S. Computer Science
University of Colorado

BARRON'S

About the Author

Gary Rubinstein has been teaching math for 25 years. He is a three-time recipient of the Math for America Master Teacher Fellowship. Gary lives with his wife Erica and his two children, Sarah and Sam. He has a YouTube channel at nymathteacher where students can find extra test tips and strategies for learning Algebra I.

Dedication

To Erica, Sarah, and Sam

© Copyright 2020, 2018, 2017, 2015 by Kaplan, Inc., d/b/a Barron's Educational Series

All rights reserved.
No part of this publication may be reproduced in any form or by
any means without the written permission of the copyright owner.

Published by Kaplan, Inc., d/b/a Barron's Educational Series
750 Third Avenue
New York, NY 10017
www.barronseduc.com

ISBN: 978-1-5062-5382-4

Printed in Canada

10 9 8 7 6 5 4 3 2 1

Kaplan, Inc., d/b/a Barron's Educational Series print books are available at special quantity discounts to use for sales promotions, employee premiums, or educational purposes. For more information or to purchase books, please call the Simon & Schuster special sales department at 866-506-1949.

TABLE OF CONTENTS

PREFACE		**vii**
CHAPTER 1	**SOLVING LINEAR EQUATIONS WITH ALGEBRA**	**1**
	1.1 Properties of Algebra	1
	1.2 Solving One-Step Algebra Equations	7
	1.3 Solving Multi-Step Algebra Equations	11
	1.4 Isolating Variables in Equations with Multiple Variables	14
CHAPTER 2	**POLYNOMIAL ARITHMETIC**	**19**
	2.1 Classifying Monomials, Binomials, and Trinomials	19
	2.2 Multiplying and Dividing Monomials	22
	2.3 Combining Like Terms	26
	2.4 Multiplying Monomials and Polynomials	29
	2.5 Adding and Subtracting Polynomials	32
	2.6 Multi-Step Algebra Equations Involving Polynomial Arithmetic	36
	2.7 Multiplying Polynomials by Polynomials	40
	2.8 Factoring Polynomials	45
	2.9 More Complicated Factoring	51
CHAPTER 3	**QUADRATIC EQUATIONS**	**55**
	3.1 Solving Quadratic Equations by Taking the Square Root of Both Sides of the Equation	55
	3.2 Solving Quadratic Equations by Guess and Check	63
	3.3 Solving Quadratic Equations by Completing the Square	66
	3.4 Solving Quadratic Equations by Factoring	71
	3.5 The Relationship Between Factors and Roots	76
	3.6 Solving Quadratic Equations with the Quadratic Formula	79
	3.7 Word Problems Involving Quadratic Equations	83

CHAPTER 4	**SYSTEMS OF LINEAR EQUATIONS**	**89**
	4.1 Solving Systems with Guess and Check	89
	4.2 Solving Simpler Systems of Equations with Algebra	95
	4.3 Solving More Complicated Systems of Equations with Algebra	102
	4.4 Solving Word Problems with Systems of Equations	111
CHAPTER 5	**GRAPHS OF SOLUTION SETS OF LINEAR EQUATIONS**	**117**
	5.1 Producing a Graph by Identifying Two or More Points	117
	5.2 Calculating and Interpreting Slope	129
	5.3 Slope-Intercept Form	138
	5.4 Graphing Solution Sets to Linear Equations with a Graphing Calculator	146
	5.5 Determining an Equation for a Given Graph	158
	5.6 Word Problems Involving Finding the Equation of a Line	165
CHAPTER 6	**GRAPHING SOLUTION SETS FOR QUADRATIC EQUATIONS**	**171**
	6.1 Graphing Solution Sets to Quadratic Equations	171
	6.2 Using the Graphing Calculator to Solve Quadratic Equations	188
	6.3 Solving Linear-Quadratic Systems of Equations	196
	6.4 Graphing Quadratic Equations for Real-World Applications	205
CHAPTER 7	**LINEAR INEQUALITIES**	**213**
	7.1 One-Variable Linear Inequalities	213
	7.2 Graphing Two-Variable Linear Inequalities	218
	7.3 Graphing Systems of Linear Inequalities	230
CHAPTER 8	**EXPONENTIAL EQUATIONS**	**243**
	8.1 Solving Exponential Equations	243
	8.2 Graphing Solution Sets to Two-Variable Exponential Equations	246
	8.3 Distinguishing Between Linear, Quadratic, and Exponential Equations	254
	8.4 Real-World Problems Involving Exponential Equations	268

CHAPTER 9   **CREATING AND INTERPRETING EQUATIONS FROM REAL-WORLD SCENARIOS**   **273**
    9.1 Creating and Interpreting Linear Equations   273
    9.2 Creating and Interpreting Exponential Equations   279

CHAPTER 10   **FUNCTIONS**   **287**
    10.1 Describing Functions   287
    10.2 Function Graphs   292
    10.3 Defining a Function with an Equation   304
    10.4 Function Transformations   316

CHAPTER 11   **SEQUENCES**   **329**
    11.1 Types of Sequences   329
    11.2 Recursively Defined Sequences   331
    11.3 Closed Form Defined Sequences   338

CHAPTER 12   **REGRESSION CURVES**   **345**
    12.1 Line of Best Fit   345
    12.2 The Correlation Coefficient   356
    12.3 Parabolas and Exponential Curves of Best Fit   367

CHAPTER 13   **STATISTICS**   **381**
    13.1 Measures of Central Tendency   381
    13.2 Graphically Representing Data   391

CHAPTER 14   **TEST-TAKING STRATEGIES**   **399**
    14.1 Time Management   399
    14.2 Know How to Get Partial Credit   400
    14.3 Know Your Calculator   401
    14.4 Use the Reference Sheet   407
    14.5 How Many Points Do You Need to Pass?   408

ANSWERS AND SOLUTION HINTS TO PRACTICE EXERCISES   **409**

GLOSSARY OF ALGEBRA I TERMS   **439**

THE ALGEBRA I REGENTS EXAMINATION   **447**
    June 2018 Examination   450
    Answer Key   471
    June 2019 Examination   472
    Answer Key   494

INDEX   **495**

PREFACE

In June of 2014, New York State adopted a more difficult Algebra I course based on the Common Core curriculum and changed the Regents exam to reflect this new course. On the new Algebra I Regents exam, students are expected not only to solve algebra problems and answer algebra-related questions but also to explain their reasoning. This is a major shift in the testing of algebra concepts, so a thorough review is even more critical for success.

This book will introduce you to the concepts covered throughout the new Algebra I Regents course as well as to the different types of problems you will encounter on the Regents exam at the end of the year.

Who Should Use This Book?

Students and teachers alike can use this book as a resource for preparing for the Algebra I Regents exam.

Students will find this book to be a great study tool because it contains a review of all Algebra I concepts, useful examples, and practice problems of varying difficulty that can be practiced throughout the school year to reinforce what they are learning in class. The most ideal way to prepare for the Algebra Common Core Regents exam is to work through the practice problems in the review sections and then the recently administered Regents exams at the end of the book.

Teachers can use this book as a tool to help structure an Algebra I course that will culminate with the Regents exam. The topics in the book are arranged by priority, so the sections in the beginning of the book are the ones from which more of the questions on the Regents exam are drawn. There are 60 sections in the book, spread out over 14 chapters, dedicated to all topics for Algebra I, each with practice exercises and solutions.

Why Is This Book a Helpful Resource?

Becoming familiar with the specific types of questions on the Algebra I Regents exam is crucial to performing well on this test. There are questions in which the math may be fairly easy but the way in which the question is asked makes the question seem much more difficult. For example, the question "Find all zeros of the function $f(x) = 2x + 6$" is a fancy way of asking

Preface

the much simpler sounding "Solve for x if $2x + 6 = 0$." <u>Knowing exactly what the questions are asking is a big part of success on this test.</u>

Algebra has been around for thousands of years, and its basic concepts have never changed. So fundamentally, the algebra curriculum today is not very different from the algebra taught in schools two years ago, ten years ago, or twenty years ago. But the exam that follows this Common Core-based course, with more complicated ways of asking questions and presenting problems, requires a specifically presented study plan that's more important than ever.

Gary Rubinstein

Chapter One: SOLVING LINEAR EQUATIONS WITH ALGEBRA

1.1 PROPERTIES OF ALGEBRA

KEY IDEAS

Algebra has a collection of tools that, when used appropriately, can be used to solve equations. These tools have names like the *addition property of equality*. Knowing the names and how to use these properties enables you to not only solve equations but also explain the logic behind your reasoning

Algebra is what you are doing when you solve equations for unknown *variables*. The most basic type of algebra problem is something like

$$x - 1 = 3$$

Of all the possible numbers that can be substituted into the variable x, the only one that makes the equation true is the number 4. Any other number would make the equation false. Another way of saying this is that the number 4 *satisfies* the equation $x - 1 = 3$ while other numbers do not satisfy that equation. All the numbers that satisfy an equation form the *solution set* of the equation. In this example, since only one number satisfies the equation, the solution set is just the set with the number 4 in it, or {4}.

If the equation is simple enough to *solve* the equation for the unknown variable x, you might be able to just see the answer, like in the $x - 1 = 3$ example. But for more involved equations, you will use the *properties of algebra*.

The Addition Property of Equality

For an equation like, $x - 1 = 3$ the way to solve this equation with algebra is to *eliminate* the -1 from the left-hand side of the equation. You do this by adding 1 to each side of the equation.

$$\begin{aligned} x - 1 &= 3 \\ +1 &= +1 \\ x &= 4 \end{aligned}$$

Solving Linear Equations with Algebra

It is the *addition property of equality* that lets you add something to both sides of an equation to create a new equation that has the same solution set as the original equation.

The Subtraction Property of Equality

If the equation, instead, is $x + 1 = 3$ it can be solved by subtracting 1 from both sides of the equation. When you subtract something from both sides of an equation, you are using the *subtraction property of equality*.

$$x + 1 = 3$$
$$-1 = -1$$
$$x = 2$$

The Division Property of Equality

For an equation like $2x = 10$, the 2 can be eliminated by dividing both sides of the equation by 2.

$$2x = 10$$
$$\frac{2x}{2} = \frac{10}{2}$$
$$x = 5$$

The *division property of equality* is the reason this is allowed.

The Multiplication Property of Equality

Sometimes when fractions are involved, a way to solve the equation is to multiply both sides of the equation by some number. For example,

$$\frac{x}{2} = 10$$

By multiplying both sides by 2, the 2 in the denominator of the left-hand side of the equation will be eliminated.

$$\frac{x}{2} = 10$$
$$2 \cdot \frac{x}{2} = 2 \cdot 10$$
$$x = 20$$

This property is also useful for questions like $\frac{2}{3}x = 10$, which you continue by multiplying both sides of the equation by the *reciprocal* (also known as the *multiplicative inverse*) of $\frac{2}{3}$, which is $\frac{3}{2}$.

$$\frac{2}{3}x = 10$$
$$\frac{3}{2} \cdot \frac{2}{3}x = \frac{3}{2} \cdot \frac{10}{1}$$
$$x = 15$$

The Distributive Property of Multiplication Over Addition

When there is an expression like 3(2 + 5), there are two ways of simplifying it. If you simplify the parentheses first, it becomes 3(2 + 5) = 3 · 7 = 21. If, instead, you use *the distributive property of multiplication over addition*, it becomes

$$3(2 + 5) = 3 \cdot 2 + 3 \cdot 5 = 6 + 15 = 21$$

You get the same answer!
Even though this isn't a very practical way to simplify 3(2 + 5), the distributive property of multiplication over addition is useful for progressing on certain algebra problems.

$$3(2x + 5) = 27$$
$$3 \cdot 2x + 3 \cdot 5 = 27$$
$$6x + 15 = 27$$
$$-15 = -15$$
$$6x = 12$$
$$\frac{6x}{6} = \frac{12}{6}$$
$$x = 2$$

The Commutative Property of Addition or Multiplication

In both addition and multiplication, switching the order of the numbers does not change the value of the expression. So 5 + 3 = 3 + 5 and 5 · 3 = 3 · 5. There is not a commutative property for subtraction or division.

Solving Linear Equations with Algebra

Example 1

When solving the equation $2x - 5 = 17$, Giovanny wrote $2x = 22$ as his first step. Which property justifies this step?

Solution: He must have added 5 to both sides of the equation. This is the addition property of equality.

Example 2

Brooklynn solved the equation $5(x - 3) = 20$ by writing, as her first step, $x - 3 = 4$. Which property of algebra justifies her first step?

Solution: Though this could have been done by beginning with the distributive property, she seems to have divided both sides of the equation by 5 to eliminate the 5 from the left-hand side. This is the division property of equality.

Check Your Understanding of Section 1.1

A. *Multiple-Choice*

1. Antonio started the question $2x + 1 = 11$ by writing $2x = 10$. Which property justifies this step?
   (1) Commutative property of addition
   (2) Distributive property of multiplication over addition
   (3) Addition property of equality
   (4) Subtraction property of equality

2. Mila used the multiplication property to justify the first step in solving an equation. The original equation was $\frac{x}{2} + 4 = 10$. What could the equation have been transformed into after this step?
   (1) $\frac{x}{2} = 14$
   (2) $\frac{x}{2} + 6 = 12$
   (3) $x + 8 = 20$
   (4) $x + 2 = 5$

3. To more easily simplify an expression, Michael changed $2x + 3 + 5x + 2$ into $2x + 5x + 3 + 2$. What property justifies this step?
   (1) Commutative property of addition
   (2) Commutative property of multiplication
   (3) Addition property of equality
   (4) Distributive property of multiplication over addition

4. When solving the equation $3x - 2 = 13$, Jocelyn used a property of algebra to transform the equation into $3x = 15$. Which property did she use?
   (1) Subtraction property of equality
   (2) Addition property of equality
   (3) Multiplication property of equality
   (4) Division property of equality

5. When solving the equation $\frac{x}{3} = 8$, Devin uses a property of algebra and transforms the equation into $x = 24$. Which property did he use?
   (1) Subtraction property of equality
   (2) Addition property of equality
   (3) Multiplication property of equality
   (4) Division property of equality

6. Elijah used two different properties from algebra to solve this problem. What were the two steps and in what order did he do them?

$$3x - 4 = 17$$
$$+4 = +4$$
$$3x = 21$$
$$\frac{1}{3} \cdot 3x = \frac{1}{3} \cdot 21$$
$$x = 7$$

   (1) Addition property of equality followed by multiplication property of equality
   (2) Addition property of equality followed by division property of equality
   (3) Subtraction property of equality followed by multiplication property of equality
   (4) Subtraction property of equality followed by division property of equality

Solving Linear Equations with Algebra

7. Which sentence uses the multiplication property of equality?
   (1) $1 + 1 = 2$; therefore, $1 + 1 + 5 = 2 + 5$
   (2) $1 + 1 = 2$; therefore, $\dfrac{1+1}{6} = \dfrac{2}{6}$
   (3) $1 + 1 = 2$; therefore, $1 + 1 - 2 = 2 - 2$
   (4) $1 + 1 = 2$; therefore, $5(1 + 1) = 5 \cdot 2$

8. Which sentence uses the subtraction property of equality?
   (1) $5 + 2 = 7$; therefore, $5 + 2 - 2 = 7 - 2$
   (2) $5 + 2 = 7$; therefore, $5 + 2 + 3 = 7 + 3$
   (3) $5 + 2 = 7$; therefore, $2(5 + 2) = 2 \cdot 7$
   (4) $5 + 2 = 7$; therefore, $\dfrac{5+2}{7} = \dfrac{7}{7}$

9. Which sentence illustrates the addition property of equality?
   (1) $a - 2 = 3$; therefore, $a - 2 - 2 = 2 - 3$
   (2) $a - 2 = 3$; therefore, $a - 2 + 2 = 3 + 2$
   (3) $a - 2 = 3$; therefore, $2(a - 2) = 2 \cdot 3$
   (4) $a - 2 = 3$; therefore, $\dfrac{a-2}{5} = \dfrac{3}{5}$

10. Which number sentence illustrates the commutative property of multiplication?
    (1) $3 \cdot x \cdot 2 = 3 \cdot 2 \cdot x$
    (2) $3 + x + 2 = 3 + 2 + x$
    (3) $3 - x - 2 = 3 - 2 - x$
    (4) $\dfrac{\left(\dfrac{3}{x}\right)}{2} = \dfrac{\left(\dfrac{3}{2}\right)}{x}$

B. Show how you arrived at your answers.

1. Malachi and Wesley each solved the equation $2(3x - 1) = 16$. Malachi used the distributive property of multiplication over addition to justify his first step. Wesley used the division property of equality to justify his first step. Write what equation each student would have after the first step is completed?

2. Justify each step of this solution to the equation $3(x + 3) - 5 = 22$.

$$3(x + 3) - 5 = 22$$
$$3(x + 3) = 27$$
$$3x + 9 = 27$$
$$3x = 18$$
$$x = 6$$

3. Sandra calculates 5(3 + 7) by writing 5 · 3 + 7 = 15 + 7 = 22. Solomon does it by writing 5(10) = 50. Which person is correct, and why?

4. Joel rewrites 6 − x as x − 6 and justifies it with the reason "The commutative property of subtraction." What is wrong with this reasoning?

5. Natalie simplified 3(5 · 2) as 3 · 5 · 3 · 2 to get 90. Lucy says the answer is 30. Who is correct, and what is wrong with the other person's logic?

1.2 SOLVING ONE-STEP ALGEBRA EQUATIONS

KEY IDEAS

When an equation has a variable by itself on one side of the equals sign, we say that that variable has been *isolated*. For example, in the equation $x = 5 - 2$, the *x* is isolated. In order to *solve* an equation with algebra, use the properties described earlier in this chapter to isolate the variable until the *solution set* becomes clear.

Zero-Step Solutions

In an equation like $x = 3$, there is no further work to be done since the variable is already isolated. The only number you can replace the *x* with to make this a true equation is the number 3. Sometimes the elements of the solution set are written inside curly brackets like {3}.

One-Step Solutions

In the equation $x + 2 = 5$, the variable *x* is not isolated yet. In order to isolate the *x*, the *constant* +2 must be *eliminated* from the left-hand side of the equals sign. To eliminate the +2, use the subtraction property of equality by subtracting 2 from both sides of the equals sign.

$$x + 2 = 5$$
$$-2 = -2$$
$$x + 2 - 2 = 5 - 2$$
$$x + 0 = 3$$
$$x = 3$$

Solving Linear Equations with Algebra

The third and fourth lines of this solution are optional. It could have looked like this:

$$x + 2 = 5$$
$$-2 = -2$$
$$x = 3$$

If the constant is negative, you eliminate it with the addition property of equality.

$$x - 2 = 5$$
$$+2 = +2$$
$$x = 7$$

Some equations like $3x = 12$ have a number being multiplied by the variable. This number, which is 3 in this example, is called the *coefficient*. Coefficients can be eliminated with the division property of equality.

$$3x = 12$$
$$\frac{3x}{3} = \frac{12}{3}$$
$$1x = 4$$
$$x = 4$$

The third line of this solution is optional.

When the coefficient is a fraction, the multiplicative property of equality can be used. Multiply both sides of the equals sign by the *reciprocal* (also known as *the multiplicative inverse*) to eliminate the coefficient.

$$\frac{2}{3}x = 18$$
$$\frac{3}{2} \cdot \frac{2}{3}x = \frac{3}{2} \cdot \frac{18}{1}$$
$$1x = \frac{54}{2}$$
$$x = 27$$

The third line of this solution is optional.

MATH FACTS

When a fraction is multiplied by its reciprocal, the product is the number 1. So if the coefficient of a variable is a fraction, as in $\frac{2}{3}x = 18$, multiplying both sides of the equation by $\frac{3}{2}$ results in $1x = 27$. Then since anything multiplied by 1 is itself, $1x$ can be replaced with x.

Example 1

Solve the equation $\frac{x}{3} = 12$ for x.

Solution: Just like the equation $\frac{1}{3}x = 12$, eliminate the 3 by multiplying both sides of the equals sign by 3.

$$\frac{x}{3} = 12$$

$$3 \cdot \frac{x}{3} = 3 \cdot 12$$

$$x = 36$$

Example 2

Solve the equation $-2 + x = 7$ for x.

Solution: The constant does not have to be on the right-hand side of the variable. You still eliminate a -2 by using the addition property of equality and adding 2 to both sides of the equals sign.

$$-2 + x = 7$$
$$+2 = +2$$
$$x = 9$$

Check Your Understanding of Section 1.2

A. Multiple-Choice

1. What value of x makes the equation $x + 4 = 12$ true?
   (1) 2 (2) 4 (3) 8 (4) 16

2. What is the solution set for the equation $x - 6 = 7$?
   (1) {1} (2) {6} (3) {11} (4) {13}

3. What value of x satisfies the equation $4x = 8$?
   (1) 1 (2) 2 (3) 8 (4) 32

4. Solve for x in the equation $\frac{3}{4}x = 12$.
   (1) 9 (2) 16 (3) 36 (4) 48

Solving Linear Equations with Algebra

5. What value of x makes the equation $x + 8 = 3$ true?
   (1) –5   (2) 5   (3) 11   (4) –11

6. What is the solution to the equation $\frac{4}{5}x = 10$?
   (1) $\frac{46}{5}$ (2) 8   (3) $\frac{25}{2}$ (4) 20

7. Find the solution to $a - 5 = -2$.
   (1) –7   (2) –3   (3) 3   (4) 7

8. Solve for t in the equation $-3t = -18$.
   (1) –54   (2) 54   (3) –6   (4) 6

9. Find the solution for v in the equation $-\frac{3}{4}v = -12$.
   (1) 9   (2) –9   (3) 16   (4) –16

10. What is the solution set to the equation $x = 7$?
    (1) {–7}
    (2) {7}
    (3) {0}
    (4) It doesn't have a solution since it is not an equation.

B. Show how you arrived at your answers.

1. Logan decides to solve the equation $\frac{5}{6}x = 10$ by multiplying both sides of the equals sign by 6. How might he still get the correct answer?

2. Andrew says that to solve the equation $\frac{3}{5}x = 6$, you have to use the multiplication property of equality. Vanessa says that it can be done with the division property of equality. Both are correct. Explain.

3. Noah says he can solve the equation $x - 2 = 5$ by using the subtraction property of equality. He is right. How can he justify this?

4. What property of algebra could be used to solve the equation $-3 + x = 5$?

5. The equation $0x = 10$ does not have any solutions. Why?

10

1.3 SOLVING MULTI-STEP ALGEBRA EQUATIONS

=== **KEY IDEAS** ===

More complicated equations require more than one step to isolate the variable. Often there is more than one possible order in which to apply the algebra properties, though some orders are quicker and easier than others.

Two-Step Algebra Equations

When both the addition property of equality and the division property of equality are used in an equation, it is simpler to use the addition property of equality first. Similarly, when the subtraction property of equality and the division property of equality are involved in the solution, use the subtraction property of equality first.

Example 1

Solve for x in the equation $2x - 4 = 6$.

Solution: Generally, it is simpler to eliminate the constant (in this case the -4) before eliminating the coefficient, which is 2.

$$2x - 4 = 6$$
$$+4 = +4$$
$$\frac{2x}{2} = \frac{10}{2}$$
$$x = 5$$

It is possible to do these steps in the opposite order, as long as you do it properly.

$$\frac{2x-4}{2} = \frac{6}{2}$$
$$x - 2 = 3$$
$$+2 = +2$$
$$x = 5$$

Solving Linear Equations with Algebra

The first way is the most straightforward, but you could be asked to explain the second way for one of the part II, part III, or part IV questions.

Even if the constant is to the left of the variable, it is generally easier to eliminate the constant before eliminating the coefficient.

Example 2

Solve for x in the equation $5 - \frac{2}{3}x = -1$

Solution:

$$5 - \frac{2}{3}x = -1$$
$$-5 = -5$$
$$-\frac{2}{3}x = -6$$
$$\left(-\frac{3}{2}\right) \cdot \left(-\frac{2}{3}\right)x = \left(-\frac{3}{2}\right) \cdot \left(-\frac{6}{1}\right)$$
$$x = 9$$

Three-Step Algebra Equations

When the distributive property of multiplication over addition is involved, algebra equations usually take three steps to solve.

Example 3

What is the solution set for the equation $2(x + 3) = 14$?

Solution:

$$2(x+3) = 14$$
$$2x + 6 = 14$$
$$-6 = -6$$
$$\frac{2x}{2} = \frac{8}{2}$$
$$x = 4$$
$$\{4\}$$

Check Your Understanding of Section 1.3

A. Multiple-Choice

1. What value of x makes the equation $3x + 7 = 22$ true?
   (1) 1   (2) 3   (3) 5   (4) 7

2. Solve the equation $\frac{3}{4}x - 6 = 3$ for x.
   (1) 12   (2) 14   (3) 16   (4) 18

3. Find the solution set for the equation $5(x + 4) = 35$.
   (1) {1}   (2) {2}   (3) {3}   (4) {4}

4. Solve the equation $-3x + 7 = -5$ for x.
   (1) 4   (2) –4   (3) 6   (4) –6

5. Solve for x. $3 + 5x = 43$.
   (1) 2   (2) 4   (3) 6   (4) 8

6. What value for x makes the equation $5 - 2x = -9$ true?
   (1) 7   (2) 8   (3) 9   (4) 10

7. Solve for x. $\frac{3}{5}(x - 2) = 6$.
   (1) 10   (2) 12   (3) 14   (4) 16

8. Find the solution set for the equation $-9 + 7x = 19$.
   (1) {4}   (2) {5}   (3) {6}   (4) {7}

9. Solve for x in the equation $\frac{x+4}{3} = 6$.
   (1) 14   (2) 16   (3) 18   (4) 20

10. Solve for x. $-\frac{2}{7}(2 - x) = -2$.
    (1) –2   (2) –3   (3) –4   (4) –5

Solving Linear Equations with Algebra

B. *Show how you arrived at your answers.*

1. Solve for *x*. Show all work and justify each step with a property of algebra.

$$\frac{5}{6}(x-2) = \frac{5}{2}$$

2. Solve for *x*. Justify each step with a property of algebra.

$$5x - 4 = 31$$

3. Solve for *x*. Justify each step with a property of algebra.

$$2(x - 5) = -8$$

4. Cristiano says he can solve the equation $5(x + 2) = 35$ with three steps, distributive property, subtraction property of equality, and division property of equality. Kayla says she can solve the same equation in two steps, division property of equality then subtraction property of equality. Are they both right? Explain.

5. To figure out what price to charge for a pair of sneakers in order to make $44,000 profit, the manufacturer must solve the equation $44 = -0.3p + 50$. Use properties of algebra to solve for the variable *p*.

1.4 ISOLATING VARIABLES IN EQUATIONS WITH MULTIPLE VARIABLES

=== KEY IDEAS ===

Whether an equation has just one variable or multiple variables, the process of using the properties of algebra remain the same. The goal is to eliminate all the numbers and variables from one side of the equation except for the one that you are trying to isolate.

Algebra Equations with Multiple Variables

Solving for x in an equation like $2x + 1 = 11$ requires two steps.

$$2x + 1 = 11$$
$$-1 = -1$$
$$\frac{2x}{2} = \frac{10}{2}$$
$$x = 5$$

If the equation was, instead, to isolate the x in the equation $ax + b = c$, the process would be very similar. You have to recognize that the b is like the constant and the a is like the coefficient.

Just as you subtract 1 from each side as the first step in solving $2x + 1 = 11$, for this equation, you subtract b from both sides.

$$ax + b = c$$
$$-b = -b$$
$$ax = c - b$$

Now, eliminate the coefficient a the same way that you eliminate the 2 in the $2x + 1 = 11$ question, by dividing both sides of the equation by it.

$$\frac{ax}{a} = \frac{c-b}{a}$$
$$x = \frac{c-b}{a}$$

Example

Solve for d, in the equation $abd - e = f$

Solution: First eliminate the e, since it is like the constant in the problem. Then eliminate the a and the b, since they are like the coefficient.

$$abd - e = f$$
$$+e = +e$$
$$\frac{abd}{ab} = \frac{f+e}{ab}$$
$$d = \frac{f+e}{ab}$$

Solving Linear Equations with Algebra

Check Your Understanding of Section 1.4

A. *Multiple-Choice*

1. Solve for d in terms of c, e, and f.
   $cd - e = f$

   (1) $\dfrac{f-e}{c}$ (3) $\dfrac{f}{c} + e$

   (2) $\dfrac{f+e}{c}$ (4) $\dfrac{f}{c} - e$

2. Solve for m in terms of a, b, and c.
   $b - ma = c$

   (1) $\dfrac{b-c}{-a}$ (3) $\dfrac{c-b}{-a}$

   (2) $\dfrac{b+c}{a}$ (4) $\dfrac{c+b}{-a}$

3. Solve for r in terms of c and π.
   $c = 2\pi r$

   (1) $\dfrac{2c}{\pi}$ (3) $\dfrac{c}{2\pi}$

   (2) $\dfrac{2\pi}{c}$ (4) $\dfrac{2}{c\pi}$

4. Solve for c in terms of π and d.
   $\pi = \dfrac{c}{d}$

   (1) πd (3) $\dfrac{d}{\pi}$

   (2) $\dfrac{\pi}{d}$ (4) $\pi + d$

5. Solve for x in terms of y, m, and b.
$y = mx + b$

   (1) $\dfrac{y-b}{m}$ (3) $\dfrac{y}{m} - b$

   (2) $\dfrac{y+b}{m}$ (4) $\dfrac{y}{m} + b$

6. Solve for m in terms of x, y, and b.
$y = mx + b$

   (1) $\dfrac{y-b}{x}$ (3) $\dfrac{y}{x} + b$

   (2) $\dfrac{y+b}{x}$ (4) $\dfrac{y}{x} - b$

7. Solve for b in terms of x, y, and m.
$y = mx + b$

   (1) $\dfrac{y}{mx}$ (3) $y - mx$

   (2) ymx (4) $y + mx$

8. Solve for c in terms of a, b, and d.
$abc = d$

   (1) $\dfrac{ab}{d}$ (3) $\dfrac{d}{ab}$

   (2) $\dfrac{db}{a}$ (4) $\dfrac{da}{b}$

9. Solve for b in terms of a, c, and d.
$abc = d$

   (1) $\dfrac{ac}{d}$ (3) $\dfrac{da}{c}$

   (2) adc (4) $\dfrac{d}{ac}$

Solving Linear Equations with Algebra

10. Solve for w in terms of l and p.
$2l + 2w = p$

(1) $\dfrac{p+2l}{2}$

(2) $\dfrac{p-2l}{2}$

(3) $\dfrac{2}{p+2l}$

(4) $\dfrac{2}{p-2l}$

B. *Show how you arrived at your answers.*

1. If an equation that relates profit earned (p), quantity sold (q), price (r), and fixed expenses (f), is $p = qr - f$, rewrite the equation with q solved in terms of p, r, and f.

2. The volume of a rectangular prism is $v = lwh$, where v is the volume, l is the length, w is the width, and h is the height. Rewrite this equation with w solved in terms of v, l, and h.

3. The area of a triangle is related to the base and height of the triangle by the formula $a = bh/2$. Rewrite this equation with b solved in terms of a and h.

4. The surface area of a cylinder is $s = 2\pi rh$, where s is the surface area, r is the radius, and h is the height. Rewrite this equation with r solved in terms of s, π, and h.

5. The nth term of an arithmetic sequence can be calculated with the formula $a_n = a_1 + (n-1)d$. Rewrite this equation with n solved in terms of a_n, a_1, and d.

Chapter Two

POLYNOMIAL ARITHMETIC

2.1 CLASSIFYING MONOMIALS, BINOMIALS, AND TRINOMIALS

Key Ideas

Terms in mathematical expressions are separated by plus or minus signs. An expression like $2x + 5$ has two terms while $x^2 + 2x + 5$ has three terms. Expressions with only one term are called *monomials*. Expressions with two terms are called *binomials*. Expressions with three terms are called *trinomials*. They are all special types of *polynomials*.

Determining the Type of a Polynomial by the Number of Terms

By counting the number of terms, you can identify which type of polynomial it is. Terms are separated by either plus or minus signs.

Example 1

Is the polynomial $6x - 2$ a monomial, a binomial, or a trinomial?

Solution: There are two terms, $6x$ and 2, so this is a binomial.

Example 2

Is the polynomial $5x^2$ a monomial, a binomial, or a trinomial?

Solution: Since there are not plus or minus signs, this is just one term, and this is a monomial.

The Degree of a Polynomial

When the terms of a polynomial have exponents on the variables, the *degree* of a polynomial is the largest exponent that a variable is raised to. The binomial $2x^5 + 3x$ is degree 5 since the largest exponent that a variable is raised to is 5. The trinomial $3x^2 - 7x + 2$ is degree 2. In a monomial like 7 the degree is 0 since it is equivalent to $7x^0$.

Polynomial Arithmetic

MATH FACTS

A polynomial with a degree of 0 is called a *constant polynomial*. A polynomial with a degree of 1 is called a *linear polynomial*. A polynomial with a degree of 2 is called a *quadratic polynomial*. A polynomial with a degree of 3 is called a *cubic polynomial*.

Check Your Understanding of Section 2.1

A. Multiple-Choice

1. Classify this polynomial $5x^2 + 3$.
   - (1) Monomial
   - (2) Binomial
   - (3) Trinomial
   - (4) None of the above

2. Classify this polynomial $7x^2 - 3x + 2$.
   - (1) Monomial
   - (2) Binomial
   - (3) Trinomial
   - (4) None of the above

3. Classify this polynomial $3x^3$.
   - (1) Monomial
   - (2) Binomial
   - (3) Trinomial
   - (4) None of the above

4. Classify this polynomial $3x^2y + 5xy^2$.
   - (1) Monomial
   - (2) Binomial
   - (3) Trinomial
   - (4) None of the above

5. Classify this polynomial $3x^3 + 5x^2 + 7x - 3$.
   - (1) Monomial
   - (2) Binomial
   - (3) Trinomial
   - (4) None of the above

6. Classify this polynomial $3x^2 - 7x + 5$.
   - (1) Monomial
   - (2) Binomial
   - (3) Trinomial
   - (4) None of the above

7. Which of the following is a binomial?
   (1) $2 \cdot 5x^2$
   (2) $5x^2 + 7x + 2$
   (3) $5x^2 + 2$
   (4) $5x^2 + 2 + 3x$

8. Which of the following is a trinomial?
   (1) $2x^2$
   (2) $2x^2 + 5x$
   (3) $2x^2 + 5x + 3$
   (4) $(2x^2)(5x)(3)$

9. What is the degree of the polynomial $5x^2 + 7x^3 - 2x$.
   (1) 0
   (2) 1
   (3) 2
   (4) 3

10. Which of the following is a quadratic trinomial?
    (1) $5x^2 + 8$
    (2) $5x^3 - 3x + 8$
    (3) $5x^2 - 3x + 8$
    (4) $5x^2$

B. *Show how you arrived at your answers.*

1. An expression for the height of a projectile is $-16t^2 + 72t + 5$. What kind of polynomial is this and what is its degree?

2. An equation for the amount of profit a company makes is $5q - 800$. What kind of polynomial is that and what is its degree?

3. Cross out one of the terms in this polynomial to make it into a third degree trinomial. $5x^3 + 2x^2 - 7x + 9$.

4. Create a fourth degree monomial.

5. Mark says that the expression $2 + 3$ is a binomial since it has two terms. Layla says that it is a monomial since $2 + 3 = 5$, and 5 is just one term. Who is correct?

Polynomial Arithmetic

2.2 MULTIPLYING AND DIVIDING MONOMIALS

KEY IDEAS

A monomial, like $5x^3$, has a coefficient of 5 and a variable part of x^3. Just as numbers can be multiplied together, monomials can be multiplied. When multiplying monomials, you have to use the rules for exponents.

Multiplying Expressions Involving Exponents

If you multiply a number, like 5, by itself you get $5 \cdot 5 = 25$, which can also be written as 5^2. When you multiply the variable x by itself, you get $x \cdot x$, which can be written as x^2. Multiplying monomials with larger exponents can be done with a shortcut.

Example 1

Simplify $x^2 \cdot x^3$.

Solution: One way to do this is to expand each of the expressions.

$$x^2 \cdot x^3$$
$$(x \cdot x) \cdot (x \cdot x \cdot x) = x \cdot x \cdot x \cdot x \cdot x = x^5$$

A shorter way is to use the rule that when multiplying two exponential expressions with the same base, the solution has that same base and the exponent will be the sum of the two exponents.

$$x^2 \cdot x^3 = x^{2+3} = x^5$$

Multiplying Monomials That Have Coefficients and Variable Parts

To multiply two monomials, multiply the two coefficients and then multiply the two variable parts using the exponent multiplication shortcut of adding the exponents.

Example 2

Simplify $3x^2 \cdot 5x^3$.

Solution: Since 3 · 5 = 15, the coefficient of the solution is 15. Since $x^2 \cdot x^3 = x^5$, the variable part of the solution is x^5. The complete answer is $15x^5$.

Example 3
Simplify $3x^4 \cdot 4x$.

Solution: 3 · 4 = 12, which is the coefficient of the solution. $x^4 \cdot x$ is like $x^4 \cdot x^1$ so the exponent multiplication rule says that $x^4 \cdot x = x^5$. The solution is $12x^5$.

Example 4
Simplify $2x^5 \cdot 3x^4 \cdot 5x^3$.

Solution: Since 2 · 3 · 5 = 30 and $x^5 \cdot x^4 \cdot x^3 = x^{12}$, the solution is $30x^{12}$.

Example 5
Simplify $4x^2y \cdot 7xy^3$.

Solution: The product of the two coefficients is 28. The product of x^2 and x is x^3. The product of y and y^3 is y^4. So the simplified expression is $28x^3y^4$.

MATH FACTS

When raising something to a negative power, like 3^{-2}, this is not the same as $-(3^2) = -9$. Instead, the rule is that $x^{-a} = \dfrac{1}{x^a}$ so $3^{-2} = \dfrac{1}{3^2} = \dfrac{1}{9}$. In scientific notation, there are often expressions like $5 \cdot 10^{-3}$, which is equivalent to $5 \cdot \dfrac{1}{10^3} = \dfrac{5}{1,000} = 0.005$.

Polynomial Arithmetic

Dividing Monomials

To divide two monomials, divide the two coefficients and subtract the exponent of the divisor from the exponent of the dividend. For example $8x^{12} \div 4x^3 = 2x^9$ because 8 divided by 4 is 2 and $12 - 3 = 9$. This is also true when the division is written as a fraction $\frac{8x^{12}}{4x^3} = 2x^9$.

Example 6

Simplify $18x^8 \div 2x^2$.

Solution: $\frac{18}{2} = 9$ and $8 - 2 = 6$ so the answer is $9x^6$.

Check Your Understanding of Section 2.2

A. Multiple-Choice

1. Multiply $3x^3 \cdot 4x^5$.
   - (1) $7x^8$
   - (2) $7x^{15}$
   - (3) $12x^8$
   - (4) $12x^{15}$

2. Multiply $2x^5 \cdot 5x$.
   - (1) $10x^6$
   - (2) $10x^5$
   - (3) $7x^6$
   - (4) $7x^5$

3. Which of the following would *not* simplify to $12x^6$?
   - (1) $2x \cdot 6x^5$
   - (2) $3x^2 \cdot 4x^4$
   - (3) $12x^3 \cdot x^3$
   - (4) $4x^3 \cdot 3x^2$

4. Which of the following would *not* simplify to $24x^5$?
   - (1) $3x^2 \cdot 8x^3$
   - (2) $4x \cdot 6x^4$
   - (3) $2x^3 \cdot 12x^2$
   - (4) $6x^5 \cdot 4x$

5. Multiply $x^6 \cdot x$.
   - (1) x^5
   - (2) x^6
   - (3) x^7
   - (4) x^8

6. Multiply $5 \cdot 3x^5$.
   - (1) $15x^4$
   - (2) $15x^5$
   - (3) $15x^6$
   - (4) $15x^7$

7. If $2x^3 \cdot y = 10x^6$, which expression is equivalent to y?
   (1) $5x^2$
   (2) $5x^3$
   (3) $5x^4$
   (4) $5x^5$

8. Multiply $3x^5 \cdot 2x^{-2}$.
   (1) $6x^7$
   (2) $6x^5$
   (3) $6x^3$
   (4) $6x$

9. If $5x^7 \cdot z = 20x^4$, which expression is equivalent to z?
   (1) $4x^5$
   (2) $4x^7$
   (3) $4x^3$
   (4) $4x^{-3}$

10. Simplify $(4x^3)^2$.
    (1) $16x^6$
    (2) $16x^9$
    (3) $8x^6$
    (4) $8x^9$

B. *Show how you arrived at your answers.*

1. Alexa multiplies $4x^2 \cdot 5x^3$ and says the answer is $20x^6$ since $4 \cdot 5$ is 20 and $2 \cdot 3$ is 6. Is she correct? Explain your reasoning.

2. What does $12x^6 \div 3x^2$ simplify to?

3. Joseph says that $x^0 = 0$ since there are no xs multiplied together. Sawyer says that this is not correct. Who is right?

4. What is $5x^2 \cdot 2x^{-2}$? Explain your reasoning?

5. When multiplying a quadratic monomial by a cubic monomial, what type of polynomial will be the result?

Chapter Two POLYNOMIAL ARITHMETIC

25

Polynomial Arithmetic

2.3 COMBINING LIKE TERMS

=== KEY IDEAS ===

Two terms are called *like terms* if they have the same variable part. For example, $3x^2$ and $5x^2$ are like terms since they both have an x^2 as the variable part. $3x^2$ and $5x^3$ are not like terms because the x^2 is different from the x^3. Like terms can be combined with addition or subtraction, whereas unlike terms cannot be combined.

Combining Two Like Terms

When you add two dogs to three dogs, you get five dogs (and a lot of dog hair!). Likewise, $2x + 3x = 5x$, $2x^2 + 3x^2 = 5x^2$, and $2x^{10} + 3x^{10} = 5x^{10}$. Unlike terms cannot be combined. For example, $2x + 3y$ cannot be simplified any further. Even though $2x$ and $3x^2$ each have a variable part containing x, they are not like terms since the variable part must be identical, Therefore, they cannot be combined.

For terms to be like terms, the variable parts must be equivalent. So $5xy^2$ cannot be combined with $5x^2y$ since the variable parts are not equivalent.

If There Is No Coefficient, It Is Really a Coefficient of 1

When a term seems to have no coefficient, it is really a coefficient of 1. To combine the like terms $5x^2 + x^2$, think of it as $5x^2 + 1x^2 = 6x^2$.

Example 1

Simplify, if it is possible, the expression $4x^3 + 6x^3$.

Solution: Since each term has a variable part of x^3, these can be combined. Just as $4 + 6 = 10$, $4x^3 + 6x^3 = 10x^3$.

Example 2

Simplify, if it is possible, the expression $6x^3 - 2x^2$.

Solution: These are not like terms since one has a variable part of x^3 and the other has a variable part of x^2, so they cannot be combined.

Example 3

Simplify, if it is possible, the expression $5x^2 y^3 + 4x^2 y^3$.

Solution: These are like terms so they can be combined into $9x^2 y^3$.

Check Your Understanding of Section 2.3

A. *Multiple-Choice*

1. Which expression is equivalent to $2x^2 + 5x^2$?
   (1) $10x^2$
   (2) $7x^2$
   (3) $7x^4$
   (4) The expression cannot be simplified any further.

2. Which expression is equivalent to $3x^4 + 5x$?
   (1) $8x^4$
   (2) $8x^5$
   (3) $8x^6$
   (4) The expression cannot be simplified any further.

3. Which expression is equivalent to $-5x^3 + 2x^3$?
   (1) $-7x^3$
   (2) $3x^3$
   (3) $-3x^3$
   (4) The expression cannot be simplified any further.

4. Which expression is equivalent to $8y^2 - 10y^2 + 5y^2$?
   (1) $7y^2$
   (2) $-7y^2$
   (3) $3y^2$
   (4) The expression cannot be simplified any further.

5. Which expression is equivalent to $5xy^2 + 3xy^2 - 4x^2y$?
   (1) $4x^2y$
   (2) $4xy^2$
   (3) $8x^2y - 4xy^2$
   (4) $8xy^2 - 4x^2y$

6. Which expression is *not* equivalent to $7x^5$?
   (1) $5x^2 + 2x^3$
   (2) $2x^5 + 5x^5$
   (3) $9x^5 - 2x^5$
   (4) $7x^2 \cdot x^3$

Polynomial Arithmetic

7. Which expression is equivalent to $3x^2$?
   (1) $1 + 2x^2$
   (2) $x^2 + 2x^2$
   (3) $x + 2x$
   (4) $x + 2x^2$

8. Which expression is *not* equivalent to $7x^2y$?
   (1) $2x^2y + 5x^2y$
   (2) $3x^2y + 4x^2y$
   (3) $7x \cdot xy$
   (4) $2 + 5x^2y$

9. Which expression cannot be simplified any further?
   (1) $5x + 3x$
   (2) $5x^2 + 3x^2$
   (3) $5x + 3x^2$
   (4) $5x^2 - 3x^2$

10. Which expression is equivalent to $x^2 + x^2 + x^2 + x^2$?
    (1) x^8
    (2) $4x^8$
    (3) $8x^4$
    (4) $4x^2$

B. *Show how you arrived at your answers.*

1. Katherine simplifies the expression $3x^2 + 5x^2$ by noticing that $x^2(3 + 5)$ would become $3x^2 + 5x^2$ if she used the distributive property of multiplication over addition, but would become $x^2(8) = 8x^2$ if she simplified the parentheses first. Was this a valid way to arrive at the correct answer $8x^2$? Explain your reasoning.

2. Chelsea says that the expression $3xy^2 + 5y^2x$ cannot be simplified any further since the terms are not like terms. Her friend Josie says that it is possible. Who is right? Explain your reasoning.

3. If $5xy^2z + a = 7xy^2z$, what expression in terms of x, y, and z must the variable a be equivalent to?

4. Combine all like terms in the expression
$$x^3 + x^2y + xy^2 + y^3 + x^2y + xy^2 + xy^2 + x^2y$$

5. Combine all like terms in the expression $6 - 3i + 2i - i^2$.

2.4 MULTIPLYING MONOMIALS AND POLYNOMIALS

> **KEY IDEAS**
>
> To multiply a monomial by a binomial, trinomial, or polynomial with more than three terms, use the distributive property of multiplication over addition. Just as you can use the distributive property to simplify an expression like $2 \cdot (3 + 5)$ to become $2 \cdot 3 + 2 \cdot 5 = 6 + 10 = 16$, you can multiply monomials by polynomials the same way.

Multiplying Monomials by Binomials

To multiply the monomial $2x^2$ by the binomial $3x + 5$, put the $3x + 5$ in parentheses, and write the problem as $2x^2(3x + 5)$. Now *distribute* the $2x^2$ to each of the terms inside the parentheses to get $2x^2 \cdot 3x + 2x^2 \cdot 5$. Each of these terms is now a monomial times a monomial and can be simplified separately to become $6x^3 + 10x^2$.

Multiplying Monomials by Polynomials with More than Two Terms

The distributive property applies even when there are more than two terms in the parentheses. So $5(1 + 2 + 3) = 5 \cdot 1 + 5 \cdot 2 + 5 \cdot 3 = 30$. The same thing happens when multiplying a monomial by a polynomial with more than two terms.

To multiply $5x$ by $2x^2 + 7x - 3$, multiply the $5x$ by each of the three terms in the second expression and then simplify the terms if possible.

$$5x(2x^2 + 7x - 3) = 5x \cdot 2x^2 + 5x \cdot 7x + 5x(-3) = 10x^3 + 35x^2 - 15x$$

Example 1

Multiply $4x^2$ by $5x^3 + 7x^5$.

Solution: Write as

$$4x^2(5x^3 + 7x^5)$$

Distribute the $4x^2$ through each term in the parentheses.

$$4x^2 \cdot 5x^3 + 4x^2 \cdot 7x^5$$
$$20x^5 + 28x^7$$

Polynomial Arithmetic

Example 2

Multiply −3 by $2x - 3$.

Solution: First, write as $-3(2x - 3)$. Then distribute the −3 through each term in the parentheses.

$$-3(2x) = -6x$$
$$-3(-3) = +9$$

So the answer is $-6x + 9$. Because $-3(-3) = 9$, you put a plus sign between the two terms in the answer.

Example 3

Multiply −4 by $3x + 5$.

Solution:

$$-4(3x + 5)$$
$$-4(3x) = -12x$$
$$-4(5) = -20$$

The answer is $-12x - 20$. Since $-4(5) = -20$, put a minus sign between the two terms in the answer.

Example 4

Multiply $2x^2 - 5x + 2$ by $3x$.

Solution: It is simpler if you write the $3x$ on the left of the parentheses.

$$3x(2x^2 - 5x + 2)$$

Distribute the $3x$ through each term in the parentheses.

$$6x^3 - 15x^2 + 6x$$

Check Your Understanding of Section 2.4

A. *Multiple-Choice*

1. Multiply 5 by $2x + 3$.
   - (1) $10x + 3$
   - (2) $10x + 15$
   - (3) $2x + 15$
   - (4) $7x + 15$

2. Multiply –2 by $5x^2 - 2x + 6$.
   (1) $-10x^2 - 4x - 12$
   (2) $-10x^2 - 2x + 6$
   (3) $-10x^2 - 4x + 12$
   (4) $-10x^2 + 4x - 12$

3. Multiply $3x^2 + 7x - 5$ by $4x^2$.
   (1) $-17x^2 + 7x$
   (2) $12x^4 + 7x - 5$
   (3) $12x^4 + 28x^3 - 20x^2$
   (4) $12x^4 + 28x^3 + 20x^2$

4. Simplify $6x(2x + 3)$.
   (1) $8x^2 + 18x$
   (2) $12x^2 + 18x$
   (3) $12x^2 + 3$
   (4) $20x$

5. Simplify $-2x(3x - 5)$.
   (1) $-6x - 5$
   (2) $-6x^2 - 5$
   (3) $-6x^2 - 10x$
   (4) $-6x^2 + 10x$

6. Simplify $-4x(-2x - 7)$.
   (1) $-8x^2 - 28x$
   (2) $-8x^2 + 28x$
   (3) $8x^2 - 28x$
   (4) $8x^2 + 28x$

7. Simplify $3x(5x^2 - 2x + 3)$.
   (1) $15x^3 - 2x + 3$
   (2) $15x^3 - 6x^2 + 9x$
   (3) $5x^2 + x + 3$
   (4) $15x^3 + 6x^2 - 9x$

8. Simplify $4x^2y(3xy - 2xy^2)$.
   (1) $12x^2y^2 - 8x^2y^3$
   (2) $12x^3y^2 - 6x^3y^3$
   (3) $12x^3y^2 - 8x^3y^3$
   (4) $12x^3y^3 - 6x^2y^3$

9. If $a(2x + 5) = 8x^2 + 20x$, which expression is a equivalent to?
   (1) $4x$
   (2) $6x$
   (3) $4x^2$
   (4) $6x^2$

10. Which expression is *not* equivalent to $2x^3 + 10x^2 - 6x$?
    (1) $2(x^3 + 5x^2 - 3x)$
    (2) $2x^2(x^2 + 5x - 6)$
    (3) $x(2x^2 + 10x - 6)$
    (4) $2x(x^2 + 5x - 3)$

B. Show how you arrived at your answers.

1. Tucker does the question $5(2x + 3x)$ by first combining the like terms and gets $5(5x) = 25x$. Alexander does the same question by applying the distributive property. Show how Tucker completed the question, and determine if Alexander gets the same answer as Tucker.

Polynomial Arithmetic

2. Jack says that $(3x^2 + 2x + 1)5x = 15x^3 + 10x^2 + 5x$. His friend Izabella says that the monomial has to be on the left of the parentheses to do it this way. Who is correct and why?

3. The polynomial $10x^3 - 6x^2 + 14x$ can be written as $a(5x^2 - 3x + 7)$. What expression must the variable a represent? Explain your reasoning.

4. Simplify and write as a trinomial $4x(2x + 3) - 2(2x + 3)$.

5. Simplify and write as a trinomial $2x(3x - 5) - 6(3x - 5)$.

2.5 ADDING AND SUBTRACTING POLYNOMIALS

KEY IDEAS

Polynomials are composed of one or more monomials. When you add or subtract polynomials, you locate the like terms and combine them separately. Trinomials often have a constant term, an x term, and an x^2 term. If two trinomials like this are combined, the result is also a trinomial.

Adding Polynomials

When two polynomials have like terms, those terms can be combined the way you combined monomials.

Example 1

Simplify the expression $(2x + 3) + (3x + 5)$.

Solution: When adding polynomials, you can disregard the parentheses and get $2x + 3 + 3x + 5$. The $2x$ and the $3x$ are like terms that can be combined to become $+5x$. The 3 and the 5 are also like terms that can be combined to become $+8$, so the answer is $5x + 8$.

32

Example 2

Simplify the expression $(x^2 + 3x - 2) + (x^2 - 5x + 7)$.

Solution:

$$x^2 + 3x - 2 + x^2 - 5x + 7$$
$$x^2 + x^2 = 2x^2$$
$$+3x - 5x = -2x$$
$$-2 + 7 = +5$$

The answer is $2x^2 - 2x + 5$.

Subtracting Polynomials

When subtracting polynomials, you have to beware of one of the most common errors in all of high school math: not distributing the negative sign through the parentheses.

Example 3

Simplify the expression $(8x + 5) - 3(3x + 2)$.

Solution: You can disregard the parentheses of the expression on the left side of the minus sign.

$$8x + 5 - 3(3x + 2)$$

Now be sure to distribute the -3 through, as if the $8x + 5$ were not there. Since $-3(3x + 2)$ is $-9x - 6$, this becomes

$$8x + 5 - 9x - 6$$

Now the like terms can be combined.

$$8x - 9x = -1x = -x$$
$$+5 - 6 = -1$$

The simplified answer is $-x - 1$.

MATH FACTS

One of the most common mistakes in algebra is to not distribute the minus sign through all the terms in a polynomial subtraction problem. If the problem is $(2x + 5) - (3x + 2)$, the correct thing to do is write it as $2x + 5 - 3x - 2$, not as $2x + 5 - 3x + 2$.

Polynomial Arithmetic

Example 4

Simplify the expression $(5x - 2) - (2x + 7)$.

Solution: Think of this as $(5x - 2) - 1(2x + 7)$. Remove the parentheses from the first expression and distribute the -1 through the second expression. Since $-1(2x + 7)$ is $-2x - 7$, this becomes

$$5x - 2 - 1(2x + 7)$$
$$5x - 2 - 2x - 7$$
$$5x - 2x = 3x$$
$$-2 - 7 = -9$$

So the answer is $3x - 9$.

Warning! The most common mistake students make is to just remove both sets of parentheses and get $5x - 2 - 2x + 7$ with a $+7$ instead of the correct -7 before combining like terms.

Example 5

Simplify the expression $(7x + 3) - (4x - 5)$.

Solution:

$$7x + 3 - 1(4x - 5)$$
$$7x + 3 - 4x + 5 \text{ (notice the +5 is correct, not } -5.)$$
$$3x + 8$$

Check Your Understanding of Section 2.5

A. *Multiple-Choice*

1. Simplify $(2x + 3) + (4x + 5)$.
   (1) $6x + 8$ (3) $6x + 15$
   (2) $8x + 15$ (4) $8x + 8$

2. Simplify $(3x - 2) + (6x + 5)$.
   (1) $9x + 3$ (3) $9x + 7$
   (2) $9x - 7$ (4) $9x - 3$

3. Simplify $(x^2 + 3x + 5) + (x^2 + 4x - 2)$.
   (1) $2x^4 + 7x^2 + 3$ (3) $2x^2 + 7x - 7$
   (2) $2x^2 + 7x + 3$ (4) $2x^2 + 7x + 7$

34

4. Simplify $(x^2 - 6x + 4) + 3(x^2 + 2x - 5)$.
   (1) $4x^4 - 4x - 1$
   (2) $4x^4 - 12x - 11$
   (3) $4x^2 - 11$
   (4) $4x^2 - 12x + 11$

5. Simplify $(5x + 3) - (3x + 2)$.
   (1) $2x - 5$
   (2) $2x - 1$
   (3) $2x + 5$
   (4) $2x + 1$

6. Simplify $(3x - 4) - (5x - 3)$.
   (1) $-2x - 7$
   (2) $-2x - 1$
   (3) $2x - 1$
   (4) $2x - 7$

7. Simplify $2x(3x + 4) - 4x(2x - 5)$.
   (1) $-2x^2 + 28x$
   (2) $-2x^2 + 8x - 5$
   (3) $-2x^2 - 12x$
   (4) $-2x^2 - 20x$

8. If $C = 2x^2 + 3x - 2$ and $D = -4x^2 - 3x + 5$, then $C - D$ equals
   (1) $-2x^2 + 3$
   (2) $6x^2 + 3$
   (3) $6x^2 + 6x - 7$
   (4) $6x^2 - 7$

9. Simplify $(3x^2 - 4x + 2) - 2(x^2 - 2x + 1)$.
   (1) x^2
   (2) $x^2 - 2x + 1$
   (3) $x^2 + 2x - 1$
   (4) $x^2 - 4x + 5$

10. Simplify $(x^2 - 5x + 6) - (x^2 - 2x + 1)$.
    (1) $-7x + 7$
    (2) $-7x + 5$
    (3) $-3x + 5$
    (4) $-3x + 7$

B. Show how you arrived at your answers.

1. Nathan tried to simplify the expression $(2x + 3) - (4x + 2)$ by first writing $2x + 3 - 4x + 2$ and then simplifying to $-2x + 5$. Did he do this correctly? Explain your reasoning.

2. If $C = 3x + 5$ and $D = 2x - 5$, determine $C - 3D$ and simplify as much as possible.

3. To solve $49 - 2(13)$, Kaleigh changed it into $(40 + 9) - 2(10 + 3)$ and then simplified to $40 + 9 - 20 - 6 = 20 + 3 = 23$. Explain why this produced the right answer.

4. If $F = -2x + 5$ and $G = 3x - 7$, express $2F - 5G$ in terms of x and y simplified as much as possible.

5. If $P = x^2 + 5x + 2$, $Q = x^2 - 3x + 4$, and $R = x^2 - 2x + 5$, what is $P + Q - R$ in terms of x simplified as much as possible?

Polynomial Arithmetic

2.6 MULTI-STEP ALGEBRA EQUATIONS INVOLVING POLYNOMIAL ARITHMETIC

=== Key Ideas ===

More complicated algebra equations often require combining like terms before solving. These equations can take three or more steps. When variables are on both sides of the equals sign, they can be eliminated from one side, which will require combining like terms on the other side.

Simplifying One Side of the Equation Before Solving

Linear equations of the form $ax + b = c$ are the simplest to solve. If there is a linear equation that is not in this form, often it can first be converted into an equation of this form and then solved. This could require using the distributive property and/or combining like terms.

Example 1

Solve for x in the equation $5(x + 3) - 2x = 27$.

Solution: First, simplify the left-hand side, using the distributive property and then combining like terms.

$$5x + 15 - 2x = 27$$
$$3x + 15 = 27$$

Then solve the two-step equation.

$$3x + 15 = 27$$
$$-15 = -15$$
$$3x = 12$$
$$\frac{3x}{3} = \frac{12}{3}$$
$$x = 4$$

Solving an Equation with Variables on Both Sides of the Equals Sign

Variables can be eliminated from one side of an equation by using the addition property of equality or the subtraction property of equality.

Example 2

Solve for x in the equation $5x + 3 = 3x + 15$.

Solution: In order to make this equation so that there are only variables on one side of the equation, it is simplest to subtract $3x$ from both sides of the equation

$$5x + 3 = 3x + 15$$
$$-3x = -3x$$

Now combine the like terms $5x$ and $-3x$, and complete the two-step equation.

$$2x + 3 = 15$$
$$-3 = -3$$
$$2x = 12$$
$$\frac{2x}{2} = \frac{12}{2}$$
$$x = 6$$

Word Problems Involving Combining Like Terms

Some word problems can be converted into algebra equations where there are like terms that have to be combined or variables on both sides of the equation. It is important to set these equation up properly.

Example 3

Angelina has x dollars. Alexander has five more than three times the amount of money that Angelina has. Together they have 41 dollars. How much does each have?

Polynomial Arithmetic

Solution: Alexander has $3x + 5$ dollars. The equation to solve this problem, then, is

$$x + (3x + 5) = 41$$
$$x + 3x + 5 = 41$$
$$4x + 5 = 41$$
$$-5 = -5$$
$$\frac{4x}{4} = \frac{36}{4}$$
$$x = 9$$
$$3x + 5 = 3 \cdot 9 + 5 = 27 + 5 = 32$$

So Angelina has $9 and Alexander has $32.

Example 4

In five years Gabriel will be three years less than twice his current age. How old is he now?

Solution: Call his current age x. In five years, his age will be $x + 5$. The equation becomes

$$x + 5 = 2x - 3$$
$$-x = -x$$
$$5 = x - 3$$
$$+3 = +3$$
$$8 = x$$

Check Your Understanding of Section 2.6

A. Multiple-Choice

1. Solve for x: $(2x + 3) + (4x + 5) = 32$.
   (1) 4   (2) 5   (3) 6   (4) 7

2. Solve for x: $(5x + 7) - (2x + 3) = 25$.
   (1) 5   (2) 6   (3) 7   (4) 8

3. Solve for x: $x + 3x - 5 = 195$.
   (1) 30   (2) 40   (3) 50   (4) 60

4. Solve for y: $y + (y + 1) + (y + 2) = 45$.
   (1) 14   (2) 15   (3) 16   (4) 17

5. Solve for x: $5x + 50 = 7x$.
   (1) 22   (2) 23   (3) 24   (4) 25

6. Solve for z: $30z = 300 - 20z$.
   (1) 3   (2) 4   (3) 5   (4) 6

7. Solve for a: $2a + 5 = 4a - 1$.
   (1) 2   (2) 3   (3) 4   (4) 5

8. Solve for x. $x^2 + 4x + 6 = x^2 - 2x + 18$
   (1) 0   (2) 1   (3) 2   (4) 3

9. The width of a rectangle is 5 inches more than its length. The perimeter of the rectangle is 58 inches. Which equation could be used to find the length of the rectangle?
   (1) $2x + 2(x + 5) = 58$ (3) $x + (x - 5) = 58$
   (2) $x + 2(x - 5) = 58$ (4) $x + (x + 5) = 58$

10. Jonathan weighs thirty pounds more than three times his son Aidan's weight. If their combined weight is 210 pounds, which equation could be used to find the weight of Aidan?
    (1) $x + 3x + 30 = 210$ (3) $3x - 30 - x = 210$
    (2) $x + 3x - 30 = 210$ (4) $x(3x + 30) = 210$

B. *Show how you arrived at your answers.*

1. In 20 years Lillian will be three years older than twice her current age. Set up an equation that can be used to solve for x, Lillian's current age.

2. The equation $2x + 5 = 6x - 7$ can be solved several ways. Makayla solves it by first subtracting $2x$ from both sides of the equation. David solves it by first subtracting $6x$ from both sides. Which method would you choose and why?

3. Solve for x in $2(x + 3) - 5(x - 2) = 4$.

4. Two trains are 300 miles away from each other on the same set of tracks. The first train goes east at a speed of 30 miles per hour. The second train goes west at a speed of 20 miles per hour. An equation that can be used to determine how long it will take for the trains to pass is $30x = 300 - 20x$. Explain how this equation was formed, and solve for x to determine when the trains will pass.

Polynomial Arithmetic

5. On Tuesday an electronics store sells five items less than double the amount it sold on Monday. On Wednesday the store sells ten items more than triple the amount it sold on Monday. If it sold 125 items altogether on the three days, create an equation and use it to determine the number of items sold on each day.

2.7 MULTIPLYING POLYNOMIALS BY POLYNOMIALS

=== KEY IDEAS ===

Multiplying polynomials can be done by applying the distributive property several times and then combining like terms. When multiplying a binomial by another binomial, some teachers teach the FOIL shortcut.

Multiplying Binomials with the Distributive Property

Just as $2(x + 5)$ can be first simplified with the distributive property into $2 \cdot x + 2 \cdot 5$, a more complicated multiplication like $(x + 2)(x + 5)$ can be first simplified to

$$(x + 2) \cdot x + (x + 2) \cdot 5.$$

To make this more familiar, change this to $x \cdot (x + 2) + 5 \cdot (x + 2)$. Now each of the terms can be simplified with the distributive property.

$$x^2 + 2x + 5x + 10$$

After combining like terms, this becomes $x^2 + 7x + 10$.

Example 1

Multiply $(x + 4) \cdot (x - 3)$.

Solution:

$$(x + 4)x + (x + 4)(-3)$$
$$x(x + 4) - 3(x + 4)$$
$$x^2 + 4x - 3x - 12$$
$$x^2 + 1x - 12$$
$$x^2 + x - 12$$

40

Multiplying Binomials with the FOIL Shortcut

Four different multiplications occur when multiplying binomials with the distributive property method. The FOIL shortcut is a way to skip two steps of the longer way and accomplish all four of these multiplications before combining like terms.

$$(x-6)(x+2) = \overbrace{x \cdot x}^{F} + \overbrace{2 \cdot x}^{O} + \overbrace{(-6)(x)}^{I} + \overbrace{(-6)(+2)}^{L}$$
$$= x^2 + [2x - 6x] - 12$$
$$= x^2 - 4x - 12$$

Example 2

Multiply $(x + 2)(x + 5)$.

Solution: The F stands for "Firsts." The first term in the first binomial is x. The first term in the second binomial is also x. When you multiply these firsts together, you get x^2.

$$x^2$$

The O stands for "Outers." The x term in the first binomial and the +5 in the second binomial are the outermost terms. Multiply them together to get +5x.

$$x^2 + 5x$$

The I stands for "Inners." The +2 in the first binomial and the x in the second binomial are the two innermost terms. Multiply them together to get +2x.

$$x^2 + 5x + 2x$$

The L stands for "Lasts." The last term in the first binomial is +2. The last term in the second binomial is +5. Multiply them together to get +10.

$$x^2 + 5x + 2x + 10$$

Now combine like terms to get $x^2 + 7x + 10$, which is the product.

Example 3

Multiply $(x + 4)(x - 2)$.

Solution: Using the FOIL shortcut, the product of the Firsts is x^2, the product of the Outers is $-2x$, the product of the Inners is +4x, and the product of the Lasts is –8. $x^2 - 2x + 4x - 8$ becomes $x^2 + 2x - 8$ after combining like terms.

Polynomial Arithmetic

Example 4

Multiply $(2x - 3)(5x + 2)$.

Solution: Using FOIL, the four terms become $10x^2$, $+4x$, $-15x$, and -6. Combine the like terms of $10x^2 + 4x - 15x - 6$ to get $10x^2 - 11x - 6$.

Example 5

Multiply $(x + 5)(x + 5)$.

Solution: The four terms you get with FOIL are x^2, $+5x$, $+5x$, and $+25$, which combine to get $x^2 + 10x + 25$.

Squaring Binomials

To square a binomial, like $(x + 5)^2$, write it as $(x + 5)(x + 5)$ and then use FOIL to get $x^2 + 10x + 25$. The answer is *not* $x^2 + 25$ as there is no distributive property for exponents. In general, the square of a binomial $(x + a)$ is $x^2 + 2ax + a^2$.

Example 6

Simplify $(x + 6)^2$.

Solution: Rewrite as $(x + 6)(x + 6)$ and use FOIL.

$$(x + 6)(x + 6) = x^2 + 6x + 6x + 36 = x^2 + 12x + 36.$$

Note that it did not become $x^2 + 36$, as some students get when they try to distribute the exponent through the parentheses, which is not a valid way to calculate this.

Multiplying the (a – b)(a + b) Pattern

When two binomials are the same except that one has a minus sign between the two terms and the other has a plus sign between the two terms, like $(a - b)(a + b)$, the answer will be $a^2 - b^2$. If you do it with FOIL, the middle two terms, the OI, will cancel each other out.

Example 7

Multiply $(x + 5)(x - 5)$.

Solution: The four terms you get with FOIL are x^2, $-5x$, $+5x$, and -25. When you combine them you get $x^2 - 5x + 5x - 25 = x^2 + 0x - 25 = x^2 - 25$, which is a binomial.

This can also be done with the $(a - b)(a + b)$ shortcut. In this case the a is the x and the b is the 5 so the solution is $x^2 - 5^2 = x^2 - 25$.

Multiplying Polynomials by Polynomials

If one of the polynomials has more than two terms, the FOIL shortcut cannot be used and the problem must be done the long way with the distributive property.

Example 8

Multiply $(x + 3)(x^2 + 5x + 6)$.

Solution: Using the distributive property, first get

$$(x + 3)x^2 + (x + 3)5x + (x + 3)6$$

Using the distributive property three more times, get

$$x^3 + 3x^2 + 5x^2 + 15x + 6x + 18$$

Combine like terms

$$x^3 + 8x^2 + 21x + 18$$

Check Your Understanding of Section 2.7

A. Multiple-Choice

1. $(x + 2)(x + 7) =$
   (1) $x^2 + 14$
   (2) $x^2 + 5x + 14$
   (3) $x^2 + 9x + 14$
   (4) $x^2 + 14x + 9$

2. $(x + 2)(x - 7) =$
   (1) $x^2 - 5x - 14$
   (2) $x^2 - 17$
   (3) $x^2 + 5x + 14$
   (4) $x^2 - 5x + 14$

Polynomial Arithmetic

3. $(2x + 3)(3x - 1) =$
   - (1) $6x^2 - 3$
   - (2) $6x^2 + 11x - 3$
   - (3) $6x^2 + 7x - 3$
   - (4) $6x^2 - 7x - 3$

4. $(2a + 5b)(4a + 3b) =$
   - (1) $8a^2 + 15b^2$
   - (2) $8a^2 + 30ab + 15b^2$
   - (3) $8a^2 + 28ab + 15b^2$
   - (4) $8a^2 + 26ab + 15b^2$

5. $(x + 5)^2 =$
   - (1) $x^2 + 25$
   - (2) $x^2 - 25$
   - (3) $x^2 + 10x - 25$
   - (4) $x^2 + 10x + 25$

6. $(x - 7)(x + 7) =$
   - (1) $x^2 + 49$
   - (2) $x^2 - 49$
   - (3) $x^2 - 14x - 49$
   - (4) $x^2 + 14x - 49$

7. $(x^2 - 2)(x^2 + 6) =$
   - (1) $x^4 + 4x^2 - 12$
   - (2) $x^4 - 12$
   - (3) $x^2 + 4x - 12$
   - (4) $x^4 - 4x^2 - 12$

8. $(x^2 - 4)(x^2 - 9) =$
   - (1) $x^4 + 36$
   - (2) $x^4 - 13x^2 + 36$
   - (3) $x^4 + 13x^2 + 36$
   - (4) $x^4 - 13x^2 - 36$

9. If $(x + 4)(x + a) = x^2 + 7x + 12$. What is the value of a?
   - (1) 1
   - (2) 2
   - (3) 3
   - (4) 4

10. A 4 foot by 6 foot patio has a 2 foot border around it. The area of the combined patio and border can be expressed as $(2x + 4)(2x + 6)$. Which expression is equivalent to this?
    - (1) $4x^2 - 20x + 24$
    - (2) $4x^2 + 20x + 24$
    - (3) $4x^2 + 20x - 24$
    - (4) $4x^2 + 24$

B. *Show how you arrived at your answers.*

1. If $C = x + 4$ and $D = x - 1$, express $CD + 2C$ as a trinomial.

2. Express $(2 + 3i)^2 + (1 + 2i)$ as a trinomial.

3. Maria says that $(x + 4)^2 = x^2 + 16$ because of the distributive property. Is she correct? Explain your reasoning.

4. What is $(x + 3)(x^2 + 7x + 10)$ expressed as a four-term polynomial?

5. Noah says that he has a shortcut for multiplying $48 \cdot 52$ in his head. He starts by writing 48 as $50 - 2$ and 52 as $50 + 2$. Explain how he can complete this question.

2.8 FACTORING POLYNOMIALS

> **KEY IDEAS**
>
> Factoring a number is when you find two numbers whose product is that number. For example, the number 15 can be *factored* into $3 \cdot 5$. Some polynomials can also be factored by reversing the multiplication processes from the previous sections.

Greatest Common Factor Factoring

When you use the distributive property to multiply an expression like $4(3x + 5)$, you get $12x + 20$. *Factoring* $12x + 20$ requires undoing the distributive property. Examine the terms $12x$ and 20 and determine if they have a common factor. In this case since 4 is a factor of both $12x$ and of 20, it is a *common factor*. There are other common factors, such as 2, but of all the common factors, 4 is the largest one, or the *greatest common factor*. To factor $12x + 20$, first write the greatest common factor, 4, and then put an empty set of parentheses, which will be filled in later. Since the goal is to find what must be multiplied by 4 to get $12x + 20$, put the $12x + 20$ on the right-hand side of the equals sign.

$$4(\quad\quad) = 12x + 20$$

Now think, "Four times what is $12x$?" The answer is $3x$. Then think, "Four times what is 20?" The answer is $+5$. These are the two terms that go inside the parentheses.

$$4(3x + 5)$$

You can check your answer by using the distributive property.

$$4(3x + 5) = 4 \cdot 3x + 4 \cdot 5 = 12x + 20$$

Polynomial Arithmetic

Example 1

Factor the trinomial $2x^2 - 8x + 10$.

Solution: The greatest common factor is 2.

$$2(\qquad) = 2x^2 - 8x + 10$$
$$2(x^2 - 4x + 5) = 2x^2 - 8x + 10$$

Example 2

Factor the trinomial $12x^3 + 18x^2 - 24x$.

Solution: In this case, the greatest common factor is not just 6, but $6x$ since all of the terms have xs in them.

$$6x(\qquad) = 12x^3 + 18x^2 - 24x$$
$$6x(2x^2 + 3x - 4) = 12x^3 + 18x^2 - 24x$$

Factoring a Trinomial into the Product of Two Binomials

When you multiply the binomials $(x + 2)$ and $(x + 5)$, you get $x^2 + 7x + 10$. The answer has three terms, an x^2 term, an x term, and a constant. The coefficient of the x term is 7 and the constant is 10. Notice that 5 and 2 are the two constants of the binomials and that $7 = 5 + 2$ and $10 = 5 \cdot 2$. This is the key to factoring trinomials that have the form $x^2 + bx + c$. You will need to find two numbers that add up to the b coefficient and whose product is the c constant as the first step.

Example 3

Factor $x^2 + 8x + 15$ into the product of two binomials.

Solution: You need to find two numbers whose product is +15 and whose sum is +8. There are a lot of pairs of numbers whose sum is +8, but not as many whose product is +15. The list of numbers whose product is +15 is $+1 \cdot +15$, $+3 \cdot +5$, $-1 \cdot -15$, and $-3 \cdot -5$.

Of these four pairs, +3 and +5 are the two numbers that also have a sum of +8. The solution, then, is $(x + 3)(x + 5)$.

Check your answer by multiplying with FOIL.

$$(x + 3)(x + 5) = x^2 + 5x + 3x + 15 = x^2 + 8x + 15$$

Example 4

Factor $x^2 - 3x - 18$ into the product of two binomials.

Solution: The six factor pairs of –18 are –1 · +18, +1 · –18, –2 · +9, +2 · 9, –3 · +6, and +3 · –6. From this list, the pair whose sum is –3 is +3 and –6, so the solution is $(x + 3)(x - 6)$. Multiply to check that the product of these is $x^2 - 3x - 18$.

Example 5

Factor $3x^2 + 30x + 63$.

Solution: If possible, first look for a common factor. In this case all three terms have a factor of 3.

$$3x^2 + 30x + 63 = 3(x^2 + 10x + 21)$$

Now factor the $x^2 + 10x + 21$ by finding two numbers with a product of 21 and a sum of 10. The two numbers are +3 and +7, so it factors into $(x + 3)(x + 7)$. The answer is $3(x + 3)(x + 7)$.

Math Facts

There are four different combinations of plus and minus signs that can separate the terms in a quadratic trinomial. $x^2 + bx + c$, $x^2 - bx + c$, $x^2 + bx - c$, and $x^2 - bx - c$. The signs between the terms give hints of what the factors could be.

Case 1: $x^2 + bx + c$. The factors will be of the form $(x + p)(x + q)$. For example, $x^2 + 7x + 10 = (x + 2)(x + 5)$.

Case 2: $x^2 - bx + c$. The factors will be of the form $(x - p)(x - q)$. For example $x^2 - 7x + 10 = (x - 2)(x - 5)$.

For both cases 1 and 2, the rule is that when the c term is positive, the signs between the terms in both factors will be the same as the sign of the b term.

Case 3: $x^2 + bx - c$. The factors will be of the form $(x - p)(x + q)$ where q is the larger factor. For example, $x^2 + 3x - 10 = (x - 2)(x + 5)$.

Case 4: $x^2 - bx - c$. The factors will be of the form $(x - p)(x + q)$ where p is the larger factor. For example, $x^2 - 3x - 10 = (x - 5)(x + 2)$.

For both cases 3 and 4, the rule is that when the c term is negative, one of the factors will have a plus sign between the two terms and the other will have a minus sign between the two terms. The larger number will go after whatever the sign of the b term is.

Factoring a Perfect Square Trinomial into the Square of a Binomial

When you multiply a number by itself, like 3 times 3, the result is 9, which is known as a *square* number because it can be factored into 3^2. When you use FOIL to square a binomial, like $(x + 5)^2$, you get $x^2 + 10x + 25$. This is known as a *perfect square trinomial* because it can be factored back into $(x + 5)^2$. By recognizing that a trinomial is a perfect square trinomial, it is easier to factor. All perfect square trinomials $x^2 + bx + c$ have the property that $c = \left(\frac{b}{2}\right)^2$, and factor into $\left(x+\frac{b}{2}\right)^2$.

Example 6

Is $x^2 + 12x + 36$ a perfect square trinomial?

Solution: Yes. If you calculate $\left(\frac{12}{2}\right)^2$ you get 36, which is equal to the constant term.

Example 7

Is $x^2 - 6x + 8$ a perfect square trinomial?

Solution: No. $\left(-\frac{6}{2}\right)^2 = 9$, which is not equal to the constant term, 8.

Example 8

If $x^2 - 14x + 49$ is a perfect square trinomial, factor it.

Solution: Since $\left(-\frac{14}{2}\right)^2$ does equal 49, it is a perfect square trinomial. It factors into $(x - 7)^2$. You can check by calculating $(x - 7)^2$. Note that $(x - 7)^2$ is NOT $x^2 - 49$. Instead, write $(x - 7)$ twice and use FOIL.

$$(x - 7)^2 = (x - 7)(x - 7) = x^2 - 7x - 7x + 49 = x^2 - 14x + 49.$$

Difference of Perfect Squares Factoring

When you multiply the binomials $(x - 5)$ and $(x + 5)$ with FOIL, you get $x^2 + 5x - 5x - 25$. When combining like terms, the $+5x$ and the $-5x$ become $0x$, which is the same thing as 0. So $(x - 5)(x + 5) = x^2 - 25$. Notice that x^2 and 25 are both perfect squares. When you have a binomial like $x^2 - 25$ where there is a perfect square subtracted from another perfect square, there is a shortcut for factoring it. The rule is that $(x^2 - b^2) = (x - b)(x + b)$.

Example 9

If possible, factor $x^2 - 16$.

Solution: Since both x^2 and 16 are perfect squares and one is being subtracted from the other, this difference of perfect squares pattern applies. Rewrite as $x^2 - 4^2$ and follow the pattern.

$$(x^2 - b^2) = (x - b)(x + b)$$
$$(x^2 - 4^2) = (x - 4)(x + 4)$$

Example 10

If possible, factor $x^2 + 49$.

Solution: Though this does resemble the $x^2 - b^2$ pattern, there is a plus sign between the two terms. The difference between perfect squares rule cannot be applied here.

Example 11

If possible, factor $3x^2 - 48$.

Solution: Though this is not a difference of perfect squares, you can first factor out a 3.

$$3x^2 - 48 = 3(x^2 - 16)$$

Now the $x^2 - 16$ can be written as $x^2 - 4^2$, which can be factored with the difference of perfect squares rule.

$$3(x^2 - 4^2) = 3(x - 4)(x + 4)$$

Test-taking Strategy for Multiple-Choice Factoring Questions

If a factoring question is a part A multiple-choice question, the correct answer can be found by multiplying all the answer choices and seeing which one is equivalent to the expression that needs to be factored.

Example 12

Which choice is a factorization of $x^2 - 3x - 180$?
   (1) $(x - 18)(x + 10)$ (3) $(x - 12)(x + 15)$
   (2) $(x + 12)(x - 15)$ (4) $(x - 20)(x + 9)$

Polynomial Arithmetic

Solution: Multiply the answer choices until one matches the question.

Testing choice 1: $(x - 18)(x + 10) = x^2 + 10x - 18x - 180 = x^2 - 8x - 180$. No.

Testing choice 2: $(x + 12)(x - 15) = x^2 - 15x + 12x - 180 = x^2 - 3x - 180$. Yes!

Check Your Understanding of Section 2.8

A. Multiple-Choice

1. Factor $3x^2 + 6x + 15$.
   (1) $3(x + 5)(x + 1)$
   (2) $3(x^2 + 6x + 15)$
   (3) $(3x + 5)(x + 3)$
   (4) $3(x^2 + 2x + 5)$

2. Factor completely $2x^3 - 8x^2 + 14x$.
   (1) $2(x^3 - 4x^2 + 7x)$
   (2) $x(2x^2 - 8x + 14)$
   (3) $2x(x^2 - 4x + 7)$
   (4) $(2x^2 + 7)(x + 2)$

3. Factor $x^2 + 10x + 16$.
   (1) $(x + 2)(x + 8)$
   (2) $(x + 4)(x + 4)$
   (3) $(x + 16)(x + 1)$
   (4) $(x + 3)(x + 7)$

4. Factor $x^2 - 2x - 15$.
   (1) $(x - 3)(x - 5)$
   (2) $(x + 3)(x - 5)$
   (3) $(x - 15)(x + 1)$
   (4) $(x + 15)(x - 1)$

5. Factor $x^2 + x - 42$.
   (1) $(x - 21)(x + 2)$
   (2) $(x - 6)(x + 7)$
   (3) $(x - 14)(x + 3)$
   (4) $(x + 6)(x - 7)$

6. Factor $x^2 + 8x + 16$.
   (1) $(x + 4)^2$
   (2) $(x + 8)(x + 2)$
   (3) $(x - 4)^2$
   (4) $(x - 4)(x + 4)$

7. Factor $x^2 - 12x + 36$.
   (1) $(x - 6)^2$
   (2) $(x + 6)^2$
   (3) $(x - 6)(x + 6)$
   (4) $(x + 12)(x + 3)$

8. Factor $x^2 - 100$.
   (1) $(x - 10)(x + 10)$
   (2) $(x - 10)^2$
   (3) $(x + 10)^2$
   (4) $(x - 20)(x + 5)$

9. Factor completely $5x^2 - 20$.
   (1) $5(x^2 - 4)$
   (2) $5(x - 2)^2$
   (3) $5(x + 2)^2$
   (4) $5(x - 2)(x + 2)$

10. Factor completely $6x^2 + 12x - 144$.
    (1) $6(x^2 + 2x - 24)$
    (2) $6(x - 4)(x + 6)$
    (3) $6(x + 4)(x - 6)$
    (4) $(3x - 4)(2x + 36)$

B. *Show how you arrived at your answers.*

1. Kenzie says that $x^2 + 25$ can be factored into $(x + 5)^2$. Ariana says that it cannot. Who is correct? Explain your answer.

2. Factor completely $x^3 + 4x^2 - 21x$.

3. Factor completely $5x^2 + 20x + 20$.

4. Korbin says that there are two different ways to factor $x^2 - 5x + 6$. One is $(x - 2)(x - 3)$ and the other is $(x - 6)(x - 1)$. Is he correct? Explain your answer.

5. The polynomial $2x^2 + 13x + 15$ can be factored into $(2x + 3)$ multiplied by another factor. What is the other factor? Explain your reasoning.

2.9 MORE COMPLICATED FACTORING

Key Ideas

Some polynomials don't, at first, appear to match any of the factoring patterns. When this happens, it is sometimes possible to rewrite it as something that does match one of the patterns. The three most frequent patterns are the factoring of a trinomial into two binomials, the difference of perfect squares, and the perfect square trinomial pattern.

Recognizing the Difference of Perfect Squares Pattern

If there are only two terms and they are separated by a minus sign, there is a chance that it will be possible to eventually use the difference of perfect squares pattern.

Example 1

Factor $9x^2 - 4$.

Solution: First, look for a common factor. In this case there is no number other than 1 that is a factor of both $9x^2$ and 4. Since there are two terms separated by a minus sign, see if each of the terms can be written as the square of other terms.
Since $9x^2 = (3x)^2$ and $4 = 2^2$,
$$9x^2 - 4 = (3x)^2 - 2^2 = (3x - 2)(3x + 2)$$

Example 2

Factor $x^4 - 9$.

Solution: x^4 can be written as $(x^2)^2$, and 9 can be written as 3^2. This is another difference of perfect squares situation.
$$x^4 - 9 = (x^2)^2 - 3^2 = (x^2 - 3)(x^2 + 3)$$

Example 3

Completely factor $x^4 - 81$.

Solution: This is like the previous question with an additional step.
$$(x^2)^2 - 9^2 = (x^2 - 9)(x^2 + 9)$$
Notice that the $x^2 - 9$ factor is, itself, a difference of perfect squares and can be factored further.
$$(x^2 - 9)(x^2 + 9) = (x - 3)(x + 3)(x^2 + 9)$$

Recognizing the Factoring a Trinomial into Two Binomials Pattern

If a polynomial has three terms where the exponent of one of the terms is half the largest exponent and one of the terms is a constant, it can sometimes be factored.

Example 4

Factor $x^4 + 4x^2 - 12$.

Solution: This can be written as $(x^2)^2 + 4(x^2) - 12$. Notice that this resembles the equation $y^2 + 4y - 12$. Just as $y^2 + 4y - 12$ would factor to $(y + 6)(y - 2)$, this polynomial can be factored into $(x^2 + 6)(x^2 - 2)$.
Multiply to check your answer.

Example 5

Factor the expression $x^4 + 8x^2 - 9$ completely.

Solution: This is a trinomial that can be rewritten as $(x^2)^2 + 8(x^2) - 9$. It has the same structure as the trinomial $y^2 + 8y - 9$, which would factor as $(y - 1)(y + 9)$. So this factors as $(x^2 - 1)(x^2 + 9)$.

Since $(x^2 - 1)$ can be written as $(x^2 - 1^2)$ the first factor can be further factored by the difference of perfect squares rule.
The answer, then, is $(x - 1)(x + 1)(x^2 + 9)$.

Recognizing the Perfect Square Trinomial Pattern

If there is a trinomial where one of the exponents is half of another exponent, and one term is a constant, it will be a perfect square trinomial when the square of half of the coefficient of the term with the smaller exponent is equal to the constant.

Example 6

Factor $x^6 + 8x^3 + 16$.

Solution: This can be written as $(x^3)^2 + 8(x^3) + 16$, which is similar in structure to $y^2 + 8y + 16$. Since $\left(\frac{8}{2}\right)^2 = 16$, this is a perfect square and can be factored into $(x^3 + 4)^2$.

Check Your Understanding of Section 2.9

A. *Multiple-Choice*

1. Factor $(2x)^2 - 25$.
   (1) $(2x - 5)(2x - 5)$
   (2) $(2x + 5)(2x + 5)$
   (3) $(2x - 5)(2x + 5)$
   (4) This cannot be factored.

2. Factor $x^4 - 49$.
   (1) $(x^2 - 7)(x^2 + 7)$
   (2) $(x^2 - 7)(x^2 - 7)$
   (3) $(x^2 + 7)(x^2 + 7)$
   (4) This cannot be factored.

3. Factor $x^4 + 7x^2 + 12$.
   (1) $(x^2 + 6)(x^2 + 2)$
   (2) $(x^2 - 3)(x^2 - 4)$
   (3) $(x^2 + 3)(x^2 + 4)$
   (4) This cannot be factored.

Polynomial Arithmetic

4. Factor $x^4 - x^2 - 20$.
   (1) $(x^2 - 5)(x^2 + 4)$
   (2) $(x^2 + 5)(x^2 + 4)$
   (3) $(x^2 - 5)(x^2 - 4)$
   (4) This cannot be factored.

5. Factor $9x^2 - 16y^2$.
   (1) $(3x - 4y)(3x - 4y)$
   (2) $(3x - 4y)(3x + 4y)$
   (3) $(3x + 4y)(3x + 4y)$
   (4) This cannot be factored.

6. Factor $a^4 + 10a^2 + 25$.
   (1) $(a^2 - 5)(a^2 + 5)$
   (2) $(a^2 - 5)^2$
   (3) $(a^2 + 5)^2$
   (4) This cannot be factored.

7. Factor $x^6 - 4$.
   (1) $(x^3 - 2)^2$
   (2) $(x^3 + 2)^2$
   (3) $(x^3 - 2)(x^3 + 2)$
   (4) This cannot be factored.

8. Factor $(3x)^2 + 8(3x) + 12$.
   (1) $(3x + 4)(3x + 3)$
   (2) $(3x + 6)(3x + 2)$
   (3) $(3x + 1)(3x + 12)$
   (4) This cannot be factored.

9. Factor $x^4 - 10x^2 + 16$.
   (1) $(x^2 - 8)(x^2 - 2)$
   (2) $(x^2 - 8)(x^2 + 2)$
   (3) $(x^2 + 8)(x^2 + 2)$
   (4) This cannot be factored.

10. Factor $(2x + 1)^2 - 9$.
    (1) $(2x + 1 + 3)^2$
    (2) $(2x + 1 - 3)(2x + 1 + 3)$
    (3) $(2x + 1 - 3)^2$
    (4) This cannot be factored.

B. *Show how you arrived at your answers.*

1. Factor completely $x^4 - 81$.

2. Factor completely $x^4 + 4x^2 - 5$.

3. Factor completely $x^4 - 13x^2 + 36$.

4. Factor $(3x - 5)^2 + 2(3x - 5) - 3$.

5. The polynomial $a^3 - b^3$ can factor into $(a - b)(a^2 + ab + b^2)$. Use this factoring pattern to factor the polynomial $x^6 - 8$.

Chapter Three

QUADRATIC EQUATIONS

3.1 SOLVING QUADRATIC EQUATIONS BY TAKING THE SQUARE ROOT OF BOTH SIDES OF THE EQUATION

=== KEY IDEAS ===

An equation that has a variable raised to the second power is called a quadratic equation. Examples of quadratic equations are $x^2 = 9$, $x^2 + 5x = 24$, and $x^2 + 5x + 6 = 0$. Quadratic equations require more advanced techniques to solve. Most quadratic equations have two solutions. The simplest quadratic equations to solve are the ones where there is no x term, like $x^2 = 25$ or $x^2 + 2 = 38$.

One-Step Quadratic Equations

If a quadratic equation is in the form $x^2 = c$, the exponent can be eliminated by taking the square root of both sides. There are two answers, generally, since positive times positive equals a positive, whereas negative times negative also equals a positive.

Example 1

Solve for all values of x that satisfy the equation $x^2 = 25$.

Solution: By looking at the equation and thinking through the multiplication tables, you may notice that 5 is one of the solutions since $5^2 = 5 \cdot 5 = 25$. Less obvious is the other answer, -5, since $-5 \cdot -5$ also equals positive 25.

This problem can also be solved with algebra. Using algebra, you could take the square root of both sides of the equation.

$$x^2 = 25$$
$$\sqrt{x^2} = \pm\sqrt{25}$$
$$x = \pm 5$$
$$x = 5 \text{ or } x = -5$$

Quadratic Equations

Since +5 or −5 can be squared to become +25, when you take the square root of both sides, put the ± symbol in front of the square root sign on the right side of the equation. You can leave out the second line of this solution and still get full credit.

MATH FACTS

Just as $\sqrt{3^2}$ is 3 and $\sqrt{5^2}$ is 5 (Quick! What is $\sqrt{173^2}$?), when x is a positive number $\sqrt{x^2}$ is x. This is the property that allows you to do the algebraic process of eliminating the exponent by taking the square root of both sides of the equation.

Example 2

Solve for all values of x that satisfy the equation $x^2 = 26$.

Solution:

$$x^2 = 26$$
$$\sqrt{x^2} = \pm\sqrt{26}$$
$$x = \pm\sqrt{26}$$
$$x = \sqrt{26} \text{ or } x = -\sqrt{26}$$

Since 26 is not a perfect square, the answer should be left in this form unless the question says to round the answer. In that case, you would use your calculator to find $\sqrt{26} \approx 5.09901951$ and round appropriately.

Two-Step Quadratic Equations with Constants

More complicated quadratic equations sometimes require the addition property of equality or the subtraction property of equality in order to solve.

Example 3

Solve for all values of x that satisfy the equation $x^2 + 5 = 54$.

Solution: First isolate the radical by subtracting 5 from both sides of the equation. After that, the equation will be just like the one step quadratic equations.

$$x^2 + 5 = 54$$
$$-5 = -5$$
$$x^2 = 49$$
$$x = \pm 7$$

Example 4

Solve for all values of x that satisfy the equation $x^2 + 9 = 36$.

Solution: It is tempting to try to take the square root of both sides and get $x + 3 = 6$, but it is not true that $\sqrt{x^2 + 9} = x + 3$. Instead, solve the same way as in the previous example.

$$x^2 + 9 = 36$$
$$-9 = -9$$
$$x^2 = 27$$
$$x = \pm\sqrt{27}$$

Example 5

Solve for all values of x that satisfy the equation $(x + 1)^2 = 25$.

Solution: Do not multiply out the left-hand side. Instead, look at how similar this question is to the question $x^2 = 25$, which had two solutions $+5$ and -5. Begin by taking the square root of both sides of the equation.

$$(x + 1)^2 = 25$$
$$\sqrt{(x+1)^2} = \pm\sqrt{25}$$
$$x + 1 = \pm 5$$

Complete the question by subtracting 1 from each side.

$$x + 1 = \pm 5$$
$$-1 = -1$$
$$x = \pm 5 - 1$$
$$x = +5 - 1 = +4 \text{ or } x = -5 - 1 = -6$$

You can check that these two answers are correct by substituting them back into the original equation.

$$(4 + 1)^2 \stackrel{?}{=} 25 \qquad (-6 + 1)^2 \stackrel{?}{=} 25$$
$$5^2 \stackrel{?}{=} 25 \qquad (-5)^2 \stackrel{?}{=} 25$$
$$25 \stackrel{\checkmark}{=} 25 \qquad 25 \stackrel{\checkmark}{=} 25$$

Quadratic Equations

Example 6

Solve for all values of x that satisfy the equation $(x + 3)^2 = 17$.

Solution:
$$(x + 3)^2 = 17$$
$$\sqrt{(x+3)^2} = \pm\sqrt{17}$$
$$x + 3 = \pm\sqrt{17}$$
$$-3 = -3$$
$$x = -3 \pm \sqrt{17}$$
$$x = -3 + \sqrt{17} \text{ or } x = -3 - \sqrt{17}$$

MATH FACTS

When combining something like -3 with $\sqrt{17}$, it is customary to put the $\sqrt{17}$ on the right as $-3 + \sqrt{17}$ rather than as $\sqrt{17} - 3$ which might get confused with $\sqrt{17-3} = \sqrt{14}$.

Example 7

Solve for all values of x that satisfy the equation $x^2 + 6x + 9 = 17$.

Solution: There are some more advanced techniques for a problem like this, but if you notice that $x^2 + 6x + 9$ is a perfect square trinomial and can be factored into $(x + 3)^2$, the equation becomes $(x + 3)^2 = 17$, which is the same as in Example 6.

Two-Step Quadratic Equations with Coefficients

To solve the equation $2x^2 = 50$ with a coefficient of 2 in front of the x^2 term, first divide both sides of the equation by 2.

$$2x^2 = 50$$
$$\frac{2x^2}{2} = \frac{50}{2}$$
$$x^2 = 25$$
$$x = \pm\sqrt{25} = \pm 5$$

Two-Step Quadratic Equations with Multiple Variables

If an equation has more than one variable and the goal is to isolate one of the variables, follow the same steps you would if the other variables were numbers.

Example 8

Solve for x in terms of a and b in the equation $ax^2 = b$.

Solution:

$$ax^2 = b$$

$$\frac{ax^2}{a} = \frac{b}{a}$$

$$x^2 = \frac{b}{a}$$

$$x = \pm\sqrt{\frac{b}{a}}$$

Example 9

The formula for the volume of a cylinder is $V = \pi r^2 h$. Solve this equation for r in terms of V, h, and π.

Solution: First eliminate the π and the h from the right-hand side.

$$\frac{V}{\pi h} = \frac{\pi r^2 h}{\pi h}$$

$$\frac{V}{\pi h} = r^2$$

Eliminate the exponent by taking the square root of both sides of the equation:

$$\frac{V}{\pi h} = \sqrt{r^2}$$

$$\sqrt{\frac{V}{\pi h}} = r$$

Since the r represents the radius of the cylinder, the negative answer can be disregarded.

Quadratic Equations

Solving Square Root Equations

Related to quadratic equations are equations like $\sqrt{x} = 3$ and $\sqrt{x-2} = 5$. To eliminate a square root sign, raise each side of the equation to the second power.

$$\sqrt{x} = 3$$
$$(\sqrt{x})^2 = 3^2$$
$$x = 9$$

$$\sqrt{x-2} = 5$$
$$(\sqrt{x-2})^2 = 5$$
$$x - 2 = 25$$
$$+2 = +2$$
$$x = 27$$

If the square root sign is not already isolated, first isolate it with algebra.

$$\sqrt{x} + 3 = 8$$
$$-3 = -3$$
$$\sqrt{x} = 5$$
$$(\sqrt{x})^2 = 5^2$$
$$x = 25$$

Check Your Understanding of Section 3.1

A. Multiple-Choice

1. Find all solutions to $x^2 = 36$.
   (1) 6   (2) –6   (3) 36, –36   (4) 6, –6

2. Find all solutions to $x^2 = 37$.
   (1) $\sqrt{37}, -\sqrt{37}$
   (2) $\sqrt{37}$
   (3) $-\sqrt{37}$
   (4) There are no solutions since 37 is not a perfect square.

3. Find all solutions to $x^2 - 9 = 40$.
   (1) 7
   (2) –7
   (3) 7, –7
   (4) $3 + \sqrt{40}$, $3 - \sqrt{40}$

4. Find all solutions to $(x + 2)^2 = 64$.
   (1) $\sqrt{62}$, $-\sqrt{62}$
   (2) 6, –10
   (3) 6
   (4) –10

5. Find all solutions to $(x - 3)^2 = 13$.
   (1) $3 + \sqrt{13}$
   (2) $3 - \sqrt{13}$
   (3) $3 + \sqrt{13}$, $3 - \sqrt{13}$
   (4) $\sqrt{10}$, $-\sqrt{10}$

6. Find all solutions to $(x + 2)^2 = 17$.
   (1) $-2 + \sqrt{17}$
   (2) $-2 - \sqrt{17}$
   (3) $\sqrt{15}$, $-\sqrt{15}$
   (4) $-2 + \sqrt{17}$, $-2 - \sqrt{17}$

7. Find all solutions to $3x^2 = 108$.
   (1) $\sqrt{\dfrac{108}{3}}$, $-\sqrt{\dfrac{108}{3}}$
   (2) 6
   (3) –6
   (4) 6, –6

8. Find all solutions to $4x^2 + 3 = 103$.
   (1) 5   (2) –5   (3) 5, –5   (4) 10, –10

9. Solve for x in terms of c and d.
   $cx^2 = d$.
   (1) $\dfrac{d}{c}$
   (2) $\sqrt{\dfrac{c}{d}}$, $-\sqrt{\dfrac{c}{d}}$
   (3) $\sqrt{\dfrac{d}{c}}$, $-\sqrt{\dfrac{d}{c}}$
   (4) $\dfrac{\sqrt{d}}{c}$, $-\dfrac{\sqrt{d}}{c}$

10. Solve for x in terms of g and h.
    $(x + g)^2 = h$
    (1) $-g + \sqrt{h}$, $-g - \sqrt{h}$
    (2) $-g + \sqrt{h}$
    (3) $-g - \sqrt{h}$
    (4) $\sqrt{h-g}$, $-\sqrt{h-g}$

Quadratic Equations

B. *Show how you arrived at your answers.*

1. Find all solutions to the equation $x^2 = 15$, rounded to the nearest hundredth.

2. The left-hand side of the equation $x^2 + 10x + 25 = 64$ can be factored into $(x + 5)^2$. Show how you can find the two solutions to this equation by first factoring the left-hand side of the equation.

3. If the area of the largest square in this diagram (composed of the two smaller squares and the two rectangles) is 49 square units, what is the length of segment x?

4. Diana solves the equation $(x + 4)^2 = 97$ as follows:

$$(x + 4)^2 = 97$$
$$x^2 + 16 = 97$$
$$-16 = -16$$
$$x^2 = 81$$
$$x = 9, -9$$

What is wrong with this reasoning?

5. Find the solution to the equation $(x + 1)^3 = 8$.

3.2 SOLVING QUADRATIC EQUATIONS BY GUESS AND CHECK

KEY IDEAS

There are several ways to solve a quadratic equation like $x^2 + 10x = 39$, which has the $10x$ term involved. For some examples, when the answer is an integer, it is possible to find the answer through guess and check.

Guess = 1	Guess = 2	Guess = 3
$1^2 + 10 \cdot 1 \stackrel{?}{=} 39$	$2^2 + 10 \cdot 2 \stackrel{?}{=} 39$	$3^2 + 10 \cdot 3 \stackrel{?}{=} 39$
$1 + 10 \stackrel{?}{=} 39$	$4 + 20 \stackrel{?}{=} 39$	$9 + 30 \stackrel{?}{=} 39$
$11 \neq 39$	$24 \neq 39$	$39 \stackrel{\checkmark}{=} 39$

Guess and check is useful if the quadratic equation is a multiple-choice question so there are only four things to check.

Example 1

Which of the four choices is a solution to the equation $x^2 + 5x - 3 = 33$?
(1) 2   (2) 3   (3) 4   (4) 5

Solution: Since $4^2 + 5 \cdot 4 - 3 = 16 + 20 - 3 = 33$, 4 is a solution to the equation, which is choice (3).

MATH FACTS

The solutions to a quadratic equation are sometimes called the *roots* or the *zeros* of the equation.

Example 2

Which of the four choices is a root of the equation $x^2 - 6x + 4 = 0$?
(1) $3 + \sqrt{2}$ (2) $3 + \sqrt{3}$ (3) $3 + \sqrt{5}$ (4) $3 + \sqrt{6}$

Solution: To test the four answer choices, there is a useful feature on the graphing calculator.

Quadratic Equations

For the TI–84:

Store the value from choice (1) into the x variable by pressing [3], [+], [2ND], [x^2], [2], [ )], [STO>], [x]. The [STO>] key is right above the [ON] key and the [x] key is next to the [ALPHA] key in the second row.

```
3+√(2)→X
         4.414213562
```

Then check the answer by entering [x], [x^2], [–], [6], [x], [+], [4], and [ENTER].

```
3+√(2)→X
         4.414213562
X^2-6X+4
                  -3
```

Since it did not evaluate to 0, choice (1) is not the answer.
When using this method to check choice (3), it becomes

```
3+√(5)→X
         5.236067977
X^2-6X+4
                   0
```

For the TI-Nspire:

Store the value from choice (1) into the x variable by pressing [3], [+], [ctrl], [x^2], [2], [right arrow], [ctrl], [var], [x].

```
         Scratchpad
3+√2 →x              4.41421

                        1/99
```

64

Check if this solves the quadratic by entering [x], [x²], [−], [6], [x], [+], [4], and [enter].

When checking choice (3), it will look like this:

Either calculator will confirm that the answer is choice (3).

Check Your Understanding of Section 3.2

A. Multiple-Choice

For each question, use your calculator to check which answer satisfies the quadratic equation.

1. $x^2 - 8x + 15 = 0$
   (1) 3   (2) 4   (3) 6   (4) 7

2. $x^2 - 2x - 8 = 0$
   (1) 1   (2) 2   (3) 3   (4) 4

3. $x^2 + 5x - 6 = 0$
   (1) 6   (2) −6   (3) −4   (4) −2

4. $2x^2 + 5x - 3 = 0$
   (1) $\frac{1}{4}$ (2) $\frac{1}{3}$ (3) $\frac{1}{2}$ (4) 1

Quadratic Equations

5. $x^2 + 7x = 30$
   (1) 1 (2) 3 (3) 5 (4) 7

6. $x^2 - 2x - 2 = 0$
   (1) $1 + \sqrt{6}$
   (2) $1 + \sqrt{5}$
   (3) $1 + \sqrt{4}$
   (4) $1 + \sqrt{3}$

7. $x^2 + 5x = 6$
   (1) –7 (2) –6 (3) –5 (4) –4

8. $x^2 + 10x + 13 = 0$
   (1) $-5 + 2\sqrt{3}$
   (2) $-4 + 2\sqrt{3}$
   (3) $-3 + 2\sqrt{3}$
   (4) $-2 + 2\sqrt{3}$

9. $x^2 + 4x = 25$
   (1) $-1 - \sqrt{29}$
   (2) $-2 - \sqrt{29}$
   (3) $-3 - \sqrt{29}$
   (4) $-4 - \sqrt{29}$

10. $x^2 - 12x - 14 = 0$
    (1) $6 + 2\sqrt{2}$
    (2) $6 + 3\sqrt{2}$
    (3) $6 + 4\sqrt{2}$
    (4) $6 + 5\sqrt{2}$

3.3 SOLVING QUADRATIC EQUATIONS BY COMPLETING THE SQUARE

=== KEY IDEAS ===

Completing the square is a technique for solving quadratic equations that relies on the concept of a perfect square trinomial. It is not the quickest method to solve certain quadratic equations, but it can be used for all quadratic equations.

A perfect square trinomial, like $x^2 + 6x + 9$ is one where the constant term, the 9 in this case, is equal to the square of half the coefficient of the x, in this case the 6. Since $\left(\frac{6}{2}\right)^2$ does equal 9, this is a perfect square trinomial and is equal to $\left(x + \frac{6}{2}\right)^2 = (x+3)^2$.

If the left-hand side of a quadratic equation is already a perfect square, you can factor it and solve using the technique in Section 3.1, example 7.

If the left-hand side is not a perfect square trinomial, it is possible to make it into one by adding to both sides of the equation. For the equation $x^2 + 10x = 39$, you look at the left-hand side and ask "what constant would make the left-hand side of the equation into a perfect square trinomial." Since $\left(\frac{10}{2}\right)^2$ is 25, $x^2 + 10x + 25$ would be a perfect square trinomial.

Add 25 to both sides and complete the question:

$$x^2 + 10x = 39$$
$$+25 = +25$$
$$x^2 + 10x + 25 = 64$$
$$(x + 5)^2 = 64$$
$$x + 5 = \pm 8$$
$$-5 = -5$$
$$x = -5 \pm 8$$
$$x = -5 + 8 = 3 \text{ or } x = -5 - 8 = -13$$

You can check these answers by first substituting 3 for both xs in the equation and then by substituting -13 for both xs in the equation and making sure that they evaluate to 39.

Example 1

What are the roots of the equation $x^2 + 12x = 28$.

Solution:

$$x^2 + 12x = 28$$
$$x^2 + 12x + 36 = 28 + 36$$
$$(x + 6)^2 = 64$$
$$x + 6 = \pm 8$$
$$x = -6 + 8 = 2 \text{ or } x = -6 - 8 = -14$$

Example 2

What are the roots of the equation $x^2 + 4x - 21 = 0$.

Solution: Completing the square is simplest when there is no constant term on the left-hand side of the equation. Eliminate the -21 by adding 21 to both sides. Then continue as in the previous example.

$$x^2 + 4x - 21 = 0$$
$$+21 = +21$$
$$x^2 + 4x = 21$$
$$x^2 + 4x + 4 = 21 + 4$$
$$(x + 2)^2 = 25$$
$$x + 2 = \pm 5$$
$$x = -2 + 5 = 3 \text{ or } x = -2 - 5 = -7$$

Quadratic Equations

Example 3

What are the roots of the equation $x^2 + 6x - 3 = 0$?

Solution: At the end of the completing the square process, something unusual happens.

$$x^2 + 6x - 3 = 0$$
$$+3 = +3$$
$$x^2 + 6x = 3$$
$$x^2 + 6x + 9 = 3 + 9$$
$$(x + 3)^2 = 12$$
$$x + 3 = \pm\sqrt{12}$$
$$x = -3 \pm \sqrt{12}$$

Since 12 is not a perfect square, it does not become a whole number as it did in previous examples. There is something further you can do, though. By factoring 12 into $4 \cdot 3$, $\sqrt{12}$ can be simplified to $\sqrt{4 \cdot 3} = \sqrt{4} \cdot \sqrt{3} = 2\sqrt{3}$. If the question asks for *simplest radical form*, the solution would be written as $x = -3 \pm 2\sqrt{3}$.

Example 4

Which equation has the same roots as $x^2 - 8x - 19 = 0$.
(1) $(x + 4)^2 = 35$ (3) $(x + 4)^2 = 3$
(2) $(x - 4)^2 = 35$ (4) $(x - 4)^2 = 3$

Solution: Notice that the answer choices all look like one of the steps of the completing the square process. Begin the question as if you are solving by completing the square, but you can stop when the equation resembles the structure of the answer choices.

$$x^2 - 8x - 19 = 0$$
$$+19 = +19$$
$$x^2 - 8x = 19$$
$$x^2 - 8x + 16 = 19 + 16$$
$$(x - 4)^2 = 35$$

You do not have to go any further. The answer is choice (2).

The Nature of the Roots

The quadratic equation $x^2 = 5$ has two solutions $x = \sqrt{5}$ and $x = -\sqrt{5}$. Since 5 is not a perfect square, these two solutions are known as *irrational* numbers. The equation $x^2 = 5$ has two irrational solutions.

Since 4 is a perfect square, the quadratic equation $x^2 = 4$ has two solutions, $x = 2$ and $x = -2$. These are known as *rational* numbers. A rational number is any number that can be written as a fraction with integers in both the numerator and the denominator. Numbers like $\frac{2}{5}$, $\frac{-7}{3}$, and $\frac{8}{1}$ are all rational. Integers like 2 and –2 are rational because they can be expressed as $\frac{2}{1}$ and $\frac{-2}{1}$. The equation $x^2 = 4$ has two rational solutions.

The quadratic equation $x^2 - 4x - 1 = 0$ has solutions $x = 2 + \sqrt{5}$ and $x = 2 - \sqrt{5}$. When a rational number is added to an irrational number, the result is an irrational number. The equation $x^2 - 4x - 1 = 0$ has two irrational solutions.

MATH FACTS

When a rational number is added to or subtracted from a rational number, the result is always a rational number. When a rational number is added to or subtracted from an irrational number, the result is always an irrational number. When an irrational number is added to or subtracted from another irrational number, the result is often irrational, but not always. For example, $3 - \sqrt{5}$ is an irrational number, but $\sqrt{5} + (3 - \sqrt{5}) = 3$, which is rational. The product of a rational and an irrational number is always irrational, but the product of two irrationals can be either irrational or rational. For example, $\sqrt{3} \cdot \sqrt{2} = \sqrt{6}$ which is irrational, but $\sqrt{5} \cdot \sqrt{5} = 5$ is rational.

Example 5

Which of the following expressions is a rational number?

(1) $\sqrt{7}$ (3) $\sqrt{9} + \sqrt{4}$

(2) $\sqrt{9} + \sqrt{7}$ (4) $2\sqrt{5}$

Solution: The answer is (3) because $\sqrt{9} + \sqrt{4} = 3 + 2 = 5$

Check Your Understanding of Section 3.3

A. Multiple-Choice

1. $x^2 + 12x + c$ is a perfect square trinomial. What is the value of c?
   (1) 36   (2) 18   (3) 12   (4) 6

Quadratic Equations

2. $x^2 + bx + 49$ is a perfect square trinomial. What is one possible value of b?
   (1) 7   (2) 14   (3) 21   (4) 49

3. Use completing the square to find both solutions for x in the equation $x^2 + 8x + 16 = 9$.
   (1) −1, −7   (2) −1, −2   (3) −2, −3   (4) −3, −4

4. Use completing the square to find both solutions for x in the equation $x^2 + 10x = 24$.
   (1) 2, −10   (2) 4, −12   (3) 3, −8   (4) 2, −12

5. Use completing the square to find both solutions for x in the equation $x^2 − 18x = 40$.
   (1) −2, 10   (2) −2, 20   (3) −4, 20   (4) −4, 10

6. What is one solution to the equation $x^2 − 6x = 8$?
   (1) $3 + \sqrt{9}$
   (2) $3 + \sqrt{13}$
   (3) $3 + \sqrt{17}$
   (4) $3 + \sqrt{21}$

7. The quadratic equation $x^2 + 2x − 24 = 0$ has the same solutions as which of the following equations?
   (1) $(x + 1)^2 = 24$
   (2) $(x + 1)^2 = 25$
   (3) $(x − 1)^2 = 24$
   (4) $(x − 1)^2 = 25$

8. Find the solutions using completing the square $x^2 + 4x − 5 = 0$.
   (1) 1, −5   (2) 1, −6   (3) 2, −5   (4) 2, −6

9. Solve by completing the square $x^2 − 12x + 32 = 0$.
   (1) 4, 8   (2) 4, 6   (3) 3, 8   (4) 3, 6

10. Which of the following is not a rational number?
    (1) $\frac{7}{5}$
    (2) $\sqrt{7} - \sqrt{7}$
    (3) $\sqrt{7} \cdot \sqrt{7}$
    (4) $7 + \sqrt{7}$

B. Show how you arrived at your answers.

1. What are all possible integer solutions for b and p in the equation $x^2 + bx + 64 = (x + p)^2$?

2. Use the completing the square method to find the two solutions to the equation $x^2 − 16x − 7 = 0$.

3. Ramon says that π is irrational and that $\pi = \frac{22}{7}$; therefore, $\frac{22}{7}$ is irrational also. What is wrong with his reasoning process? Explain.

4. Use the completing the square method to find the two solutions to the equation $x^2 - 5x = 14$.

5. Use the completing the square method to find the two solutions to the equation $x^2 + 2ax = b$ in terms of a and b.

3.4 SOLVING QUADRATIC EQUATIONS BY FACTORING

=== KEY IDEAS ===

When the quadratic equation contains a trinomial that can be factored, the equation can be solved very quickly. This requires first getting the equation in a form with all the terms on one side of the equation and a zero on the other side of the equation.

Solving a Quadratic Equation That Is Already in Factored Form

The only way for the product of two numbers to be zero is if one, or both, is the number zero. In the equation $(x - 2)(x - 3) = 0$, there are two expressions $(x - 2)$ and $(x - 3)$ that have a product of zero. This means that you need to find x values that make either of these expressions equal to zero.

$$(x - 2)(x - 3) = 0$$

This equation will be satisfied if either $(x - 2)$ or $(x - 3)$ is equal to 0.
This becomes two separate questions

$$x - 2 = 0 \quad \text{or} \quad x - 3 = 0$$
$$x - 2 + 2 = 0 + 2 \quad \quad x - 3 + 3 = 0 + 3$$
$$x = +2 \quad \text{or} \quad x = +3$$

Either of these two values, 2 or 3, would make the expression $(x - 2)(x - 3)$ equal to 0.

$$(2 - 2)(2 - 3) = 0 \cdot -1 = 0$$
$$(3 - 2)(3 - 3) = 1 \cdot 0 = 0$$

Quadratic Equations

Example 1

Solve for both values of x that satisfy the equation $(x - 4)(x - 5) = 0$.

Solution: This is true for x values that make either $x - 4$ or $x - 5$ equal to 0.

$$\begin{array}{lcl} x - 4 = 0 & \text{or} & x - 5 = 0 \\ x - 4 + 4 = 0 + 4 & & x - 5 + 5 = 0 + 5 \\ x = +4 & \text{or} & x = +5 \end{array}$$

So the solutions are 4 and 5.

Example 2

Determine the two roots of the equation $(x - 2)(x + 6) = 0$.

Solution:

$$\begin{array}{lcl} x - 2 = 0 & \text{or} & x + 6 = 0 \\ x - 2 + 2 = 0 + 2 & & x + 6 - 6 = 0 - 6 \\ x = +2 & \text{or} & x = -6 \end{array}$$

The solutions are 2 and –6.

Solving Equations by First Factoring the Quadratic Trinomial

One of the most common types of quadratic equations in the course is when you have to first factor and then solve the equation. This technique is likely to be applied several times on each test.

Example 3

Determine the roots of the equation $x^2 + 8x + 15 = 0$.

Solution: Though this question can be solved by completing the square, it is much quicker if you can first factor the left-hand side and use the techniques from this section to complete the question. In Section 2.8, the trinomial $x^2 + 8x + 15$ was factored into $(x + 3)(x + 5)$.

$$x^2 + 8x + 15 = 0$$
$$(x + 3)(x + 5) = 0$$

$$\begin{array}{lcl} x + 3 = 0 & \text{or} & x + 5 = 0 \\ x + 3 - 3 = 0 - 3 & & x + 5 - 5 = 0 - 5 \\ x = -3 & \text{or} & x = -5 \end{array}$$

Example 4

Determine the roots of the equation $x^2 + 3x - 10 = 0$.

Solution:

$$x^2 + 3x - 10 = 0$$
$$(x + 5)(x - 2) = 0$$

$x + 5 = 0$	or	$x - 2 = 0$
$x + 5 - 5 = 0 - 5$		$x - 2 + 2 = 0 + 2$
$x = -5$	or	$x = 2$

Example 5

Determine the roots of the equation $x^2 - x - 6 = 14$.

Solution: Do not factor the left-hand side of the equation until you have arranged the equation so that a zero is on the right-hand side.

$$x^2 - x - 6 = 14$$
$$-14 = -14$$
$$x^2 - x - 20 = 0$$
$$x^2 - 1x - 20 = 0$$
$$(x - 5)(x + 4) = 0$$

$x - 5 = 0$	or	$x + 4 = 0$
$x - 5 + 5 = 0 + 5$		$x + 4 - 4 = 0 - 4$
$x = 5$	or	$x = -4$

Example 6

Find the values of x that satisfy the equation $x^2 - 7x = 0$.

Solution: Since this polynomial does not have a constant term, it can be factored with the greatest common factor method.

$$x(x - 7) = 0$$

| $x = 0$ | or | $x - 7 = 0$ |
| $x = 0$ | or | $x = 7$ |

Example 7

Find the values of x that satisfy the equation $2x^2 + 12x + 16 = 0$.

Quadratic Equations

Solution: Start by factoring out the common factor of 2.

$$2(x^2 + 6x + 8) = 0$$
$$2(x + 4)(x + 2) = 0$$

The only way this expression can be 0 is if either $x + 4 = 0$ or $x + 2 = 0$.

| $x + 4 = 0$ | or | $x + 2 = 0$ |
| $x = -4$ | or | $x = -2$ |

Example 8

Find all values of x that are solutions to the equation $x^2 - 25 = 0$.

Solution: One way to do this is to factor the left-hand side.

$$x^2 - 25 = 0$$
$$(x - 5)(x + 5) = 0$$

| $x - 5 = 0$ | or | $x + 5 = 0$ |
| $x = 5$ | or | $x = -5$ |

Another way is to add 25 to both sides of the equation.

$$x^2 - 25 = 0$$
$$+25 = +25$$
$$x^2 = 25$$
$$x = \pm\sqrt{25}$$
$$x = \pm 5$$

Check Your Understanding of Section 3.4

A. Multiple-Choice

For each equation, find all values of x that satisfy it.

1. $(x - 2)(x - 3) = 0$
   (1) 2, 3   (2) –2, –3   (3) –2, 3   (4) 2, –3

2. $(x - 3)(x + 4) = 0$
   (1) 3, –4   (2) –3, 4   (3) –3, –4   (4) 3, 4

3. $x(x - 5) = 0$
   (1) 5   (2) 0   (3) –5, 0   (4) 5, 0

4. $x^2 + 10x + 24 = 0$
   (1) 4, –6   (2) –4, –6   (3) –4, 6   (4) 4, 6

5. $x^2 + 2x = 15$
   (1) –3, 5   (2) 3, –5   (3) 3, 5   (4) –3, –5

6. $x^2 + 6x = 0$
   (1) 0   (2) –6   (3) 0, –6   (4) 0, 6

7. $x^2 - 6x + 9 = 0$
   (1) 3   (2) 3, –3   (3) –3   (4) 3, 0

8. $2x^2 - 16x + 14 = 0$
   (1) 1   (2) 7   (3) 1, 7   (4) –1, –7

9. $3x^2 + 15x - 108 = 0$
   (1) –4, 9   (2) 4, –9   (3) 4, 9   (4) –4, –9

10. $x^2 - 9 = 0$
    (1) 3   (2) –3   (3) 3, –3   (4) 3, 0

B. Show how you arrived at your answers.

1. Faith tries to solve the equation $x^2 - 3x = 0$ by first dividing both sides by x to get $x - 3 = 0$. She concludes that the only solution is $x = 3$. Is she correct? Explain.

2. The cubic polynomial $x^3 - 6x^2 + 11x - 6$ can be factored into $(x - 1)(x - 2)(x - 3)$. How can you use this fact to find all solutions to the equation $x^3 - 6x^2 + 11x - 6 = 0$?

3. If $ab = 0$ and it is known that $a \neq 0$, what conclusion can you make about b? Explain.

4. The equation $x^4 - 13x^2 + 36 = 0$ has four solutions. What are they?

5. Stephanie tries to solve the equation $x^2 + 6x = 40$ by first factoring the left-hand side into $x(x + 6) = 40$ and then concludes that either $x = 40$ or $x + 6 = 40$ giving two solutions 40 and 34. What is wrong with this reasoning?

Quadratic Equations

3.5 THE RELATIONSHIP BETWEEN FACTORS AND ROOTS

=== KEY IDEAS ===

When a quadratic trinomial $x^2 + bx + c = 0$ factors into $(x + r)(x + s) = 0$, the roots of the polynomial $x^2 + bx + c$ are $x = -r$ and $x = -s$. These are also sometimes called the roots of the equation $x^2 + bx + c = 0$. The roots, $-r$ and $-s$, are very closely related to the factors of the quadratic expression $(x + r)(x + s)$. It is possible to very quickly translate between the factors and the roots.

The quadratic trinomial $x^2 + 8x + 15$ has *factors* of $(x + 3)$ and $(x + 5)$. The quadratic equation $x^2 + 8x + 15 = 0$ has two solutions $x = -3$ and $x = -5$. These two solutions are called the *roots* of the polynomial $x^2 + 8x + 15$. In general, the two roots are the opposites of the constants in the two factors.

In the equation $x^2 + 3x - 10 = 0$, the factors of the trinomial on the left-hand side of the equation are $(x + 5)$ and $(x - 2)$, and the roots of the polynomial are $x = -5$ and $x = 2$.

Example 1

What are the roots of the equation $(x + 6)(x - 4) = 0$?

Solution: The first root is the opposite of the constant of the first factor. That constant is $+6$ so the root is $x = -6$. The second root is the opposite of the constant of the second factor. That constant is -4 so the root is $x = -(-4) = +4$.

Example 2

If the roots of a quadratic equation are $+2$ and $+7$, what could the equation be?

Solution: Since the roots are the opposites of the constants of the factors, the constants of the factors must be $(x - 2)$ and $(x - 7)$. So the equation is $(x - 2)(x - 7) = 0$.

Example 3

If the roots of a quadratic equation are –5 and 3, what could the equation be?

Solution: Since the opposite of –5 is +5 and the opposite of +3 is –3, the factors are $(x + 5)$ and $(x - 3)$ so the equation could be $(x + 5)(x - 3) = 0$.

Example 4

If the roots of a quadratic equation are +2 and +8, what could the equation be?

    (1) $x^2 + 10x + 16 = 0$ (3) $x^2 + 10x - 16 = 0$
    (2) $x^2 - 10x + 16 = 0$ (4) $x^2 - 10x - 16 = 0$

Solution: Since the roots are +2 and +8, the factors are $(x - 2)$ and $(x - 8)$ and the equation is $(x - 2)(x - 8) = 0$. Use FOIL to multiply the left-hand side of the equation to make it look more like the answer choices and get $(x - 2)(x - 8) = x^2 - 8x - 2x + 16 = x^2 - 10x + 16$, which is choice (2).

Check Your Understanding of Section 3.5

A. *Multiple-Choice*

1. What are the roots of the equation $(x - 2)(x + 5) = 0$?
   (1) 2 and 5   (3) –2 and –5
   (2) 2 and –5   (4) –2 and 5

2. What are the roots of the polynomial $(x + 4)(x - 7)$?
   (1) 4 and 7   (3) –4 and 7
   (2) 4 and –7   (4) –4 and –7

3. If the roots of an equation are 3 and –6, what could the equation be?
   (1) $(x - 3)(x - 6) = 0$ (3) $(x + 3)(x - 6) = 0$
   (2) $(x - 3)(x + 6) = 0$ (4) $(x + 3)(x + 6) = 0$

4. If the roots of a polynomial are 4 and –2, what could the polynomial be?
   (1) $(x - 4)(x - 2)$ (3) $(x - 4)(x + 2)$
   (2) $(x + 4)(x + 2)$ (4) $(x + 4)(x - 2)$

Quadratic Equations

5. If the factors of a polynomial are $(x - 5)$ and $(x - 2)$, what are the roots of that polynomial?
   (1) 5 and 2
   (2) 5 and -2
   (3) -5 and 2
   (4) -5 and -2

6. If the roots of a polynomial are 1 and -8, what could be the factors?
   (1) $(x - 1)$ and $(x + 8)$
   (2) $(x - 1)$ and $(x - 8)$
   (3) $(x + 1)$ and $(x + 8)$
   (4) $(x + 1)$ and $(x - 8)$

7. If a polynomial has factors of $(x - p)$ and $(x + q)$, what are the roots of the polynomial?
   (1) p and q
   (2) $-p$ and q
   (3) $-p$ and $-q$
   (4) p and $-q$

8. If a cubic polynomial has factors of $(x + 2)$, $(x - 3)$, and $(x - 7)$, what are the roots of the polynomial?
   (1) 2, -3, and -7
   (2) -2, 3, and 7
   (3) 2, 3, and 7
   (4) -2, -3, and -7

9. If a cubic polynomial has roots of 5, 3, and -1, what could the factors of the polynomial be?
   (1) $(x + 5)$, $(x + 3)$, and $(x - 1)$
   (2) $(x + 5)$, $(x + 3)$, and $(x + 1)$
   (3) $(x - 5)$, $(x - 3)$, and $(x + 1)$
   (4) $(x - 5)$, $(x - 3)$, and $(x - 1)$

10. Which is a root of the equation $x^3 - 6x^2 + 13x - 120 = 0$?
    (1) 1   (2) 3   (3) 5   (4) 7

B. *Show how you arrived at your answers.*

1. A cubic equation has three roots, 2, -2, and 4. What could the equation be?

2. If an equation has two roots, -4 and -7, what could the equation be?

3. Camila says the roots of a polynomial are just the factors with the signs changed. Is this accurate? Explain.

4. If a polynomial has factors $(2x + 3)$ and $(2x + 5)$, what are the two roots?

5. The polynomial $x^2 - 4x - 3$ does not seem to factor into $(x - p)(x - q)$ with p and q as integers, but it might factor if p and q don't have to be integers. But by solving the equation $x^2 - 4x - 3 = 0$ by completing the square, it is possible to find the roots. Find the roots, and then use them to find the factors.

3.6 SOLVING QUADRATIC EQUATIONS WITH THE QUADRATIC FORMULA

=== KEY IDEAS ===

Many quadratic polynomials do not factor. Though completing the square is a technique that will work when factoring is not possible, it is a lengthy process, with many opportunities for careless errors. An efficient way to solve for the roots of a quadratic equation is to use the quadratic formula.

=== MATH FACTS ===

The two solutions to the quadratic equation $ax^2 + bx + c = 0$ are

$x = \dfrac{-b \pm \sqrt{b^2 - 4ac}}{2a}$. These are the same two answers you would get if you did the completing the square process with the equation

$x^2 + \dfrac{b}{a} x + \dfrac{c}{a} = 0$. If the terms of the quadratic are in a different order, like $bx + c + ax^2$, the a value is still the coefficient of the x^2, the b value is still the coefficient of the x, and the c value is the constant.

Applying the Quadratic Formula

The equation $x^2 + 5x + 2 = 0$ cannot be solved with factoring since $x^2 + 5x + 2$ does not factor easily. It would also be difficult to use completing the square for this equation since the 5 is odd so the constant would need to be $\left(\dfrac{5}{2}\right)^2 = \dfrac{25}{4}$. This is a job for the quadratic formula!

Since the equation could be written as $1x^2 + 5x + 2 = 0$, the a coefficient is 1, the b coefficient is 5, and the c is 2.

Substituting these three values into the quadratic formula lead to

$$x = \dfrac{-5 \pm \sqrt{5^2 - 4 \cdot 1 \cdot 2}}{2 \cdot 1}$$

Quadratic Equations

Rather than type this entire expression into the calculator, evaluate the number inside the radical sign, and also the denominator to get the expression

$$x = \frac{-5 \pm \sqrt{25-8}}{2} = \frac{-5 \pm \sqrt{17}}{2}$$

If you get a decimal approximation of this, it is safest to first calculate $-5 + \sqrt{17}$ and then divide by 2 to get approximately -0.44. Then calculate $-5 - \sqrt{17}$ and divide that by 2 to get approximately -4.56.

If this is a multiple-choice question, the answer might also look like $-\frac{5}{2} \pm \frac{\sqrt{17}}{2}$.

Applying the Quadratic Formula When Some of the Coefficients Are Negative

Be careful when you see negatives in the quadratic equation. When b is negative, then the $-b$ becomes positive while the b^2 is still positive. If c is negative while a is positive, the expression $-4ac$ becomes positive when the product ac is negative.

For the equation $x^2 + 5x - 2 = 0$, the quadratic equation becomes

$$x = \frac{-5 \pm \sqrt{5^2 - 4 \cdot 1 \cdot (-2)}}{2 \cdot 1} = \frac{-5 \pm \sqrt{25+8}}{2} = \frac{-5 \pm \sqrt{33}}{2}$$

Notice that the 8 got added to the 25 because the expression became positive.

Example 1

Use the quadratic formula to solve for all values of x that satisfy the equation $x^2 + 8x + 15 = 0$.

Solution: Though this one can be done by factoring, the quadratic formula can be used on any equation like this.

$$a = 1, b = 8, \text{ and } c = 15.$$

$$x = \frac{-8 \pm \sqrt{8^2 - 4 \cdot 1 \cdot 15}}{2 \cdot 1} = \frac{-8 \pm \sqrt{64-60}}{2} = \frac{-8 \pm \sqrt{4}}{2} = \frac{-8 \pm 2}{2}$$

$$x = \frac{-8+2}{2} = \frac{-6}{2} = -3 \text{ or } x = \frac{-8-2}{2} = \frac{-10}{2} = -5$$

Example 2

Which is a solution to the quadratic equation $x^2 + 6x + 2 = 0$?

(1) $3 + \sqrt{7}$ (2) $3 - \sqrt{7}$ (3) $-3 + \sqrt{7}$ (4) $-3 + \sqrt{5}$

Solution:

$$a = 1, b = 6, c = 2$$

$$x = \frac{-6 \pm \sqrt{6^2 - 4 \cdot 1 \cdot 2}}{2 \cdot 1} = \frac{-6 \pm \sqrt{36 - 8}}{2} = \frac{-6 \pm \sqrt{28}}{2}$$

This doesn't exactly match any of the four answer choices. But $\sqrt{28}$ can be written as $\sqrt{4 \cdot 7} = \sqrt{4} \cdot \sqrt{7} = 2\sqrt{7}$, which allows you to simplify the answer to $\frac{-6 \pm 2\sqrt{7}}{2} = -3 \pm \sqrt{7}$ so the answer is choice (3).

Example 3

Use the quadratic formula to solve for all values of x that satisfy the equation $3x^2 - 4x - 5 = 0$

Solution: $a = 3$, $b = -4$, and $c = -5$.

$$x = \frac{-(-4) \pm \sqrt{(-4)^2 - 4 \cdot 3 \cdot (-5)}}{2 \cdot 3} = \frac{4 \pm \sqrt{16 + 60}}{6} = \frac{4 \pm \sqrt{76}}{6}$$

The $\sqrt{76}$ can be further simplified into $\sqrt{4 \cdot 19} = 2\sqrt{19}$, if needed.

Check Your Understanding of Section 3.6

A. Multiple-Choice

For each question, find all solutions for the variable using the quadratic formula.

1. $x^2 - 6x + 8 = 0$
   (1) 2, 4   (2) 3   (3) –2, –4   (4) 2, –4

2. $x^2 + 2x - 15 = 0$
   (1) –3, 5   (2) –5, 3   (3) –1, 15   (4) 3, 5

Quadratic Equations

3. $x^2 + 6x - 16 = 0$
   (1) –2, 8   (2) 2, –8   (3) 4, –4   (4) –1, 16

4. $-x^2 - 9x - 20 = 0$
   (1) 4, –5   (2) –4, 5   (3) 4, 5   (4) –4, –5

5. $2x^2 + 7x - 4 = 0$
   (1) $\frac{1}{2}$, –4   (2) 1, –4   (3) $-\frac{1}{2}$, 4   (4) –1, 4

6. $x^2 + 4x - 7 = 0$
   (1) 1.3   (2) $2 \pm \sqrt{11}$ (3) $2 \pm \sqrt{12}$ (4) $-2 \pm \sqrt{11}$

7. $6x^2 - 13x + 6 = 0$
   (1) $-\frac{2}{3}, -\frac{3}{2}$ (2) $\frac{2}{3}, -\frac{3}{2}$ (3) $\frac{2}{3}, \frac{3}{2}$ (4) $-\frac{2}{3}, \frac{3}{2}$

8. $x^2 - 4 - 2x = 0$
   (1) $1 \pm \sqrt{5}$ (2) $2 \pm 2\sqrt{6}$ (3) $-1 \pm \sqrt{5}$ (4) $-2 - 2\sqrt{6}$

9. $x^2 = x + 1$
   (1) $\frac{1 \pm \sqrt{5}}{2}$ (2) $\frac{-1 \pm \sqrt{5}}{2}$ (3) $\frac{1 \pm \sqrt{-3}}{2}$ (4) $\frac{-1 \pm \sqrt{-3}}{2}$

10. $-16t^2 + 64t + 80 = 0$
    (1) 1, –5   (2) –1, 5   (3) –1, –5   (4) 1, 5

B. Show how you arrived at your answers.

1. Solve for all values of x that satisfy the equation $x^2 - 10x + 22 = 0$. Round answers to the nearest hundredth.

2. If the solution to a quadratic equation, $ax^2 + bx + c = 0$ is $x = \frac{-5 \pm \sqrt{17}}{2}$, what could the values of a, b, and c be?

3. Use the quadratic formula to solve for both values of x in terms of c. $x^2 + 4x + c = 0$.

4. Are the solutions to the quadratic equation $x^2 + 3x - 2 = 0$ rational or irrational? Explain.

5. What are the two solutions to the equation $x^2 + 3 = 4x + 6$?

3.7 WORD PROBLEMS INVOLVING QUADRATIC EQUATIONS

=== KEY IDEAS ===

Many real-world situations, especially ones that involve products, can be modeled as quadratic equations. After the equation is set up, the techniques from the previous sections in this chapter can be used to complete the question.

Questions Involving the Area of a Rectangle

Since the area of a rectangle is equal to the length times the width, questions about areas of rectangle often become quadratic equations.

Example 1

The length of a rectangle is 4 feet more than its width. The area of the rectangle is 140 square feet. If the width is represented by x, the equation that could be used to solve for x is $x(x + 4) = 140$. Solve this equation for x to find the width of the rectangle.

Solution:

$$x(x + 4) = 140$$
$$x^2 + 4x = 140$$
$$x^2 + 4x - 140 = 0$$
$$(x + 14)(x - 10) = 0$$
$$x + 14 = 0 \text{ or } x - 10 = 0$$
$$x = -14 \text{ or } x = 10$$

Since x represents the width, it must be a positive number, so the answer is 10. The quadratic equation $x^2 + 4x - 140 = 0$ could have also been solved with completing the square or with the quadratic formula.

Questions Involving the Height of a Projectile

When a projectile is shot into the air, the relationship between the time in the air and the projectile's height above the ground can be represented with a quadratic equation.

Example 2

A projectile is shot into the air such that its height can be calculated with the expression $-16x^2 + 48x + 64$, where x is the number of seconds since the projectile has been shot. The equation $-16x^2 + 48x + 64 = 0$ can be used to find the amount of time it will take for the projectile to land on the ground. Solve for x to find this answer.

Solution:

$$-16x^2 + 48x + 64 = 0$$
$$-16(x^2 - 3x - 4) = 0$$
$$-16(x - 4)(x + 1) = 0$$
$$x = 4 \text{ or } x = -1$$

Since x represents the amount of time since the projectile was shot, only the answer $x = 4$ is possible.

Questions Involving the Price of an Item

Usually when the cost of something increases, the number of things sold decreases. Another type of word problem that becomes a quadratic equation is one that describes the effect on sales of raising the price of an item.

The price to rent a bicycle is $10 an hour, and a store rents 40 bicycles in a day. Each time the store raises the rental price $2, the number of bicycles the store rents decreases by 5. When they charge $10 and rent 40 bicycles, their total revenue per day is $10 · 40 = $400. If they increase the price to $12 an hour, they only rent 35 bicycles a day, and their revenue becomes $12 · 35 = $420.

In general, the equation for revenue is $R = (10 + 2x)(40 - 5x)$ where x is the number of $2 increases.

To find out what they should charge to make a revenue of $300, solve the equation.

$$300 = (10 + 2x)(40 - 5x)$$
$$300 = 400 - 50x + 80x - 10x^2$$
$$300 = -10x^2 + 30x + 400$$
$$-300 = -300$$
$$0 = -10x^2 + 30x + 100$$
$$0 = -10(x^2 - 3x - 10)$$
$$0 = -10(x - 5)(x + 2)$$
$$x - 5 = 0 \text{ or } x + 2 = 0$$
$$x = 5 \text{ or } x = -2$$

When $x = 5$, they charge $10 + $2 · 5 = $20.
When $x = -2$, they charge $10 + $2 · (−2) = $6.

So if they charge $20 or $6, they will make $300 in revenue.

Check Your Understanding of Section 3.7

A. Multiple-Choice

1. A printer wants to put 54 square inches of text into a rectangle on an 8 by 11 sheet of paper. She wants the text to be surrounded by a border of constant width. Which equation could be used to find the width of the border (x)?

    (1) $2(8 - 2x) + 2(11 - 2x) = 54$
    (2) $(8 - 2x) + (11 - 2x) = 54$
    (3) $(8 - x)(11 - x) = 54$
    (4) $(8 - 2x)(11 - 2x) = 54$

Quadratic Equations

2. The width of a rectangle is 10 inches longer than its length. If the area of the rectangle is 56 square inches, which equation could be used to determine its length (l)?
   (1) $l(l + 10) = 56$
   (2) $l(l - 10) = 56$
   (3) $2l + 2(l + 10) = 56$
   (4) $2l - 2(l + 10) = 56$

3. A 100-foot fence is used to enclose a rectangular land plot. If the area of the land plot is 456 square feet, which equation can be used to determine the length (x) of the land plot?
   (1) $x(100 - x) = 456$
   (2) $2x + 2(100 - x) = 456$
   (3) $x(50 - x) = 456$
   (4) $2x + 2(50 - x) = 456$

4. The height of a projectile in feet at time t is determined by the equation $h = -16t^2 + 128t + 320$. At what time will the projectile be 560 feet high?
   (1) 4 seconds
   (2) 5 seconds
   (3) 6 seconds
   (4) 7 seconds

5. The height of a projectile in feet at time t is determined by the equation $h = -16t^2 + 112t + 128$. At what time will the projectile be 0 feet high?
   (1) 5 seconds
   (2) 6 seconds
   (3) 7 seconds
   (4) 8 seconds

6. The height of a projectile in meters at time t is determined by the equation $h = -4.9t^2 + 14.7t + 88.2$. At what time will the projectile be 68.6 meters high?
   (1) 3 seconds
   (2) 4 seconds
   (3) 5 seconds
   (4) 6 seconds

7. When a textbook sells for $20, 70 copies are purchased. Every time the price is increased by $3, the number of people purchasing the book decreases by 8. Which equation can be used to determine the number of $3 increases needed to make the total revenue from the books equal $1,426?
   (1) $(20 - 3x)(70 - 8x) = 1,426$
   (2) $(20 - 3x)(70 + 8x) = 1,426$
   (3) $(20 + 3x)(70 + 8x) = 1,426$
   (4) $(20 + 3x)(70 - 8x) = 1,426$

8. When the cost of a movie is $10, 200 people buy tickets. Each time the price of the movie is increased by $3, the number of people buying tickets decreases by 20. Which equation can be used to determine the number of $3 increases needed to make the total revenue from the movie tickets equal to $2,660?
   (1) $(10 - 3x)(200 - 20x) = 2,660$
   (2) $(10 - 3x)(200 + 20x) = 2,660$
   (3) $(10 + 3x)(200 - 20x) = 2,660$
   (4) $(10 + 3x)(200 + 20x) = 2,660$

9. When a candy bar costs $0.50, 500 kids buy it. Each time the price increases by $0.05, the number of kids buying it decreases by 10. Which equation could be used to determine the number of $0.05 increases needed to make the total revenue from the candy equal to $450?
   (1) $(0.5 + 0.05x)(500 - 10x) = 450$
   (2) $(0.5 - 0.05x)(500 - 10x) = 450$
   (3) $(0.5 + 0.05x)(500 + 10x) = 450$
   (4) $(0.5 - .005x)(500 + 10x) = 450$

10. The surface area of a box with a square base is $S = 2l^2 + 4hl$ where l is the length of the side of the square base and h is the height of the box. If the height of the box is 5 feet, which equation could be used to find l to make the surface area 238 square feet?
    (1) $l^2 h = 238$
    (2) $2l^2 2 + 5l = 238$
    (3) $2l^2 + 4l = 238$
    (4) $2l^2 + 20l = 238$

B. *Show how you arrived at your answers.*

1. The height, in feet, of a projectile after t seconds is determined by the equation $h = -16t^2 + 128t + 144$. At what two times is the projectile exactly 384 feet high?

Quadratic Equations

2. A 4 inch by 6 inch photo is put into a picture frame with a border of constant width. If the area of the frame, including the picture, is 80 square inches, find an equation for determining the width of the border, and use that equation to solve for it.

3. When a song costs $0.99 it gets downloaded 200,000 times. Each time the price is increased by $0.10, the number of downloads decreases by 10,000. How may $0.10 increases need to happen for the total revenue to be $223,500? Create an equation and solve.

4. A man is 35 years older than his son. The product of their ages is 884. Write an equation that can be used to solve for the man's age. How old is the man?

5. The perimeter of a rectangle is 44 inches. The area of the rectangle is 120 square inches. If the width of the rectangle is x inches, determine the length of the rectangle using an algebraic solution.

Chapter Four

SYSTEMS OF LINEAR EQUATIONS

4.1 SOLVING SYSTEMS WITH GUESS AND CHECK

Key Ideas

An equation like $x + 3 = 10$ has just one number in its solution set, $x = 7$. When an equation has two variables, for instance $x + y = 10$, there is no longer just one solution. There are many pairs of numbers that satisfy the equation $x + y = 10$. Since $4 + 6 = 10$, one solution is $x = 4$, $y = 6$. This can be written as the *ordered pair* (4, 6) where 4 is called the *x-coordinate* and 6 is the *y-coordinate*. Other ordered pairs that satisfy the equation are (3, 7), (1, 9), (9, 1), and (–4, 14).

Finding an Ordered Pair That Satisfies an Equation with Two Variables

An equation with two variables has an infinite number of ordered pairs that satisfy it. To find one solution, try picking any number you want for the x-coordinate. Then substitute that into the equation and solve for y to find the other number in the ordered pair.

If the equation is $2x + y = 10$, you could pick 3 for the x-value and substitute that into the equation.

$$2(3) + y = 10$$
$$6 + y = 10$$
$$-6 = -6$$
$$y = 4$$

The ordered pair, then, is (3, 4). You can check by substituting 3 for x and 4 for y in the equation.

$$2(3) + 4 \stackrel{?}{=} 10$$
$$6 + 4 \stackrel{?}{=} 10$$
$$10 \stackrel{\checkmark}{=} 10$$

Checking to See If an Ordered Pair Is Part of the Solution Set of a Two-Variable Equation

By substituting the x-coordinate of an ordered pair in for the x-variable and the y-coordinate of the ordered pair in for the y-variable in an equation, you can determine if the point is in the solution set for that equation. If the resulting equation is true, then it is. If the resulting equation is not true, then the ordered pair is not part of the solution set.

Example 1

Which of the following choices is part of the solution set of the equation $2x + 3y = 14$?

(1) (1, 5)   (2) (2, 3)   (3) (3, 3)   (4) (4, 2)

Solution: There are an infinite number of solutions to the equation $2x + 3y = 14$. One of them is among the four choices. Check each choice by substituting the first coordinate into x and the second coordinate into y, and see which make the equation true.

Check (1, 5)

$$2(1) + 3(5) \stackrel{?}{=} 14$$
$$2 + 15 \stackrel{?}{=} 14$$
$$17 \neq 14$$

Check (2, 3)

$$2(2) + 3(3) \stackrel{?}{=} 14$$
$$4 + 9 \stackrel{?}{=} 14$$
$$13 \neq 14$$

Check (3, 3)

$$2(3) + 3(3) \stackrel{?}{=} 14$$
$$6 + 9 \stackrel{?}{=} 14$$
$$15 \neq 14$$

Check (4, 2)

$$2(4) + 3(2) \stackrel{?}{=} 14$$
$$8 + 6 \stackrel{?}{=} 14$$
$$14 \stackrel{\checkmark}{=} 14$$

So the answer is (4, 2).

Identifying Equations That Have Equivalent Solution Sets

The equation $x + y = 10$ has an infinite number of ordered pairs in its solution set. A few of them are (2, 8), (5, 5), and (7, 3) since $2 + 8 = 10$, $5 + 5 = 10$, and $7 + 3 = 10$. If you multiply both sides of this equation by the same number, it becomes a new equation with the same solution set as the original.

If you multiply both sides of $x + y = 10$ by 2, it becomes

$$2(x + y) = 2(10)$$
$$2x + 2y = 20$$

If you test the points (2, 8), (5, 5), and (7, 3), you will see that they are solutions of $x + y = 10$ and of $2x + 2y = 20$.

$$2(2) + 2(8) = 4 + 16 = 20$$
$$2(5) + 2(5) = 10 + 10 = 20$$
$$2(7) + 2(3) = 14 + 6 = 20$$

Every solution to $x + y = 10$ will also be a solution of $2x + 2y = 10$. When both sides of an equation are multiplied by the same number, the new equation has the same solution set as the original equation.

Example 2

Which of the following equations has the same solution set as $x + 3y = 12$?
   (1) $4x + 12y = 12$ (3) $4x + 12y = 48$
   (2) $3x + 9y = 48$ (4) $2x + 9y = 36$

Solution: If you multiply both sides of the equation $x + 3y = 12$ by 4, it becomes $4x + 12y = 48$, choice (3).

If you multiply both sides by 3, it becomes $3x + 9y = 36$, not $3x + 96 = 48$ so choice (2) is incorrect. The other two choices are also not possible to obtain by multiplying both sides of $x + 3y = 12$ by the same number.

Solving Systems of Equations by Guess and Check

A system of equations is a set of two equations that each has an x and a y. The goal is to find the ordered pairs that satisfy both equations at the same time.

An example is the system of equations:

$$x + y = 10$$
$$x - y = 2$$

By going through some of the ordered pairs that satisfy the first equation (9, 1), (8, 2), (7, 3), (6, 4), and (5, 5), you might notice that the ordered pair

(6, 4) also satisfies the second equation so the solution to the system of equations is (6, 4).

$$6 + 4 \stackrel{?}{=} 10$$
$$10 \stackrel{\checkmark}{=} 10$$
$$6 - 4 \stackrel{?}{=} 2$$
$$2 \stackrel{\checkmark}{=} 2$$

This guess and check method is not very practical unless it is a multiple-choice question where there are at most four ordered pairs to check.

Example 3

Which ordered pair is a solution to the system of equations?

$$2x + 3y = 11$$
$$4x - y = 15$$

(1) (1, 3)   (2) (2, –7)   (3) (3, 4)   (4) (4, 1)

Solution: Only if the ordered pair is a solution to both equations, is it a solution to the system. Check each of the choices in both equations.

Check (1, 3)

$$2(1) + 3(3) \stackrel{?}{=} 11$$
$$2 + 9 \stackrel{?}{=} 11$$
$$11 \stackrel{\checkmark}{=} 11$$

$$4(1) - 3 \stackrel{?}{=} 15$$
$$4 - 3 \stackrel{?}{=} 15$$
$$1 \neq 15$$

Since it satisfies only the first equation, choice (1) is not a solution to the system of equations.

Choice (2) doesn't satisfy the first equation, and choice (3) doesn't satisfy either of the equations. But when you test choice (4), it becomes

$$2(4) + 3(1) \stackrel{?}{=} 11$$
$$8 + 3 \stackrel{?}{=} 11$$
$$11 \stackrel{\checkmark}{=} 11$$

$$4(4) - 1 \stackrel{?}{=} 15$$
$$16 - 1 \stackrel{?}{=} 15$$
$$15 \stackrel{\checkmark}{=} 15$$

Choice (4) is a solution to both equations so it is a solution to the system of equations.

Check Your Understanding of Section 4.1

A. Multiple-Choice

1. (2, 5) is a solution to which equation?
   (1) $x + 2y = 9$
   (2) $2x + y = 9$
   (3) $8x - y = 9$
   (4) $3x + 4y = 25$

2. Which ordered pair is a solution to the equation $3y - 4x = 3$?
   (1) (3, 5)   (2) (5, 3)   (3) (1, 4)   (4) (8, 12)

3. Which ordered pair is *not* a solution to the equation $3y - 2x = 3$?
   (1) (6, 5)   (2) (9, 7)   (3) (12, 9)   (4) (10, 8)

4. Which equation has the same solution set as the equation $2x + 3y = 5$?
   (1) $8x + 12y = 20$
   (2) $8x + 12y = 15$
   (3) $6x + 9y = 12$
   (4) $4x + 6y = 8$

5. Which equation does *not* have the same solution set as the equation $4x - 8y = 12$?
   (1) $6x - 7y = 20$
   (2) $x - 2y = 3$
   (3) $2x - 4y = 6$
   (4) $8x - 16y = 24$

6. Which is the solution set to this system of equations?

   $$2x + 3y = 20$$
   $$5x - 2y = 31$$

   (1) (4, 4)   (2) (1, 6)   (3) (−2, 8)   (4) (7, 2)

7. Which is the solution set to this system of equations?

   $$-4x + 2y = 18$$
   $$3x - 6y = -36$$

   (1) (−2, 5)   (2) (−1, 7)   (3) (−3, 3)   (4) (0, 9)

8. The ordered pair (3, −7) is a solution to which system of equations?
   (1) $2x - 3y = 27$
       $5x - 3y = 38$
   (2) $4x + 2y = -2$
       $3x - 7y = 59$
   (3) $2x - 3y = 27$
       $4x + 2y = -2$
   (4) $5x - 3y = 38$
       $3x + 7y = -40$

Systems of Linear Equations

9. If $(a, 5)$ is a solution to the equation $3x + 6y = 42$, what is the value of a?
(1) 1 (2) 2 (3) 3 (4) 4

10. Which is *not* a solution to the system of equations?
$$x + y = 12$$
$$2x + 2y = 24$$
(1) (8, 4) (2) (3, 10) (3) (7, 5) (4) (5, 7)

B. Show how you arrived at your answers.

1. The equation $2x + 3y = 11$ has two solutions in which both coordinates are positive integers less than 6. What are those two solutions?

2. Use guess and check to find the solution to the system of equations.
$$x + y = 8$$
$$x - y = 6$$

3. The system of equations
$$2x + 5y = 25$$
$$2x + 5y = 26$$
has no solutions. Explain why?

4. The equation $x + 3y = c$ has the point (4, 7) in its solution set. What is c?

5. (6, –2) is a solution to the equation $5x - 6y = 42$. Find a solution to $15x - 18y = 126$.

4.2 SOLVING SIMPLER SYSTEMS OF EQUATIONS WITH ALGEBRA

KEY IDEAS

Certain systems of equations can be solved by combining the two equations such that one of the variables gets eliminated. Three ways of doing this are addition, when the two equations are added to each other; subtraction, when one equation is subtracted from the other; and substitution, when one of the variables in one equation is replaced by an expression involving the other variable.

Combining Two Equations to Form a New Equation

If two equations have an ordered pair that makes each of them true, then the equation formed by adding the two equations will also have that ordered pair as one of its solutions.

The equations $2x + 5y = 26$ and $3x - 2y = 1$ both have the solution $(3, 4)$. If you add the two equations together, you get

$$\begin{array}{rrrrr} & 2x & + & 5y & = 26 \\ + & 3x & - & 2y & = 1 \\ \hline & 5x & + & 3y & = 27 \end{array}$$

This new equation also is satisfied by the ordered pair $(3, 4)$.

$$5(3) + 3(4) \stackrel{?}{=} 27$$
$$15 + 12 \stackrel{?}{=} 27$$
$$27 \stackrel{\checkmark}{=} 27$$

MATH FACTS

If in a system of equations, the coefficient of one of the variables is the opposite sign of the coefficient of the same variable in the other equation, the system can be quickly solved by adding the two equations together to form an equation with just one variable.

Systems of Linear Equations

Example 1

Which equation has, as one of its solutions, the solution of the system of equations?

$$3x - 2y = 16$$
$$4x + 3y = -7$$

(1) $7x + y = 9$ (3) $7x + y = 7$
(2) $7x + y = 8$ (4) $7x + y = 6$

Solution: If you add the two equations, the new equation has as one of its solutions, the solution to the system of equations. In this case,

$$\begin{array}{rrrrr} & 3x & - & 2y & = 16 \\ + & 4x & + & 3y & = -7 \\ \hline & 7x & + & y & = 9 \end{array}$$, choice (1)

Adding Two Equations to Eliminate a Variable

The equation $5x + 2y = 36$ has an infinite number of solutions. The equation $3x - 2y = 12$ also has an infinite number of solutions. Finding the ordered pair that makes both equations true can be done quickly in this case by adding the two equations.

$$\begin{array}{rrrrr} & 5x & + & 2y & = 36 \\ + & 3x & - & 2y & = 12 \\ \hline & \frac{8x}{8} & & & = \frac{48}{8} \\ & x & & & = 6 \end{array}$$

In this example, because the $+2y$ and the $-2y$ have a sum of $0y$, which is 0, the new equation you get by adding the two equations together is an equation with only one variable. This equation, $8x = 48$ is solved by dividing both sides of the equation by 8, to get $x = 6$.

6 is the x-coordinate of the ordered pair that solves the original system of equations. To find the y-coordinate, substitute the 6 in for x in either of the original two equations.

$$5(6) + 2y = 36$$
$$30 + 2y = 36$$
$$-30 = -30$$
$$\frac{2y}{2} = \frac{6}{2}$$
$$y = 3$$

The solution to the original system of equations is (6, 3).

To check, substitute 6 for x and 3 for y into both original equations.
$$5(6) + 2(3) = 30 + 6 = 36$$
$$3(6) - 2(3) = 18 - 6 = 12$$

Example 2

Find the ordered pair that solves the system of equations.
$$4x - 3y = 22$$
$$2x + 3y = 2$$

Solution: Since the $-3y$ in the top equation will cancel out the $+3y$ in the bottom equation, this system can be solved by adding the two equations.

$$\begin{array}{rrrrr} & 4x & - & 3y & = 22 \\ + & 2x & + & 3y & = 2 \\ \hline & 6x & & & = 24 \\ & \frac{6x}{6} & & & = \frac{24}{6} \\ & x & & & = 4 \end{array}$$

Substitute $x = 4$ into either of the original two equations.
$$4(4) - 3y = 22$$
$$16 - 3y = 22$$
$$-16 = -16$$
$$\frac{-3y}{3} = \frac{6}{-3}$$
$$y = -2$$

The solution is $(4, -2)$.

Example 3

Find the solution set for the system of equations.
$$-3x + 7y = 16$$
$$3x + 4y = -5$$

Solution: Since the coefficients of the x in the first equation is the opposite sign of the coefficient of the x in the second equation, this system can be solved by adding the two equations together.

$$\begin{array}{rrrrr} & -3x & + & 7y & = 16 \\ + & 3x & + & 4y & = -5 \\ \hline & & & 11y & = 11 \\ & & & \frac{11y}{11} & = \frac{11}{11} \\ & & & y & = 1 \end{array}$$

$$-3x + 7(1) = 16$$
$$-3x + 7 = 16$$
$$-7 = -7$$
$$\frac{-3x}{-3} = \frac{9}{-3}$$
$$x = -3$$

The ordered pair that satisfies the system is (–3, 1).

Subtracting Two Equations to Eliminate a Variable

If in the two equations in a system, one of the variables has the same coefficient as the same variable has in the other equation, that variable can be eliminated by subtracting the two equations.

In this system

$$4x + 5y = 37$$
$$2x + 5y = 31$$

the y variables in both equations have a coefficient of +5. Subtract the two equations

$$\begin{array}{rrrrrr} & 4x & + & 5y & = & 37 \\ - & 2x & + & 5y & = & 31 \\ \hline & \dfrac{2x}{2} & & & = & \dfrac{6}{2} \\ & x & & & = & 3 \end{array}$$

Substitute $x = 3$ into either equation and solve for y.

$$4(3) + 5y = 37$$
$$12 + 5y = 37$$
$$-12 = -12$$
$$\frac{5y}{5} = \frac{25}{5}$$
$$y = 5$$

The solution is (3, 5).

Solving Systems of Equations with the Substitution Method

If one of the two equations is in the form $y = mx + b$, the system of equations can be solved by replacing the y in the other equation by the expression $mx + b$ and then solving the new one-variable equation.

For the system

$$y = 2x + 1$$
$$3x + 2y = 16$$

one of the equations is in the form $y = mx + b$.

Replace the y in the second equation with the expression $2x + 1$.

$$3x + 2(2x + 1) = 16$$
$$3x + 4x + 2 = 16$$
$$7x + 2 = 16$$
$$-2 = -2$$
$$\frac{7x}{7} = \frac{14}{7}$$
$$x = 2$$

Substitute this x into either equation.

$$y = 2(2) + 1$$
$$y = 4 + 1$$
$$y = 5$$

The solution is (2, 5).

Example 4

Solve this system of equations with the substitution method.

$$y = 5x - 7$$
$$3x - 4y = -6$$

Solution: Replace the y in the second equation with the expression $5x - 7$.

$$3x - 4(5x - 7) = -6$$

Distribute the -4 through the parentheses.

$$3x - 20x + 28 = -6$$
$$-17x + 28 = -6$$
$$-28 = -28$$
$$\frac{-17x}{-17} = \frac{-34}{-17}$$
$$x = 2$$

Substitute the $x = 2$ into the first equation

$$y = 5x - 7 = 5(2) - 7 = 10 - 7 = 3$$

The solution is (2, 3).

Systems of Linear Equations

Check Your Understanding of Section 4.2

A. Multiple-Choice

1. Which equation has as one of its solutions, the solution to the system of equations?

$$4x + 2y = 10$$
$$3x - 5y = 14$$

   (1) $7x - 3y = 21$ (3) $7x - 3y = 23$
   (2) $7x - 3y = 22$ (4) $7x - 3y = 24$

2. Solve the system of equations.

$$3x + 2y = 17$$
$$4x - 2y = 4$$

   (1) (4, 3)   (2) (5, 1)   (3) (3, 4)   (4) (1, 6)

3. Solve the system of equations.

$$y = 5x + 3$$
$$2x + 6y = 50$$

   (1) (1, 8)   (2) (8, 1)   (3) (−1, −8)   (4) (−8, −1)

4. Solve the system of equations.

$$6x - 5y = 19$$
$$-3x + 5y = -7$$

   (1) (−4, −1)   (2) (−4, 1)   (3) (4, −1)   (4) (4, 1)

5. Solve the system of equations.

$$2x + 3y = -1$$
$$-2x + 5y = -23$$

   (1) (−4, 3)   (2) (4, 3)   (3) (4, −3)   (4) (−4, −3)

6. Solve the system of equations.

$$-3x + 7y = 1$$
$$3x + 3y = -21$$

   (1) (5, 2)   (2) (−5, 2)   (3) (−5, −2)   (4) (5, −2)

7. Solve the system of equations.

$$3x + 7y = -2$$
$$x - 7y = -10$$

(1) (3, 1)   (2) (–3, –1)   (3) (3, –1)   (4) (–3, 1)

8. Solve the system of equations.

$$5x + 3y = 19$$
$$2x + 3y = 4$$

(1) (5, –2)   (2) (5, 2)   (3) (–5, –2)   (4) (–5, 2)

9. Solve the system of equations.

$$y = 3x - 2$$
$$4x - 2y = -4$$

(1) (10, 4)   (2) (4, 10)   (3) (–10, –4)   (4) (–4, –10)

10. Solve the system of equations.

$$4x + 8y = 36$$
$$4x + 5y = 33$$

(1) (7, 1)   (2) (7, –1)   (3) (–7, 1)   (4) (–7, –1)

B. *Show how you arrived at your answers.*

1. Find an equation that has as one of its solutions, the solution to the system of equations.

$$3x + 4y = 11$$
$$2x + 6y = 14$$

2. A student likes to use the substitution method for systems of equation. How can he use it with a system that is not in the proper form for substitution? Show with this system.

$$-2x + y = 4$$
$$3x + 4y = 49$$

3. The system

$$y = 2x - 9$$
$$y = -3x + 16$$

can be solved several ways.

Student 1 wants to use subtraction. Student 2 wants to use substitution. Show how each student would do this.

4. Two numbers, x and y, have a sum of 17 but a difference of 11. Write a system of equations that can be used to solve this and then use it to find the solution.

5. The system of equations

$$5x + 2y = 8$$
$$-5x + 2y = -32$$

can be solved several ways.

Student 1 wants to use addition. Student 2 wants to use subtraction. Who is right? Explain your answer.

4.3 SOLVING MORE COMPLICATED SYSTEMS OF EQUATIONS WITH ALGEBRA

=== KEY IDEAS ===

Often when solving systems of equations, we do not get so lucky that the coefficient of one variable in the top equation is the opposite of, or the same as, the coefficient of the same variable in the bottom equation. When this happens, questions like this involve more steps.

Solving a System by Changing One of the Equations

Look at this system of equations

$$3x + 4y = 31$$
$$5x - 2y = -9$$

Adding or subtracting the two equations will not eliminate one of the variables.

	$3x$	+	$4y$	=	31
+	$5x$	−	$2y$	=	-9
	$8x$	+	$2y$	=	22

	$3x$	+	$4y$	=	31
−	$5x$	−	$2y$	=	-9
	$-2x$	−	$6y$	=	40

Fortunately there is a way to change the equations so that a variable will get eliminated when you add the changed equations together. You are permitted to multiply both sides of either equation by whatever number you want.

In the above example, If the $3x$ were a $-5x$, it could cancel out the $+5x$, but it isn't easy to turn a $3x$ into a $-5x$ by multiplying. If the $5x$ were a $-3x$, that would also work but, again, it isn't easy to do this. Now look at the ys. If the $-2y$ in the bottom equation were a $-4y$, then it would be the opposite sign of the $+4y$ in the top equation. And best of all, it is possible to make that $-2y$ into a $-4y$ by just multiplying both sides of the bottom equation by 2. Multiplying both sides of an equation by a number does not change the solution set for that equation, or the solution set for the system of equations.

$$3x + 4y = 31$$
$$2(5x - 2y) = 2 \cdot -9$$

The system becomes

$$3x + 4y = 31$$
$$10x - 4y = -18$$

This equation is now the type that can be solved by adding the two equations together to get $13x = 13$, which leads to $x = 1$.

Now you can substitute $x = 1$ into either of the original two equations.

$$3(1) + 4y = 31$$
$$3 + 4y = 31$$
$$-3 = -3$$
$$4y = 28$$
$$y = 7$$

Example 1

Solve this system of equations by first multiplying both sides of one of the equations by a number.

$$2x + 6y = 2$$
$$3x + 2y = 10$$

Solution: Since the $6y$ is a multiple of the $2y$, multiply both sides of the bottom equation by -3 so that the $2y$ becomes $-6y$.

$$2x + 6y = 2$$
$$-3(3x + 2y) = -3(10)$$

$$2x + 6y = 2$$
$$-9x - 6y = -30$$

Systems of Linear Equations

Now add the equations together to eliminate the ys.

$$\begin{array}{r} 2x + 6y = 2 \\ +\;-9x - 6y = -30 \\ \hline \dfrac{-7x}{-7} = \dfrac{-28}{-7} \\ x = 4 \end{array}$$

$$2(4) + 6y = 2$$
$$8 + 6y = 2$$
$$-8 = -8$$
$$\dfrac{6y}{6} = \dfrac{-6}{6}$$
$$y = -1$$

The solution is $(4, -1)$.

Solving a System by Changing Both Equations

Sometimes the only way to change the equations so that a variable will drop out when adding them is to change both equations. This is needed when none of the coefficients is evenly divisible by the corresponding coefficient in the other equation.

An example of a system where both equations need to be changed is

$$2x + 3y = 23$$
$$3x + 4y = 31$$

If you want to make it so the y variable drops out after adding, first take the coefficient of the y in the bottom equation, change the sign, and multiply both sides of the first equation by it.

Since 4 is the coefficient of the y in the bottom equation, multiply both sides of the top equation by -4.

$$-4(2x + 3y) = -4(23)$$
$$-8x - 12y = -92$$

For the bottom equation, multiply both sides of it by the coefficient of the y in the top equation. In this case, multiply both sides of the bottom equation by $+3$.

$$+3(3x + 4y) = 3(31)$$
$$9x + 12y = 93$$

Multiplying one or both equations in a system by numbers does not change the solution. The new system of equations has the same solution set as the original system of equations but is easier to solve.

Now add the two new equations and complete the process

$$\begin{array}{r} -8x - 12y = -92 \\ +\quad 9x + 12y = 93 \\ \hline 1x = 1 \\ x = 1 \end{array}$$

To get the y value, substitute the x value of 1 into one of the original two equations.

$$2(1) + 3y = 23$$
$$2 + 3y = 23$$
$$-2 = -2$$
$$\frac{3y}{3} = \frac{21}{3}$$
$$y = 7$$

And the solution is (1, 7).

Example 2

Determine the solution set of the system of equations.

$$2x + 3y = -11$$
$$4x - 7y = 43$$

Since $-(-7) = +7$, multiply the top equation by 7 and the bottom equation by 3.

$$7(2x + 3y) = 7(-11)$$
$$3(4x - 7y) = 3(43)$$

$$\begin{array}{r} 14x + 21y = -77 \\ +\quad 12x - 21y = 129 \\ \hline \frac{26x}{26} = \frac{52}{26} \\ x = 2 \end{array}$$

$$2(2) + 3y = -11$$
$$4 + 3y = -11$$
$$-4 = -4$$
$$\frac{3y}{3} = \frac{-15}{3}$$
$$y = -5$$

So, the solution is (2, −5).

Systems of Linear Equations

Example 3

Find the solution to the system of equations.

$$y = 2x - 1$$
$$y = 3x - 4$$

Solution: Rearrange the equations so they are in standard form $(ax + by = c)$.

$$-2x + y = -1$$
$$-3x + y = -4$$

Multiply both sides of the top equation by -1.

$-1(-2x + y) = -1(-1)$ becomes $2x - y = 1$

$$\begin{array}{rrrrr} & 2x & - & y & = 1 \\ + & -3x & + & y & = -4 \\ \hline & -x & & & = -3 \\ & x & & & = 3 \end{array}$$

$$y = 2(3) - 1$$
$$y = 5$$

The solution is $(3, 5)$.

Example 4

Which system of equations has the same solution set as the following system?

$$3x - 4y = 7$$
$$4x + y = 22$$

(1) $3x - 4y = 7$
    $16x + 4y = 22$
(2) $6x - 8y = 7$
    $4x + y = 22$
(3) $3x - 4y = 7$
    $16x + 4y = 22$
(4) $3x - 4y = 7$
    $16x + 4y = 88$

Solution: The system of equations in choice (4) can be obtained by multiplying both sides of the bottom equation by 4. This new system will have the same solution set as the original. This would also be true if both equations were multiplied by numbers. The system

$$6x - 8y = 14$$
$$16x + 4y = 88$$

would also be correct if it were a choice since it is obtained by multiplying both sides of the top equation by 2 and both sides of the bottom equation by 4.

Notice that the system in choice (4) would be the logical first step in solving the original system.

Systems of Equations That Have No Solutions or Infinite Number of Solutions

For some systems of equations, there are no ordered pairs that satisfy both equations. An example of this is

$$x + y = 10$$
$$x + y = 11$$

Any ordered pair that satisfies the first equation, like (7, 3), will not satisfy the second equation, and any ordered pair that satisfies the second equation, like (7, 4), will not satisfy the first equation.

To determine that a system of equations has no solution, do the steps for elimination.

$$x + y = 10$$
$$-1(x + y) = -1(11)$$

$$\begin{array}{rrrrr} & x & + & y & = & 10 \\ + & -x & - & y & = & -11 \\ \hline & 0 & & & = & -1 \end{array}$$

Since 0 does not equal −1, the original system had no solution.

The system of equations

$$x + y = 10$$
$$2x + 2y = 20$$

also has something unusual occur.

$$-2(x + y) = -2(10)$$
$$2x + 2y = 20$$

$$\begin{array}{rrrrr} & -2x & - & 2y & = & -20 \\ + & 2x & + & 2y & = & 20 \\ \hline & 0 & & & = & 0 \end{array}$$

Since 0 does equal 0, the original system has an infinite number of solutions. Any ordered pair that satisfies $x + y = 10$, like (2, 8) will also satisfy $2x + 2y = 20$, $2(2) + 2(8) = 4 + 16 = 20$.

Systems of Linear Equations

=== MATH FACTS ===

If when doing the process to eliminate a variable, both variables get eliminated, there are either no solutions or an infinite number of solutions. If the equation that is left is false saying that 0 is equal to something other than zero, then there are no solutions. If the equation that is left is 0 = 0, there are an infinite number of solutions—any pair that satisfies the top equation will satisfy the bottom equation too.

Check Your Understanding of Section 4.3

A. Multiple-Choice

1. The system of equations

$$5x - 4y = -3$$
$$3x + 2y = 7$$

   has the same solution as the system

   (1) $5x - 4y = -3$
       $6x + 4y = 7$
   (2) $5x - 4y = -3$
       $6x + 2y = 14$
   (3) $5x - 4y = -3$
       $6x + 4y = 14$
   (4) $5x - 4y = -3$
       $9x + 6y = 14$

2. In order to eliminate the y from this system of equations,

$$3x + 12y = 21$$
$$6x + 4y = 22$$

   you could
   (1) Multiply both sides of the first equation by -2.
   (2) Multiply both sides of the second equation by -3.
   (3) Multiply both sides of the first equation by -3.
   (4) Multiply both sides of the second equation by $1/2$.

3. In order to eliminate the x from this system of equations,

$$12x - 3y = 21$$
$$-2x + 6y = 2$$

   you could
   (1) Multiply both sides of the first equation by 2.
   (2) Multiply both sides of the second equation by 6.
   (3) Multiply both sides of the first equation by -2.
   (4) Multiply both sides of the second equation by $1/2$.

4. Which step would not cause a variable to be eliminated after adding the new equations together on this system?

$$3x - 6y = 6$$
$$9x + 2y = 38$$

(1) Multiply both sides of the first equation by –3.
(2) Multiply both sides of the second equation by +3.
(3) Multiply both sides of the first equation by +1/3.
(4) Multiply both sides of the first equation by 4.

5. One way to eliminate the y from this system of equations

$$3x + 4y = 16$$
$$2x - 5y = 3$$

is to
(1) Multiply both sides of the first equation by –2 and both sides of the second equation by +3.
(2) Multiply both sides of the first equation by +4 and both sides of the second equation by –5.
(3) Multiply both sides of the first equation by +5 and both sides of the second equation by +4.
(4) Multiply both sides of the first equation by +2 and both sides of the second equation by –3.

6. Solve the system of equations:

$$8x - 2y = 28$$
$$4x + 3y = 6$$

(1) (3, 2)   (2) (–3, 2)   (3) (–3, –2)   (4) (3, –2)

7. Solve the system of equations:

$$4x - 3y = 31$$
$$3x + 5y = 16$$

(1) (7, –1)   (2) (7, 1)   (3) (–7, –1)   (4) (–7, 1)

8. Solve the system of equations:

$$x - 3y = 14$$
$$4x + 5y = 22$$

(1) (8, 2)   (2) (–8, –2)   (3) (–8, 2)   (4) (8, –2)

Systems of Linear Equations

9. Solve the system of equations:
$$3x - 2y = -4$$
$$5x - 4y = -12$$

(1) (4, 8)   (2) (4, –8)   (3) (–4, 8)   (4) (–4, –8)

10. The system of equations
$$2x - 7y = 2$$
$$6x - 21y = 7$$
has
(1) Exactly one solution.
(2) Every ordered pair possible satisfies this equation.
(3) Every ordered pair that satisfies the first equation also satisfies the whole system.
(4) No ordered pairs that solve the system.

B. *Show how you arrived at your answers.*

1. For the system
$$2x + 5y = 17$$
$$6x + 7y = 27$$

   Samuel wants to eliminate the y, but Georgia wants to eliminate the x. Which student's choice will require less work?

2. Solve for x and y in this system
$$x + y = 40$$
$$2x + 4y = 136$$

3. Aubree solves the system of equations
$$2x + 3y = 21$$
$$4x + 6y = 42$$
and gets the solution
$$0 = 0$$
She says this means that every ordered pair is a solution to the system of equations. Is this accurate? Explain.

4. If it is possible, find an ordered pair that is a solution to the system
$$5x - 2y = 5$$
$$10x - 4y = 11$$
If not, explain why.

5. For the system of equations

$$4x - 3y = 18$$
$$2x + 12y = 36$$

Victoria starts by multiplying both sides of the second equation by –2. Porter starts by multiplying both sides of the second equation by +2. How is it that either of these methods will lead to the correct solution?

4.4 SOLVING WORD PROBLEMS WITH SYSTEMS OF EQUATIONS

=== Key Ideas ===

Word problems requiring this simultaneous equation process are common in the free response parts of this test. This is generally a two-part process: (1) setting up the system of equations and (2) solving the system of equations.

Setting Up the System of Equations

A typical word problem that can be solved with systems of equations is:
A restaurant sells hamburgers and hot dogs. One customer purchases 3 hot dogs and 5 hamburgers and the total bill is $37. Another customer purchases 5 hot dogs and 2 hamburgers and the total bill is $30. What is the cost of one hot dog and the cost of one hamburger?

Let x be the price of one hot dog and y be the price of one hamburger. The cost of 3 hotdogs is $3x$ dollars. The cost of 5 hamburgers is $5y$ dollars. So the cost of 3 hot dogs and 5 hamburgers is $3x + 5y$ dollars. The cost of 5 hot dogs and 2 hamburgers is $5x + 2y$ dollars.

Since the question says that 3 hot dogs and 5 hamburgers cost $37, the first equation is

$$3x + 5y = 37$$

Since the question says that 5 hot dogs and 2 hamburgers cost $30, the second equation is

$$5x + 2y = 30$$

Systems of Linear Equations

Solve the system of equations.

$$3x + 5y = 37$$
$$5x + 2y = 30$$

$$-2(3x + 5y) = -2(37)$$
$$5(5x + 2y) = 5(30)$$

$$\begin{array}{r} -6x - 10y = -74 \\ +25x + 10y = 150 \end{array}$$

$$\frac{19x}{19} = \frac{76}{19}$$
$$x = 4$$

$$3(4) + 5y = 37$$
$$12 + 5y = 37$$
$$-12 = -12$$
$$5y = 25$$
$$y = 5$$

So the cost of a hot dog is $4, and the cost of a hamburger is $5.

Example

The cost of a movie ticket for a child is $7, while the cost of a movie ticket for an adult is $11. A group of 23 people, some children and some adults, go to the movies. The total cost of all the tickets is $197. How many child tickets and how many adult tickets were purchased?

Solution: If there are x children and y adults, the total number of people is $x + y$. The cost for x child tickets is $7x$ dollars. The cost for y adult tickets is $11y$ dollars. The total cost is $7x + 11y$ dollars.

The two equations are

$$x + y = 23$$
$$7x + 11y = 197$$

Multiply both sides of the top equation by -11

$$-11(x + y) = -11(23)$$
$$7x + 11y = 197$$

$$\begin{array}{r} -11x - 11y = -253 \\ +7x + 11y = 197 \end{array}$$

$$\frac{-4x}{-4} = \frac{-56}{-4}$$
$$x = 14$$

112

$$14 + y = 23$$
$$-14 = -14$$
$$y = 9$$

There were 14 child tickets and 9 adult tickets purchased.

Check Your Understanding of Section 4.4

A. Multiple-Choice

1. Which system of equations can be used to model the following scenario? There are 50 animals. Some of the animals have 2 legs and the rest of them have 4 legs. In total there are 172 legs.
   (1) $x + y = 172$
   $2x + 4y = 50$
   (2) $x + 50 = y$
   $2x + 172 = 4y$
   (3) $y + 50 = x$
   $4y + 172 = 2x$
   (4) $x + y = 50$
   $2x + 4y = 172$

2. Which system of equations could be used to model the following scenario? There are 8 people in an elevator. Some are adults and the rest are children. Each adult weighs 150 pounds. Each child weighs 50 pounds. The total weight of the 8 people is 800 pounds.
   (1) $x + y = 8$
   $150x + 50y = 800$
   (2) $x + y = 800$
   $150x + 50y = 50$
   (3) $x + 50 = y$
   $150x + 800 = 50y$
   (4) $y + 50 = x$
   $50y + 800 = 150x$

3. Which system of equations could be used to model the following scenario? There are 20 coins. Some are quarters and the rest are dimes. The quarters are worth 25 cents each and the dimes are worth 10 cents each. The total value of the 20 coins is $2.90.
   (1) $x + y = 20$
   $25x + 10y = 290$
   (2) $x + y = 20$
   $25x + 10y = 2.90$
   (3) $x + y = 290$
   $25x + 10y = 20$
   (4) $x + y = 2.90$
   $25x + 10y = 20$

4. A pet store has 30 animals. Some are cats and the rest are dogs. The cats cost $50 each. The dogs cost $100 each. If the total cost for all 30 animals is $1900, how many cats are there?
   (1) 8   (2) 20   (3) 22   (4) 24

Systems of Linear Equations

5. A restaurant sells only two desserts, pie and cake. A piece of pie costs $4. A piece of cake costs $5. The restaurant sells 100 desserts, which cost a total of $473. How many pieces of pie did they sell?
   (1) 73   (2) 27   (3) 18   (4) 11

6. A fence is put around a rectangular plot of land. The perimeter of the fence is 28 feet. Two of the opposite sides of the fence cost $10 per foot. The other two sides cost $12 per foot. If the total cost of the fence is $148, what are the dimensions of the fence?
   (1) 8 by 20   (3) 3 by 11
   (2) 4 by 10   (4) 2 by 12

7. One number is five bigger than another number. When three times the larger number is added to twice the smaller number, the result is 60. What are the two numbers?
   (1) 9 and 14   (3) 11 and 16
   (2) 10 and 15   (4) 12 and 17

8. Peanuts cost $7 a pound. Cashews cost $9 a pound. There is a mixture of peanuts and cashews that weighs 10 pounds and costs $84. How many pounds of peanuts are there in the mixture?
   (1) 1   (2) 2   (3) 3   (4) 4

9. The current in a river makes boats going upstream slower by y miles per hour and boats going downstream faster by y miles per hour. Upstream boats go 3 miles per hour and downstream boats go 13 miles per hour. Which system of equations can be used to find the speed of the boat if it were in still water?
   (1) $x + y = 13$
        $x - y = 3$
   (2) $x + y = 3$
        $x - y = 13$
   (3) $xy = 13$
        $\dfrac{x}{y} = 3$
   (4) $x + 2y = 13$
        $x - 3y = 3$

10. A basketball player made some shots worth two points and the rest worth three points. If she made 16 shots for a total of 38 points, how many of each type did she make?
    (1) 5 three pointers and 11 two pointers
    (2) 6 three pointers and 10 two pointers
    (3) 7 three pointers and 9 two pointers
    (4) 8 three pointers and 8 two pointers

B. *Show how you arrived at your answers.*

1. Red roses cost $3 each. Pink roses cost $2 each. A man buys 24 flowers for his wife with some pink roses and the rest red roses. The total cost of the flowers is $68. Write a system of equations to model this situation and use the equation to determine how many red and how many pink roses he bought.

2. Ice cream cones cost $4 each. Milkshakes cost $6 each. If ten items are purchased, either ice cream cones or milkshakes: (a) What is the greatest amount of money that can be spent? (b) What is the least amount of money that can be spent? (c) How much would it cost for 5 cones and 5 shakes? (d) If $52 is spent on ten items, how many cones were purchased and how many shakes? Explain how you got your answer to part d.

3. Marshmallows cost $2 an ounce and have 500 calories in an ounce. Cookies cost $3 an ounce and have 400 calories in an ounce. A mixture of marshmallows and cookies costs $29 and has 4,800 calories. Create a system of equations to model this scenario. How many ounces of marshmallows are there in the mixture?

4. Three burgers and two orders of French fries cost $24. Five burgers and one order of French fries cost $33. What is the cost of one burger? What is the cost of one order of French fries?

5. A cycling store sells two-wheel bicycles and three-wheel tricycles. It sells 58 cycles that had a total of 134 wheels. How many of each type of cycle did they sell?

Chapter Five

GRAPHS OF SOLUTION SETS OF LINEAR EQUATIONS

5.1 PRODUCING A GRAPH BY IDENTIFYING TWO OR MORE POINTS

=== KEY IDEAS ===

A linear equation, like $x + y = 10$, is one where neither of the variables has an exponent greater than or equal to 2. The set of ordered pairs that makes this equation true include (2, 8), (3, 7), and (9, 1). If each ordered pair that satisfies an equation is plotted as a point on the *coordinate plane*, the result is the *graph of the solution set of the equation*. For linear equations, the graph is always a line so getting two points in the solution set and drawing the line that passes through those two points will create the graph of the solution set for that line.

Graphing the Solution Set of a Linear Equation by Making a Table of Values

The equation $x + y = 10$ has an infinite number of ordered pairs that satisfy it. One way to organize the information before creating a graph is to make a table of values.

This is a chart with three ordered pairs satisfying the equation $x + y = 10$. For a linear equation, only two ordered pairs are needed, but it is wise to do an extra ordered pair in case one of your first two is incorrect.

x	y
2	8
3	7
9	1

Plot the ordered pair (2, 8) on the coordinate plane by locating the point that is two units to the right of the y-axis and 8 units above the x-axis. One way to do this is to start at the origin point where the two axes intersect and move to the right two units from there and then up 8 units.

Graphs of Solution Sets of Linear Equations

Do the same for the other two ordered pairs on the chart (3, 7) and (9, 1).

Draw a line through the three points. If the three points do not all lie on the same line, one of your ordered pairs is incorrect.

MATH FACTS

The line on a graph contains an infinite number of points. Each point corresponds to an ordered pair that is part of the solution set for the equation, and each ordered pair that is part of the solution set for the equation corresponds to a point on the line.

Example 1

Fill in this chart with three ordered pairs that satisfy the equation $2x + y = 10$. Then graph the three points that correspond to those three ordered pairs and use them to create the graph of the solution set of $2x + y = 10$.

x	y
0	8
1	6
2	4

Solution: One way to determine ordered pairs that satisfy an equation is to pick small values for *x* and then substitute those values into the equation and use algebra to solve for the corresponding *y* value that would make the equation true.

For $x = 0$, the equation becomes $2(0) + y = 10$. When you solve for *y*, it becomes $y = 10$. One ordered pair is $(0, 10)$.

For $x = 1$, the equation becomes $2(1) + y = 10$. When you solve for *y*, it becomes $y = 8$. Another ordered pair is $(1, 8)$.

For $x = 2$, the equation becomes $2(2) + y = 10$. When you solve for *y*, it becomes $y = 6$. A third ordered pair is $(2, 6)$.

x	y
0	10
1	8
2	6

Graphs of Solution Sets of Linear Equations

Plot the three points and draw a line through them to complete the graph.

MATH FACTS

It is common to call the graph of the solution set of an equation simply the graph of the equation. These are considered to mean the same thing though it is the solution set that is really being graphed and not the equation itself. Saying it the first way is more accurate, but usually on a test it would just say the graph of the equation.

Example 2

Will the point (9, 2) be on the line that is the graph of the solution set of the equation $x - 2y = 5$?

Solution: If an ordered pair is a solution to the equation, then the point that represents that ordered pair will be on the graph of the solution set. By substituting 9 for x and 2 for y into the equation, it becomes

$$9 - 2(2) \stackrel{?}{=} 5$$
$$9 - 4 \stackrel{?}{=} 5$$
$$5 \stackrel{\checkmark}{=} 5$$

So, (9, 2) is in the solution set for $x - 2y = 5$ and the point (9, 2) will be on the graph of the solution set.

Example 3

Below is the graph of the solution set for an equation. Based on this graph, is the ordered pair (4, 7) a solution to the equation?

Solution: If you plot the point (4, 7) on the same coordinate plane, it is close to being on the line, which is the graph of the solution set. If it were on the line, (4, 7) would be a solution to the equation, but since it is not, (4, 7) is not a solution to the equation.

Graphing the Solution Set of Linear Equations Using the Two-Intercept Method

---
MATH FACTS

Unless a line is vertical or horizontal, it will cross both the *x*-axis and the *y*-axis. The point where the line crosses the *x*-axis is called the *x-intercept* and the point where it crosses the *y*-axis is called the *y-intercept*. Any point on the *y*-axis has an *x*-coordinate of 0 and any point on the *x*-axis has a *y*-coordinate of 0. In the graph below, the *x*-intercept of the line is (5, 0) and the *y*-intercept is (0, 8).

Graphs of Solution Sets of Linear Equations

A quick way to make a graph of the solution set of a two variable equation is to calculate the x and y-intercepts. To get the x-intercept, substitute 0 for y and solve for x. To get the y-intercept, substitute 0 for x and solve for y.

Example 4

Find the x-intercept and y-intercept of the graph of the solution set of $2x + 4y = 16$. Use the intercepts to sketch the graph.

Solution: Substituting 0 for x leads to the equation $2(0) + 4y = 16$, which has $y = 4$ as the solution. The y-intercept is $(0, 4)$. Substituting 0 for y leads to the equation $2x + 4(0) = 16$, which has $x = 8$ as the solution. The x-intercept is $(8, 0)$. Plot both points on the coordinate plane, and draw a line through them.

Equations for Horizontal or Vertical Lines

An equation with one variable can also have a solution set of ordered pairs. One ordered pair that satisfies the equation $y = 3$ is $(0, 3)$ since it is true that $3 = 3$ if you substitute the 3 for the y (there is no x to substitute with the 0!). Other ordered pairs that satisfy this equation are $(1, 3)$, $(7, 3)$, and $(29, 3)$. When all the ordered pairs that satisfy the equation $y = 3$ are graphed, it becomes a horizontal line with y-intercept of $(0, 3)$.

For an equation with only an *x* variable, like *x* = 5, some ordered pairs that satisfy it are (5, 0), (5, 1), and (5, 2). When these are graphed, they make a vertical line.

Graphing Linear Equations Involving Absolute Value

The *absolute value* of a number is defined as the distance that number is from zero on the number line. The symbol for absolute value of *a* is |*a*|. The absolute value of 5 is 5. The absolute value of –5 is also 5.

A two variable equation can have absolute values like $y = |2x - 2|$. One way to graph the solution set is to create a chart.

x	$|2x - 2|$
–2	$|2(-2) - 2| = |-4 - 2| = |-6| = 6$
–1	$|2(-1) - 2| = |-2 - 2| = |-4| = 4$
0	$|2(0) - 2| = |0 - 2| = |-2| = 2$
1	$|2(1) - 2| = |2 - 2| = |0| = 0$
2	$|2(2) - 2| = |4 - 2| = |2| = 2$
3	$|2(3) - 2| = |6 - 2| = |4| = 4$
4	$|2(4) - 2| = |8 - 2| = |6| = 6$

Graphs of Solution Sets of Linear Equations

When these five points are graphed, they create a V shape. This is typical for linear equations involving absolute values.

Solving Systems of Linear Equations by Graphing

Chapter 4 showed how to solve systems of linear equations with algebra. Systems of linear equations can also be solved by graphing.

> **MATH FACTS**
>
> The ordered pair that is the solution to a system of linear equations will be related to the point of intersection of the two lines that are the graphs of the solution sets of the two equations.

In Chapter 4, there was an example where you had to solve the system of equations.

$$x + y = 10$$
$$x - y = 2$$

Using algebra, the solution was (6, 4).

This system can also be solved by graphing the solution sets of each equation and locating the intersection point.

Find the intercepts:

$$x + y = 10$$
x-intercept: $x + 0 = 10$, $x = 10$
y-intercept: $0 + y = 10$, $y = 10$

So the intercepts for the top equation are (10, 0) and (0, 10).

$$x - y = 2$$
x-intercept: $x - 0 = 2$, $x = 2$
y-intercept: $0 - y = 2$, $y = -2$

So, the intercepts for the bottom equation are (2, 0) and (0, –2).
Plot the intercepts on graph paper and draw the lines carefully with a ruler.

The lines intersect at the point (6, 4) so the solution to the system of equations is $x = 6$ and $y = 4$.

=== **MATH FACTS** ===

Solving with algebra is generally the faster and more accurate method for solving a system of linear equations. But this is a technique you need for systems of linear inequalities, coming later.

Check Your Understanding of Section 5.1

A. Multiple-Choice

1. Which are the coordinates of a point that will be on the line that represents the solution set of the equation $2x + 3y = 12$?
   (1) (4, 1)   (2) (2, 3)   (3) (5, 0)   (4) (3, 2)

2. What is the x-intercept of the graph of the solution set of the equation $2x + 5y = 20$?
   (1) (10, 0)   (2) (4, 0)   (3) (0, 10)   (4) (0, 4)

3. What is the y-intercept of the graph of the solution set of the equation $3x - 8y = 24$?
   (1) (0, –3)   (2) (0, 3)   (3) (8, 0)   (4) (0, 8)

4. The point $(2, k)$ is on the graph of the solution set of the equation $3x + y = 15$. What is the value of k?
   (1) 7   (2) 8   (3) 9   (4) 10

Graphs of Solution Sets of Linear Equations

5. Based on the x-intercept and y-intercept, this is the graph of the solution set of which equation?

   (1) $9x + 2y = 18$
   (2) $2x + 9y = 18$
   (3) $9x - 2y = 18$
   (4) $2x - 9y = 18$

6. Which is a graph of the solution set of the equation $y = 5$?

7. This is a graph of the solution set of which equation?

(1) $x = 8$
(2) $x = -8$
(3) $y = 8$
(4) $y = -8$

8. The equations $2x - 3y = 9$ and $3x + 2y = 20$ are graphed below. What is the solution to the system of equations?

$$2x - 3y = 9$$
$$3x + 2y = 20$$

(1) (5, 2)
(2) (2, 5)
(3) (1, 6)
(4) (6, 1)

Graphs of Solution Sets of Linear Equations

9. Below is the graph of the solution set of an equation. Based on this graph, which ordered pair does not seem to be part of the solution set of the equation $y = \frac{1}{3}x + 2$?

    (1) (3, 3)   (2) (6, 4)   (3) (6, 8)   (4) (9, 5)

10. What is the equation of the x-axis?

    (1) $x = 0$ (2) $y = 0$ (3) $x + y = 0$ (4) $x - y = 0$

B. *Show how you arrived at your answers.*

1. Make a table of values to graph the solution set of the equation $y + 2x = 8$.

2. Determine the x-intercept and y-intercept for the graph of the solution set of this equation and use them to produce a sketch of the graph $4x - 6y = 24$.

3. Tenley says that this is the graph of the equation $y = 3$. Ingrid says that it is the graph of the equation $x = 3$. Who is correct? Explain your reasoning.

4. Graphically solve the system of equations

$$x - y = 4$$
$$3x + 5y = 20$$

5. Below is the graph of $2x + y = k$ with intercepts at (0, 8) and (4, 0). What must the value of k be?

5.2 CALCULATING AND INTERPRETING SLOPE

KEY IDEAS

The *slope* of a line is a number that measures how steep it is. A horizontal line has a slope of 0. A line with a positive slope goes up as it goes to the right. A line with a negative slope goes down as it goes to the right. The variable used for slope is the letter m.

$m = 0$

$m = 1$

$m = 2$

$m = 1/2$

$m = -1$

$m = -2$

$m = -1/2$

Calculating the Slope of a Line

The slope of a line is the amount that it moves up (or down) for every one unit it moves to the right. If the two points are (x_1, y_1) and (x_2, y_2), then the slope can be calculated with the formula

$$m = \frac{y_2 - y_1}{x_2 - x_1}$$

For example, the line in the graph below passes through the two points (3, 1) and (6, 8). Notice that the line goes up as it moves to the right, so you should expect a positive slope. Also, the line is steeper than the line with a slope of 1 so the answer should be greater than 1.

Substituting into the slope formula $x_1 = 3$, $y_1 = 1$, $x_2 = 6$, and $y_2 = 8$, you get $m = \frac{8-1}{6-3} = \frac{7}{3}$ so the slope is $\frac{7}{3}$.

MATH FACTS

It doesn't matter which point you choose to make (x_1, y_1). As long as the y_1 is the y-coordinate that corresponds to the x_1 you choose, you will get the same answer. In the previous example, it would have become $m = \frac{1-8}{3-6} = \frac{-7}{-3} = \frac{7}{3}$.

Example 1

What is the slope of the line that passes through the points (1, 2) and (9, –2)?

(1) $\frac{1}{2}$ (2) –2 (3) $-\frac{2}{3}$ (4) $-\frac{1}{2}$

Solution: A sketch of the graph indicates that the solution should be negative.

Using the formula, you get $m = \dfrac{-2-2}{9-1} = \dfrac{-4}{8} = -\dfrac{1}{2}$, choice (4).

=== **MATH FACTS** ===

When two lines have the same slope they are parallel. If the lines have different slopes, then they are not parallel. A special case of nonparallel lines are lines whose slopes have a product of −1. When this happens, the lines are perpendicular. When two fractions have a product of −1, one is the *negative reciprocal* of the other. For example, the negative reciprocal of $\dfrac{2}{3}$ is $-\dfrac{3}{2}$ and $\left(\dfrac{2}{3}\right)\left(-\dfrac{3}{2}\right) = -1$.

=== **MATH FACTS** ===

A horizontal line has a slope of 0, whereas the slope of a vertical line is said to be *undefined*. When using the slope formula for two points on a vertical line, like (5, 2) and (5, 3), it becomes $m = \dfrac{(3-2)}{(5-5)} = \dfrac{1}{0}$. Since division by 0 is undefined, the slope of the vertical line is undefined too.

Interpreting Slope in a Time/Distance Graph

If a car is traveling 30 miles per hour, it will have traveled 30 miles after 1 hour, 60 miles after 2 hours, and 90 miles after 3 hours. If you graph the points (1, 30), (2, 60), and (3, 90) on a coordinate axis with elapsed time as the horizontal axis and distance traveled as the vertical axis, it will look like this:

The slope of this line is positive since it goes up as it goes to the right. It isn't easy to estimate the slope since the unit size on the time axis is larger than the unit size on the distance axis. But the slope can still be calculated using any two points, like (1, 30) and (2, 60).

$$m = \frac{60-30}{2-1} = \frac{30}{1} = 30.$$ The slope is 30 and the speed is 30 miles per hour.

Math Facts

In a distance time graph, the slope of a line segment corresponds to the speed of the thing that is moving.

Distance Time Graphs for Things That Are Changing Speeds

When a car is traveling at a constant speed, the distance time graph will be a line with the same slope on every interval. If the car changes speed, the slope of the distance time graph will also change.

In the graph below, the car has traveled 30 miles after 1 hour, 60 miles after 2 hours, and 140 miles after 4 hours.

Since the line segment between (2, 60) and (4, 140) is steeper than the line segment between (1, 30) and (2, 60), the car was moving faster between hours 2 and 4 than it was between hours 1 and 2.

If necessary, you can also calculate the two different speeds by finding the slopes of the two segments.

$$m = \frac{60-30}{2-1} = \frac{30}{1} = 30$$ mph between hours 1 and 2.

$$m = \frac{140-60}{4-2} = \frac{80}{2} = 40$$ mph between hours 2 and 4.

Example 2

Below is the distance time graph for a 120 mile bicycle trip. During which interval was the average speed of the bicycle the least?

(1) hours 0 to 2   (3) hours 4 to 6
(2) hours 2 to 4   (4) hours 6 to 10

Solution: Between hours 2 and 4, the slope appears to be the least steep, which means the bicycle is going slowest in that interval. To make sure, you can calculate the slopes for each interval separately. Between hours 0 and 2, the speed was 15 mph, between hours 2 and 4 the speed was 5 mph, between hours 4 and 6 the speed was 10 mph, and between hours 6 and 10 the speed was 15 mph. The answer is choice (2).

Graphs of Solution Sets of Linear Equations

Check Your Understanding of Section 5.2

A. *Multiple-Choice*

1. Which line has a positive slope?

   (1)   (3)

   (2)   (4)

2. Which line's slope is the greatest number?

   (1)   (3)

   (2)   (4)

134

3. Which is closest to the slope of this line?

(1) $-\dfrac{2}{3}$ (2) -3 (3) $\dfrac{2}{3}$ (4) 3

4. What is the slope of the line that passes through $(-2, 1)$ and $(8, 5)$?

(1) $\dfrac{2}{5}$ (2) $-\dfrac{2}{5}$ (3) $\dfrac{5}{2}$ (4) $-\dfrac{5}{2}$

5. What is the slope of the line that passes through $(2, 8)$ and $(5, -1)$?

(1) $-\dfrac{1}{3}$ (2) $\dfrac{1}{3}$ (3) -3 (4) 3

6. A line with a slope of $\dfrac{3}{4}$ passes through the point $(2,1)$ and the point $(10, a)$. What must the value of a be?

(1) 7   (2) 8   (3) 9   (4) 10

7. Line 1 passes through the points $(-3, -2)$ and $(3, 8)$. Line 2 is parallel to line 1. What is the slope of line 2?

(1) $\dfrac{3}{5}$ (2) $\dfrac{5}{3}$ (3) $-\dfrac{3}{5}$ (4) $-\dfrac{5}{3}$

Graphs of Solution Sets of Linear Equations

8. Below is a distance–time graph for a bicycle trip. During which time interval is the cyclist going the fastest?

(1) 0 to 2 hours
(2) 2 to 3 hours
(3) 3 to 5 hours
(4) 5 to 8 hours

9. A line with a slope of 3 passes through the point (4, 1). Which point will this line *not* pass through?
(1) (5, 4)   (2) (6, 7)   (3) (7, 10)   (4) (7, 11)

10. What is the slope of the hypotenuse of this right triangle?

(1) $\dfrac{3}{4}$ (2) $\dfrac{4}{5}$ (3) $\dfrac{4}{3}$ (4) $\dfrac{3}{5}$

B. *Show how you arrived at your answers.*

1. Line 1 passes through the points (−3, 7) and (7, 1). Line 2 is perpendicular to line 1. What is the product of the slopes of line 1 and line 2?

136

2. A triangle has vertex A at (0, 0), vertex B at (2, 5), and vertex C at (4, 5). Which side of the triangle has the greatest slope?

3. The slope of a line is $\frac{2}{3}$, and it passes through the point (–5, 2). What are three other points on this line that have coordinates that are integers?

4. A line passes through the points (–7, 2) and (–3, –1). It also passes through the point $(a, -7)$. What is the value of a?

5. Find the x-intercept and y-intercept of the graph of the solution set of $4x - 6y = 12$, and use them to find the slope of the line defined by that equation.

Graphs of Solution Sets of Linear Equations

5.3 SLOPE-INTERCEPT FORM

=== KEY IDEAS ===

When a two variable equation, like $2x + 3y = 12$, is written in this form with the variables on one side of the equation and the constant on the other side of the equation, it is called *standard form*. When it is written in the form $y = mx + b$, like $y = -\frac{2}{3}x + 4$, where m and b could be fractions, it is called *slope-intercept form*. When an equation is in slope-intercept form, there is a fast way to graph the solution set. Also, when an equation is in slope intercept form, it is often quicker to solve for x or y when the other variable's value is known.

Graphing the Solution Set of a Linear Equation That Is in Slope-Intercept Form

=== MATH FACTS ===

An equation like $y = 2x + 3$ or $y = \frac{2}{3}x - 5$ is said to be in *slope-intercept form*. In general, an equation of the form $y = mx + b$ is in slope-intercept form with the m representing the slope and the b representing the y-intercept.

To graph the solution set of a linear equation that is already in slope–intercept form:

1. Plot the point (0, b), which is on the y-axis. If the equation is $y = 2x + 3$, plot the point (0, 3). This is known as the y-intercept of the graph.

2. If the coefficient of the x term is not already a fraction, turn it into a fraction by putting the coefficient in the numerator of a fraction and a 1 in the denominator. If the coefficient is a negative fraction, make the numerator negative and the denominator positive. For the example, $y = 2x + 3$, the slope, denoted by m, is 2, which gets changed into $\frac{2}{1}$.

3. Starting at the y-intercept you already plotted, move right the number in the denominator of the slope. Then, from where you stopped, move up (down if it is negative) the number in the numerator of the slope. For the $y = 2x + 3$ example, the slope is $m = \frac{2}{1}$ so from (0, 3), you move to the right 1 unit and then up 2 units to get to the point (1, 5).

Graphs of Solution Sets of Linear Equations

4. Draw a line through the y-intercept and the new point. Put arrows on both sides of the line to indicate that it continues forever on both sides.

Example 1

Make a sketch of the solution set of the graph $y = -\frac{2}{3}x + 5$ using the slope-intercept process.

Solution: Since the constant is 5, the y-intercept is (0, 5). Since the coefficient of the x term is $-\frac{2}{3}$, $m = \frac{-2}{3}$. Plot the point (0, 5). From that point, move 3 to the right and 2 down (because it is negative) and then plot a new point. That's all you need for your line, but you can make another point by again moving 3 to the right and 2 down to get a third point.

MATH FACTS

The slope-intercept process only works when the equation is in slope-intercept form, $y = mx + b$. It is possible to use algebra to change an equation that is not in slope-intercept form into one that is by isolating the y term.

Example 2

Change $2y + 6x = 4$ into slope-intercept form.

Solution: In order to isolate the *y* term, first eliminate the $6x$ term from the left-hand side of the equation by subtracting $6x$ from both sides.

$$2y + 6x = 4$$
$$-6x = -6x$$

Usually we put the $-6x$ in front of the $+4$, rather than have $4 - 6x$, as we are trying to get this to look like $y = mx + b$.

$$2y = -6x + 4$$

Now, eliminate the coefficient on the *y* by dividing both sides by 2.

$$\frac{2y}{2} = \frac{-6x}{2} + \frac{4}{2}$$
$$y = -3x + 2$$

Finding the Equation in Slope-Intercept Form When the *y*-Intercept and Another Point Are Known

The two things needed before creating the equation are the slope and the *y*-intercept. If the *y*-intercept is given, half the work is done. To calculate the slope, use the slope formula $m = \frac{y_2 - y_1}{x_2 - x_1}$.

Example 3

The graph of the solution set of a two variable equation is a line that passes through (0, 3) and (5, 6). What could the equation be?

Solution: A sketch of the graph looks like this:

Graphs of Solution Sets of Linear Equations

The b value is 3 since the y-intercept is $(0, 3)$. To calculate the slope, use the slope formula with the points $(0, 3)$ and $(5, 6)$.

$m = \dfrac{6-3}{5-0} = \dfrac{3}{5}$. The equation is $y = \dfrac{3}{5}x + 3$.

MATH FACTS

The y-intercept can be described by the point $(0, b)$ or as just the number b. For example, the y-intercept of the equation $y = 5x + 7$ could be described as $(0, 7)$ or as just 7.

Check Your Understanding of Section 5.3

A. *Multiple-Choice*

1. Which is the graph of $y = 2x - 5$?

(1)

(2)

(3)

(4)

2. Which is the graph of $y = -\frac{5}{3}x + 4$?

(1)

(3)

(2)

(4)

3. What is the slope of the line defined by the equation $y = -3x + 4$?
   (1) 3   (2) –3   (3) 4   (4) –4

4. What is the y-intercept of the line defined by the equation $y = 5x + 2$?
   (1) (2, 0)   (2) (0, 2)   (3) (5, 0)   (4) (0, 5)

5. In the equation $y = 5$, the 5 is the
   (1) y-intercept
   (2) x-intercept
   (3) slope
   (4) length

6. The line defined by the equation $y = -\frac{4}{7}x + 5$ is perpendicular to the line defined by which equation?

   (1) $y = \frac{4}{7}x + 3$ (3) $y = -\frac{7}{4}x + 3$

   (2) $y = -\frac{4}{7}x + 3$ (4) $y = \frac{7}{4}x + 3$

143

Graphs of Solution Sets of Linear Equations

7. This is the graph of the solution set of which equation?

(1) $y = 5x + \dfrac{1}{3}$ (3) $y = 3x + 5$

(2) $y = 5x + 3$ (4) $y = \dfrac{1}{3}x + 5$

8. What are two points on the graph for the equation $y = -\dfrac{4}{5}x + 7$?

(1) (0, 7) and (4, 2) (3) (0, 7) and (5, 3)
(2) (7, 0) and (5, 3) (4) (7, 0) and (4, 2)

9. What are two points on the graph for the equation $y = \dfrac{5}{6}x - 4$?

(1) (0, 4) and (6, 1) (3) (0, −4) and (6, 1)
(2) (4, 0) and (6, 1) (4) (0, −4) and (5, 2)

10. What is the slope of the graph of the solution set of the equation $y = 3 + 2x$?
    (1) −3   (2) 3   (3) 2   (4) −2

B. Show how you arrived at your answers.

1. Identify the slope and the y-intercept of the graph for the equation $y = 3x - 9$, and sketch the graph on graph paper.

2. Jaiden says the graph for the equation $2y = 3x - 8$ has a slope of 3 and a y-intercept of −8. Edwin says that this is not correct and that the slope is $\dfrac{3}{2}$ and the y-intercept is −4. Which student is correct?

3. A portion of a line that passes through (2, 4) and (4, 5) is shown below. What is the y-intercept of this line?

4. Solve the system of equations graphically. An algebraic solution will not be accepted.

$$y = -2x + 8$$
$$y = -\frac{2}{3}x + 4$$

5. What do the equations from which these five lines were determined have in common?

5.4 GRAPHING SOLUTION SETS TO LINEAR EQUATIONS WITH A GRAPHING CALCULATOR

KEY IDEAS

Though the graphing calculator can't do the real "thinking" for you, it can be a powerful tool to help with graphing solution sets. It is also possible to *graphically solve* certain equations to find solution sets without needing to use algebra. This book has instructions for the two most popular calculators, the TI-84 Plus and the TI-Nspire.

Graphing Solution Sets for Linear Equations

Any graphing calculator can graph solution sets to two variable equations. Both the TI-84 and the TI-Nspire require that the y-variable is first isolated, however. Here are instructions for graphing $y = 2x - 5$.

For the TI-84:

To graph $y = 2x - 5$ on the TI-84, first push the [$y=$] key.

Then type $2x - 5$. Be sure to use the minus sign that is two keys above [ENTER] and not the (–), which is the negative sign, and which is one key to the left of [ENTER].

Push [ZOOM] and then [6] for ZStandard. This will make the graph on a grid that has a minimum value of –10 for both x and y and a maximum value of +10 for both x and y.

For the TI-Nspire:

Press [home] to get to the home screen. Then select [B] for Graph.

At the bottom of the screen, type $2x-5$ on the entry line after $f1(x)=$ and then press [enter].

To see the entry line, press [tab].

Graphs of Solution Sets of Linear Equations

Making a Table of Values with the Calculator

The graph is useful, but since you will likely have to draw the graph in your test booklet, the image on the calculator screen is not as helpful as a table of values.

For the TI-84:

Press [2ND] and [GRAPH] to get to the TABLE function. This is a table of values for the equation $y = 2x - 5$. Use the up and down arrows to see more values. These values can be used to sketch an accurate graph on graph paper.

For the TI-Nspire:

If you are on the graphing screen, press and hold down [ctrl] and then press [t]. To hide the table, click on the graph and push [ctrl] and [t] again.

Graphing Absolute Value Equations with the Graphing Calculator

The graph of $y = |2x - 2|$ can be quickly done on the TI-84 or TI-Nspire calculators.

For the TI-84:

In the $Y=$ menu press [MATH], [Right Arrow], and [1] for the abs() function. Then type $2x - 2$.

Press [ZOOM] and [6] to graph or [2nd] and [GRAPH] to see a table of values.

For the TI-Nspire:

From the home screen, press [B] to get to the Graph scratchpad.

Press [templates] button next to the [9] key and select the absolute value template in the second row on the far left.

Graphs of Solution Sets of Linear Equations

Type $2x - 2$ inside the absolute value bars and press [enter]. To see a table of values, hold down [ctrl] and press [t].

Systems of Linear Equations with the Intersect Feature

When two solution sets are graphed on the graphing calculator, there is a feature called intersect that determines the intersection point of two graphs. Since the intersection point of two lines determines the solution set for a system of equations, the graphing calculator can be used to solve systems of equations without algebra.

To find the solution set for the system of equations

$$y = -\frac{2}{3}x + 5$$
$$y = x - 5$$

For the TI-84:

Put both equations into the $Y =$ menu.

Press [GRAPH] to see the two lines.

Using the arrow keys, you can move the cursor as close as you can to the intersection.

Because the answer is likely to have integer coordinates, the solution set appears to be (6, 1).

For a more exact answer, press [2ND] and [TRACE], which displays the CALCULATE menu.

Select [5], intersect.

Because there may be more than two curves on the screen, you have to inform the calculator which two things you are trying to find the intersection for.

There is a cursor on the line for $y = -\frac{2}{3}x + 5$. Press [ENTER] to select the first line.

Graphs of Solution Sets of Linear Equations

A + is now on the $y = -\frac{2}{3}x + 5$ line, and the calculator is asking what the second curve is. As there is now a blinking cursor on the $y = x - 5$ line, press [ENTER] to set it as the second curve.

As there may be several intersection points, the calculator is asking for a "guess" near the intersection point you are looking for. Use the right and left arrows to get near the intersection point and press [ENTER] again.

The accurate answer is (6, 1).

For the TI-Nspire:

In the entry field, enter $-\frac{2}{3}x + 5$ after $f1(x)=$ and press [enter].

Press [tab] to see the entry field again and enter $x - 5$ after $f2(x)=$ and press [enter].

Press [menu][6][4] for Intersection.

Move the pointer to the right of the intersection point and press [enter] to set the upper bound.

Graphs of Solution Sets of Linear Equations

Move the pointer to the left of the intersection and press [enter] to find the intersection point.

The intersection point (6, 1) is labeled.

Example

Use the intersect feature of the graphing calculator to find the solution to the system of equations

$$-3x + 4y = -4$$
$$3x + 2y = 16$$

Solution: Though this can be done with algebra, it can also be done with the graphing calculator. The equations must be in the form of $y=$ in order to graph it.

Isolate the y in each equation.

$$-3x + 4y = -4$$
$$+3x = +3x$$
$$\frac{4y}{4} = \frac{3x-4}{4}$$
$$y = \frac{3}{4}x - 1$$

$$3x + 2y = 16$$
$$-3x = -3x$$
$$\frac{2y}{2} = \frac{-3x+16}{2}$$
$$y = -\frac{3}{2}x + 8$$

For the TI-84:

Enter these equations on the *Y=* menu.

Graph and find the intersection point.

The solution is $x = 4$, $y = 2$.

Graphs of Solution Sets of Linear Equations

For the TI-Nspire:

Enter the two equations in the entry line and press [enter].

Press [menu], [6], and [4] and click to the right and then to the left of the intersection.

The solution is (4, 2).

Check Your Understanding of Section 5.4

B. *Show how you arrived at your answers.*

1. Use your graphing calculator to graph the equation $y = x - 3$.

2. Use your graphing calculator to graph the equation $y = -\frac{3}{4}x + 2$.

3. An algebra equation like $2x - 3 = 5$ can be solved graphically by intersecting the two graphs $y = 2x - 3$ and $y = 5$. Graph these two equations and find the *x*-coordinate of the intersection to solve the algebra problem $2x - 3 = 5$.

4. Use your graphing calculator to solve the system of equations.

$$y = -\frac{4}{5}x + 2$$
$$y = 2x + 16$$

5. Use your graphing calculator to solve the system of equations.

$$y = \frac{1}{4}x + 1$$
$$y = -x + 11$$

Graphs of Solution Sets of Linear Equations

5.5 DETERMINING AN EQUATION FOR A GIVEN GRAPH

=== **KEY IDEAS** ===

Just as it is possible to create a graph of a solution set when given a two-variable equation, it is also possible to find the two-variable equation for which a graph is the solution set. Often this equation is then used to answer other questions about the graph.

Finding the Equation When the Slope and y-Intercept Are Known

In the equation $y = mx + b$, the m represents the slope, and the b is the y-intercept. If these two values are given, substitute them into the equation.

Example 1

What is the equation for a line that has a y-intercept of $(0, -2)$ and a slope of 5?

Solution: Since m is the slope, $m = 5$. And since $(0, b)$ is the y-intercept, b is -2. The answer is $y = 5x - 2$.

Finding the Equation When the y-Intercept and Another Point Are Known

If the y-intercept is known, then the b value in the $y = mx + b$ equation is the same number. With one other point, the slope formula can be used with that point and with the y-intercept to calculate the value for m.

Example 2

If a line passes through the points $(0, 4)$ and $(6, 8)$, what is the equation for that line?

Solution: Since one of the points has an *x*-coordinate of 0, it is the *y*-intercept and *b* is 4. Then to calculate *m*, use the slope formula with points (0, 4) and (6, 8).

$$m = \frac{8-4}{6-0} = \frac{4}{6} = \frac{2}{3}$$

The equation is $y = \frac{2}{3}x + 4$.

Finding the Equation When Two Points Are Known

If two points in the solution set for a linear equation are known, neither of which is the *y*-intercept, there is a two-part process for finding the equation. First, use the two points to calculate *m*, the slope of the line. Then substitute that slope and also the *x* and *y* values from one of the points into the equation $y = mx + b$ or $b = y - mx$ to solve for *b*.

Example 3

The graph of the solution set of a linear equation contains the points (4, –1) and (8, –4). What could be the equation?

Solution: Making a sketch of the graph shows that the *y*-intercept is something positive and that the slope is negative.

To calculate *m*, use the slope formula

$$m = \frac{-4-(-1)}{8-4} = \frac{-3}{4} = -\frac{3}{4}$$

Graphs of Solution Sets of Linear Equations

Now choose just one of the known points, like (4, –1) and substitute 4 for x, –1 for y, and $-\frac{3}{4}$ for m into $y = mx + b$.

$$-1 = -\frac{3}{4} \cdot 4 + b$$
$$-1 = -3 + b$$
$$+3 = +3$$
$$2 = b$$

So, the equation is $y = -\frac{3}{4}x + 2$.

Using the Calculator to Find the Equation When Two Points Are Known

If the question is a multiple-choice question about finding the linear equation when two ordered pairs in the solution set are known, the TI calculator has the ability to do this.

Here is how you could do the last example with the calculator.

The points on the line are (4, –1) and (8, –4).

For the TI-84:

Press [STAT] and then choose [1], Edit.
In the L1 column, enter the two x-coordinates. In the L2 column, enter the two y-coordinates. Be sure that the y-coordinates are in the same row as their corresponding x-coordinates.

Press [STAT] again and then use the right arrow to move to the CALC menu and press [4] to select LinReg and then [ENTER].

The calculator uses the variable a for slope instead of m. The solution is $y = –0.75x + 2$, as above.

160

For the TI–Nspire:

Press [home] to get to the home screen, and then select the Add Lists & Spreadsheet icon

On the spreadsheet, enter *x* to name the first column and *y* to name the second column, and then enter 4 into cell A1, –1 into cell B1, 8 into cell A2, and –4 into cell B2.

Press [menu], [4], and [1] to get to Stat Calculations.

Graphs of Solution Sets of Linear Equations

Press [3] for Linear Regression (mx + b).

In the X List box, type x. In the Y List box, type y.

On the spreadsheet, underneath $m \cdot x + b$ are the numbers −0.75 and 2. These are the values of m and b.

The equation is $y = -0.75x + 2$.

Check Your Understanding of Section 5.5

A. Multiple-Choice

For each question, find the equation of the line that passes through the two given points.

1. (0, –7) and (5, 8)
   (1) $y = 3x + 7$
   (2) $y = 3x - 7$
   (3) $y = \frac{1}{3}x + 7$
   (4) $y = \frac{1}{3}x - 7$

2. (0, 5) and (–12, –1)
   (1) $y = 2x + 5$
   (2) $y = 2x - 5$
   (3) $y = \frac{1}{2}x + 5$
   (4) $y = \frac{1}{2}x - 5$

3. (0, 4) and (9, 10)
   (1) $y = \frac{3}{2}x + 4$
   (2) $y = \frac{3}{2}x - 4$
   (3) $y = \frac{2}{3}x + 4$
   (4) $y = \frac{2}{3}x - 4$

4. (2, 1) and (6, 9)
   (1) $y = 2x - 3$
   (2) $y = 2x + 3$
   (3) $y = \frac{1}{2}x - 3$
   (4) $y = \frac{1}{2}x + 3$

5. (3, 1) and (7, –3)
   (1) $y = x + 4$
   (2) $y = x - 4$
   (3) $y = -x - 4$
   (4) $y = -x + 4$

6. (4, –2) and (12, 4)
   (1) $y = \frac{4}{3}x - 5$
   (2) $y = \frac{4}{3}x + 5$
   (3) $y = \frac{3}{4}x - 5$
   (4) $y = \frac{3}{4}x + 5$

163

Graphs of Solution Sets of Linear Equations

7. (5, 1) and (15, –3)

   (1) $y = -\frac{2}{5}x - 3$ (3) $y = \frac{2}{5}x - 3$

   (2) $y = -\frac{2}{5}x + 3$ (4) $y = \frac{2}{5}x + 3$

8. (–4, 1) and (4, 3)

   (1) $y = 4x - 2$ (3) $y = \frac{1}{4}x - 2$

   (2) $y = 4x + 2$ (4) $y = \frac{1}{4}x + 2$

9. (3, 7) and (6, 7)

   (1) $y = -7$ (3) $x = -7$
   (2) $y = 7$ (4) $x = 7$

10. (3, 5) and (3, 8)

    (1) $x = -3$ (3) $y = -3$
    (2) $x = 3$ (4) $y = 3$

B. *Show how you arrived at your answers.*

1. A line passes through the points (3, 4) and (9, 8). The line also passes through the point (5, a) for what value of a, rounded to the nearest tenth?

2. A line passes through the points (4, 7) and (8, 6). The line also passes through the point (a, 3.5) for what value of a?

3. Find the x-intercept and y-intercept of the line that passes through (5, 6) and (10, 2).

4. The equation $y = \frac{1}{3}x + 5$ passes through points (6, 7) and (12, a). What is the value of a?

5. The five points (5, 4), (6, 5), (7, 5), (9, 4), and (10, 5) are plotted on a coordinate plane. A line is drawn through the points (5, 4) and (10, 5). This line does not pass through either of the other three points. What is the sum of the three vertical line segments?

5.6 WORD PROBLEMS INVOLVING FINDING THE EQUATION OF A LINE

=== KEY IDEAS ===

Usually in a scientific experiment, data are collected and then analyzed. If the graph of the collected data seems to make a line, two points on the line are enough to get an approximate equation describing the line. That equation can then be used to make predictions about the results of other experiments.

A Real-World Problem Involving Two Data Points

A party begins at 12:00 PM. At 12:21 PM, there were 56 pieces of cake remaining. At 12:36 PM, there were 46 pieces of cake remaining. Find an equation that relates the number of minutes past 12:00 PM, called T, and the number of pieces of cake left, called C. Use your equation to predict how many pieces of cake will remain after 51 minutes. Use your equation to predict how many minutes it will take before there are 10 pieces of cake remaining.

Graphs of Solution Sets of Linear Equations

A graph of the scenario might look like this:

[Graph showing a line through points (21, 56) and (36, 46) with C on vertical axis (values 20, 40, 60) and T on horizontal axis (values 10, 20, 30, 40, 50)]

It uses the same process as in Section 5.5, except that the equation will not have x and y but will instead be $C = mT + b$.

The slope of the line is

$$m = \frac{46-56}{36-21} = \frac{-10}{15} = -\frac{2}{3}$$

For b, use one of the given points for T and C and the m just calculated.

$$C = mT + b$$
$$56 = -\frac{2}{3} \cdot 21 + b$$
$$56 = -14 + b$$
$$+14 = +14$$
$$70 = b$$

So the equation is $C = -\frac{2}{3}T + 70$.

For the second part of the question, substitute 51 for T and solve for C.

$C = -\frac{2}{3} \cdot 51 + 70 = -34 + 70 = 36$ so there will be 36 pieces of cake left after 51 minutes.

For the third part of the question, substitute 10 for C and solve for T with algebra.

$$10 = -\frac{2}{3}T + 70$$
$$-70 = -70$$
$$-60 = -\frac{2}{3}T$$
$$\left(-\frac{3}{2}\right) \cdot (-60) = \left(-\frac{3}{2}\right)\left(-\frac{2}{3}\right)T$$
$$90 = T$$

So it will take 90 minutes until just 10 pieces of cake remain.

Check Your Understanding of Section 5.6

A. *Multiple-Choice*

1. A balloon is held so that it is 6 feet above the ground. The balloon is released, and 10 seconds later it is 36 feet above the ground. Which equation can be used to relate the time since the balloon was released (T) to the height the balloon is above the ground (H)?
   (1) $T = 3H - 6$
   (2) $T = 3H + 6$
   (3) $H = 3T - 6$
   (4) $H = 3T + 6$

2. Lily has $500 in the bank when the year begins. Each week she takes out the same amount of money. After 8 weeks she has $436 left in the bank. Which equation can be used to model this scenario where W is the number of weeks that has passed since the beginning of the year and M is the amount of money remaining in the bank?
   (1) $W = -8M + 500$
   (2) $W = 8M - 500$
   (3) $M = -8W + 500$
   (4) $M = 8W - 500$

3. Marcus puts the same amount of money into the bank each week. Three weeks after the year begins, he has $575 in the bank. Ten weeks after the year begins, he has $750. Which equation can be used to model the amount of money (M) he has in the bank after (W) weeks?
   (1) $M = 25W - 500$
   (2) $M = 25W + 500$
   (3) $W = 25M - 500$
   (4) $W = 25M + 500$

Graphs of Solution Sets of Linear Equations

4. A taxi costs a fixed amount to get into and then each mile is an additional fee. If a 4 mile ride costs $8 and a 12 mile ride costs $18, which equation models the cost (C) of a taxi ride compared to the number of miles (M) traveled?
   (1) $C = 1.25M + 3$
   (2) $C = 1.25M - 3$
   (3) $M = 1.25C + 3$
   (4) $M = 1.25C - 3$

5. Claire exercises each day by running around the track. She increases the amount she runs each day by the same amount. Fifteen days after she begins training, she is running six miles. Thirty-five days after she begins training, she is running ten miles. What equation relates the number of days she has been training (D) to the number of miles she runs that day (M)?
   (1) $D = \left(\frac{1}{5}\right)M - 3$
   (2) $D = \left(\frac{1}{5}\right)M + 3$
   (3) $M = \left(\frac{1}{5}\right)D - 3$
   (4) $M = \left(\frac{1}{5}\right)D + 3$

6. Miriam goes on a diet to lose weight. She loses the same amount of weight each week. If after four weeks she weighs 242 pounds and after seven weeks she weights 236 pounds, what equation relates her weight in pounds (P) to the number of weeks since she has been on the diet (W)?
   (1) $W = -2P - 250$
   (2) $W = -2P + 250$
   (3) $P = -2W - 250$
   (4) $P = -2W + 250$

7. When food is put into a special freezer, the temperature of the food decreases the same number of degrees each minute. If after six minutes the food is 67 degrees and after twenty minutes the food is 60 degrees, which equation relates the temperature of the food (T) to the number of minutes (M) the food has been in the freezer?
   (1) $M = -\left(\frac{1}{2}\right)T + 70$
   (2) $M = -\left(\frac{1}{2}\right)T - 70$
   (3) $T = -\left(\frac{1}{2}\right)M + 70$
   (4) $T = -\left(\frac{1}{2}\right)M - 70$

8. Jocelyn jumps out of an airplane with a parachute. She falls the same number of feet each second. Ten seconds after jumping she is 3,000 feet in the air. Fifteen seconds after jumping she is 2,000 feet in the air. Which equation relates her height in the air (H) to the number of seconds since she jumped (S)?
   (1) $H = -200S + 5{,}000$
   (2) $H = -200S - 5{,}000$
   (3) $S = -200H + 5{,}000$
   (4) $S = -200H - 5{,}000$

9. Isaiah and Jaxson are climbing up a mountain. Four hours after starting, they are 4,600 feet high. Nine hours after starting, they are 5,350 feet high. Which equation relates their height (A) to the number of hours they have been climbing (T)?
   (1) $T = 150A - 4,000$
   (2) $T = 150A + 4,000$
   (3) $A = 150T - 4,000$
   (4) $A = 150T + 4,000$

10. Waylon is on an elevator in the Empire State Building that is going down. After ten seconds, he is at the 104th floor. After thirty seconds he is at the 92nd floor. Which equation relates the number of seconds (S) he has been in the elevator to the floor (F) that he is at?
    (1) $F = -\frac{3}{5}S - 110$
    (2) $F = -\frac{3}{5}S + 110$
    (3) $S = -\frac{3}{5}F - 110$
    (4) $S = -\frac{3}{5}F + 110$

B. *Show how you arrived at your answers.*

1. A tree is planted in the ground. The tree grows the same amount each year. After two years the tree is 14 feet tall. After five years the tree is 26 feet tall. (a) Write an equation that relates the height of the tree (H) to the number of years since it was planted (Y). (b) Use your equation to determine how tall the tree will be after ten years?

2. Kendrick exercises by doing pull-ups. Each week he increases the number of pull-ups he can do by the same amount. After four weeks of training, he can do 9 pull-ups. After ten weeks of training, he can do 18 pull-ups. (a) Write an equation that relates the number of pull-ups he can do (P) to the number of weeks he has been training (W). (b) How many pull-ups could he do when he started his training? (c) After how many weeks will he be able to do 24 pull-ups?

3. Ava uploads to the Internet a video that goes viral. Each day the video is seen by a certain number more than the day before. On the sixth day, the video is watched 1,900 times. On the tenth day, the video is watched 3,100 times. (a) What equation relates the number of views (V) to the number of days (D) since the video was posted? (b) How many times will the video be watched on the 14th day? (c) How many times will the video be watched on the 31st day?

Chapter Five **GRAPHS OF SOLUTION SETS OF LINEAR EQUATIONS**

Graphs of Solution Sets of Linear Equations

4. The population of Mathlandia is 300 million in 2007 and 380 million in 2015. The population increases by the same amount each year. (a) What equation can be used to relate the number of years since 2000 (T) to the population in millions (P)? (b) When will the population reach 730 million?

   a) $P = 10T + 230$ b) 2050

5. At 8:03 AM there are 188 empty seats on the train. At 8:10 AM there are 160 empty seats. The number of empty seats decreases by the same amount each minute. (a) What equation can be used to relate the number of minutes after 8:00 AM (M) to the number of empty seats on the train (E)? (b) How many seats will be empty at 8:15? (c) At what time will there be 120 empty seats? (d) At what time will there be no empty seats left?

   a. $E = -4M + 200$

   b. 140

   c. 8:20

   d. 8:50

Chapter Six

GRAPHING SOLUTION SETS FOR QUADRATIC EQUATIONS

6.1 GRAPHING SOLUTION SETS TO QUADRATIC EQUATIONS

KEY IDEAS

A two-variable quadratic equation is one where at least one of the variables has an exponent of 2. An example is the equation $y = x^2 - 6x + 8$. When the solution set to a two-variable quadratic equation is graphed, it does not become a line, but instead a *parabola*.

Graphing Solution Sets to Quadratic Equations by Making a Table

The equation $y = x^2$ has an infinite number of ordered pairs that satisfy it. Since $9 = 3^2$, (3, 9) is part of the solution set. Because it is also true that $9 = (-3)^2$, (-3, 9) is also in the solution set. With linear equations, only two points were needed since the graph was a line. With quadratic equations, five points are generally needed.

For the table, you can use the x values from -2 to $+2$.

x	y
-2	$(-2)^2 = 4$
-1	$(-1)^2 = 1$
0	$0^2 = 0$
1	$1^2 = 1$
2	$2^2 = 4$

The five ordered pairs, then, are (-2, 4), (-1, 1), (0, 0), (1, 1), and (2, 4).

Graphing Solution Sets for Quadratic Equations

Graph the five points on the coordinate plane.

Connect the points with a U-shaped curve called a parabola. Make arrows on the ends to indicate that the curve continues forever in both directions.

Example 1

Make a table of values and use them to graph the solution set of the equation $y = x^2 - 2x - 3$.

Solution: Using the five x values $-2, -1, 0, 1, 2$, the table becomes

x	y
-2	$(-2)^2 - 2(-2) - 3 = 4 + 4 - 3 = 5$
-1	$(-1)^2 - 2(-1) - 3 = 1 + 2 - 3 = 0$
0	$0^2 - 2(0) - 3 = 0 - 0 - 3 = -3$
1	$1^2 - 2(1) - 3 = 1 - 2 - 3 = -4$
2	$2^2 - 2(2) - 3 = 4 - 4 - 3 = -3$

Graphing the points $(-2, 5), (-1, 0), (0, -3), (1, -4)$, and $(2, -3)$ and connecting them with a parabola results in this graph.

Graphing Solution Sets to Square Root Equations by Making a Table

The solution set to a square root equation is a sideways half-parabola. When making a table, it is easiest if x values are chosen so that what is under the square root sign is a perfect square.

To graph the solution set of $y = \sqrt{x} + 2$, the table could be

x	y
0	$\sqrt{0} + 2 = 0 + 2 = 2$
1	$\sqrt{1} + 2 = 1 + 2 = 3$
4	$\sqrt{4} + 2 = 2 + 2 = 4$
9	$\sqrt{9} + 2 = 3 + 2 = 5$

Since we can't (yet) take the square root of a negative number, and since the square root of something like 5 is not an integer, and would be more difficult to graph, the numbers 0, 1, 4, and 9 were good choices.

The Axis of Symmetry and the Vertex of a Parabola

Every parabola has an *axis of symmetry*, usually a vertical line that divides the parabola into two equal pieces. The axis of symmetry passes through the *vertex* of the parabola, which is the point where it changes from decreasing to increasing, or from increasing to decreasing.

In the parabola for $y = x^2$, the axis of symmetry is vertical line $x = 0$ and the vertex is (0, 0). In the parabola for $y = x^2 - 2x - 3$, the axis of symmetry is the vertical line $x = 1$ and the vertex is (1, –4).

Using the Axis of Symmetry to Pick x Values for a Table

Rather than always use the five x values –2, –1, 0, 1, and 2, it is better to use as the middle number the x–intercept of the axis of symmetry. To determine the x–intercept of the axis of symmetry of a quadratic equation in the form $y = ax^2 + bx + c$, use the formula

$$x = -\frac{b}{2a}$$

For the first example, $y = x^2$, it is like $y = 1x^2 + 0x + 0$, so $a = 1$ and $b = 0$. So $-\frac{b}{2a} = -\frac{0}{2} = 0$ and the axis of symmetry was $x = 0$.

For the second example, $y = x^2 - 2x - 3$, $a = 1$ (since x^2 is the same as $1x^2$) and $b = -2$ so the axis of symmetry is a vertical line with the equation

$$x = -\frac{-2}{2(1)} = \frac{2}{2} = 1.$$

Example 2

Create a table with five ordered pairs to graph the solution set to $y = x^2 + 6x + 8$.

Solution: The x-coordinate of the middle of the five points can be calculated with the formula $x = -\frac{b}{2a}$. Since $a = 1$ and $b = 6$, this is $\frac{-6}{2 \cdot 1} = -\frac{6}{2} = -3$.

The chart is

x	y
−5	$(-5)^2 + 6(-5) + 8 = 25 - 30 + 8 = 3$
−4	$(-4)^2 + 6(-4) + 8 = 16 - 24 + 8 = 0$
−3	$(-3)^2 + 6(-3) + 8 = 9 - 18 + 8 = -1$
−2	$(-2)^2 + 6(-2) + 8 = 4 - 12 + 8 = 0$
−1	$(-1)^2 + 6(-1) + 8 = 1 - 6 + 8 = 3$

Graph the five ordered pairs and use them to create a sketch of the parabola.

Determining the Coordinates of the Vertex of the Parabola

Because the vertex is on the axis of symmetry, the x-coordinate of the parabola that is the graph of the solution set of $y = ax^2 + bx + c$ is $-\frac{b}{2a}$. To determine the y-coordinate, substitute for x the value you get for $-\frac{b}{2a}$ into the equation $y = ax^2 + bx + c$.

Graphing Solution Sets for Quadratic Equations

If the equation is $y = x^2 + 6x + 8$, the x-coordinate of the vertex is $\frac{-6}{2 \cdot 1} = \frac{-6}{2} = -3$. The y-coordinate is $y = (-3)^2 + 6(-3) + 8 = 9 - 18 + 8 = -1$. So the vertex is $(-3, -1)$ as can be seen on the graph from the last example.

MATH FACTS

The x-coordinate of the vertex of the parabola defined by the equation $y = ax^2 + bx + c$ is $x = -\frac{b}{2a}$. To get the y-coordinate of the vertex, substitute the x-coordinate into the equation $y = ax^2 + bx + c$ and solve for y.

Example 3

What is the vertex of the graph of the solution set of $y = 2x^2 - 16x + 5$?

Solution: The x-coordinate is $-\frac{b}{2a} = -\frac{-16}{2 \cdot 2} = \frac{16}{4} = 4$. Substitute this 4 into the equation to determine the y-coordinate. The y-coordinate is $y = 2 \cdot 4^2 - 16 \cdot 4 + 5 = 2 \cdot 16 - 64 + 5 = 32 - 64 + 5 = -27$. The answer is $(4, -27)$.

Example 4

What is the vertex of the graph of the solution set of $y = (x - 2)^2 - 4(x - 2) + 7$?

Solution: First, get this equation into the form $y = ax^2 + bx + c$ before using the $x = -\frac{b}{2a}$ formula.

$$y = (x - 2)(x - 2) - 4(x - 2) + 7$$
$$y = x^2 - 4x + 4 - 4x + 8 + 7$$
$$y = x^2 - 8x + 19$$

The x-coordinate of the vertex is at

$$y = -\frac{-8}{2 \cdot 1} = \frac{8}{2} = 4$$

The y-coordinate of the vertex is

$$y = 4^2 - 8 \cdot 4 + 19$$
$$y = 16 - 32 + 19 = 3$$

So, the vertex is $(4, 3)$.

Graphing Parabolas with the Vertex and Intercepts Method

For the graph of a line, any two points will help produce an accurate graph. When a parabola has two x-intercepts, four points are very useful: the vertex, the y-intercept, and the two x-intercepts.

For the equation $y = x^2 + 6x + 8$ has a vertex, as calculated in the previous section, of $(-3, -1)$. Since the y-intercept has an x-coordinate of 0, the y-intercept can be determined by substituting 0 in for x.

$$y = 0^2 + 6 \cdot 0 + 8 = 8 \text{ so the } y\text{-intercept is } (0, 8).$$

The x-intercepts each have a y-coordinate of 0. To calculate them, substitute 0 for y and solve.

$$0 = x^2 + 6x + 8$$

This is a quadratic equation covered in Chapter 3. This one can be solved by the factoring method.

$$0 = (x + 4)(x + 2)$$

$$\begin{array}{ccc} x + 4 = 0 & & x + 2 = 0 \\ -4 = -4 & \text{or} & -2 = -2 \\ x = -4 & \text{or} & x = -2 \end{array}$$

So the two x-intercepts are $(-4, 0)$ and $(-2, 0)$.

The four points, then, are $(-3, -1)$, $(0, 8)$, $(-4, 0)$, and $(-2, 0)$. Plot them and draw the parabola.

Example 5

Find the vertex, the y-intercept, and the two x-intercepts of the solution set of the equation $y = -x^2 + 4x + 12$.

Graphing Solution Sets for Quadratic Equations

Solution: The x-coordinate of the vertex is $x = \dfrac{-4}{2(-1)} = \dfrac{-4}{-2} = 2$.

The y-coordinate of the vertex is $y = -2^2 + 4 \cdot 2 + 12 = -4 + 8 + 12 = 16$.
The y-intercept is $y = -0^2 + 4 \cdot 0 + 12 = 12$.
The x-intercepts are the solutions to the equation

$$0 = -x^2 + 4x + 12$$
$$0 = -1(x^2 - 4x - 12)$$
$$0 = -1(x - 6)(x + 2)$$

$x - 6 = 0$	or	$x + 2 = 0$
+6 = +6		-2 = -2
$x = 6$	or	$x = -2$

The vertex is (2, 16), the y-intercept is (0, 12), and the x-intercepts are (6, 0) and (-2, 0). Graph all four points to sketch the parabola.

=== **MATH FACTS** ===

When the coefficient of the x^2 term is negative, the parabola opens downward instead of upward. The vertex of a parabola like this is also known as a *maximum* since it is the point with the highest y-coordinate. The vertex of a parabola that opens upward is known as a *minimum*.

Graphing Solution Sets to Quadratic Equations on the Graphing Calculator

In Section 5.4 the process for graphing linear equations on the TI-84 and TI-Nspire calculators was explained. For quadratic equations, it is the same process. Below are directions for graphing $y = x^2 + 6x + 8$ on both calculators.

For the TI–84:

Press [y=] key and enter $x^2 + 6x + 8$ into the first line.

Push [ZOOM] and then [6] for a graph that has a minimum value of –10 and a maximum value of +10 for both x and y.

To obtain the coordinates of the vertex, press [2ND] and then [TRACE] for the CALCULATE menu. Then press [3] for minimum (or [4] for maximum if the vertex is the highest point on the curve.)

When the calculator asks for "Left Bound?" move the cursor to the left of the vertex and press [ENTER].

179

Graphing Solution Sets for Quadratic Equations

When the calculator asks for "Right Bound?" move the cursor to the right of the vertex and press [ENTER].

When the calculator asks for a "Guess?" move the cursor near the vertex and press [ENTER].

The calculator gives a close approximation of the vertex. Since this is very close to $(-3, -1)$, it is nearly certain that the actual answer is $(-3, -1)$. (The calculator figures out the vertex in a very complicated way that could lead to rounding errors.)

To find the x-intercepts for the parabola, use the zeros feature on the CALCULATE menu.

Just like finding the location of the vertex, the calculator will ask for a left bound, a right bound, and a guess for each x-intercept.

To make a copy of the graph onto your graph paper, press [2ND] and then [GRAPH] to see a table of values.

For the TI-Nspire:

From the home screen, press [B] for graphing on the Scratchpad. Enter $x^2 + 6x + 8$ on the entry line, and press [enter].

To find the coordinates of the vertex press [menu], [6], and [2] for Minimum (or [3] if the parabola opens downward and the vertex is the maximum point of the parabola).

Graphing Solution Sets for Quadratic Equations

Move the hand to the left of the vertex and click to set the lower bound; then move the hand to the right of the vertex and click to set the upper bound.

For the *x*-intercepts, press [menu], [6], and [1] for the Zero finder. Move the hand to the left of one of the *x*-intercepts and click to set the lower bound; then move the hand to the right of the same *x*-intercept for the upper bound and click.

To see a table of values, press and hold the [ctrl] button and then press [t].

Graphing Solution Sets to Square Root Equations on the Graphing Calculator

To get the $\sqrt{}$ symbol on either TI calculator, press [2ND] key and then the [x^2] key. This symbol can be entered into an equation for graphing.

Graphing Solution Sets for Quadratic Equations

Intervals When a Graph Is Increasing or Decreasing

A graph is *increasing* when as the x-coordinates increase, the y-coordinates increase too. Informally, the graph goes up as it moves to the right.

The graph of the line $y = 2x$ is an example of a graph that is increasing everywhere.

Graphs of the solution sets of linear equations are either always increasing or always decreasing, depending on whether the slope of the line is positive or negative.

Graphs that form parabolas will switch, at the vertex, from increasing to decreasing or from decreasing to increasing. The graph of $y = x^2 + 6x + 8$ has a vertex at $(-3, -1)$.

This curve is decreasing when $x < -3$ and is increasing when $x > -3$.

Check Your Understanding of Section 6.1

A. Multiple-Choice

1. Which is a point on the graph of the solution set of $y = x^2 + 5x - 2$?
   - (1) (3, 19)
   - (2) (3, 20)
   - (3) (3, 21)
   - (4) (3, 22)

2. The parabola defined by the equation $y = x^2 - 8x + 12$ has a y-intercept at
   - (1) (0, -12)
   - (2) (0, 12)
   - (3) (12, 0)
   - (4) (-12, 0)

3. Which is an x-intercept of the parabola defined by the equation $y = x^2 - 2x - 15$?
   - (1) (0, -15)
   - (2) (0, 15)
   - (3) (-5, 0)
   - (4) (5, 0)

4. What are the coordinates of the vertex of the parabola defined by the equation $y = x^2 - 4x - 1$?
   - (1) (-2, 5)
   - (2) (2, 5)
   - (3) (-2, -5)
   - (4) (2, -5)

5. What is the equation of the axis of symmetry of the parabola defined by the equation $y = x^2 - 6x - 2$?
   - (1) $y = 3$
   - (2) $y = -3$
   - (3) $x = 3$
   - (4) $x = -3$

6. $x = -4$ is the x-coordinate of the vertex for the parabola defined by which equation?
   - (1) $y = x^2 + 8x + 3$
   - (2) $y = x^2 - 8x + 3$
   - (3) $y = x^2 + 4x + 3$
   - (4) $y = x^2 - 4x + 3$

Graphing Solution Sets for Quadratic Equations

7. What could be the equation that determines this parabola?

(1) $y = x^2 - 6x - 8$ (3) $y = x^2 + 6x - 8$
(2) $y = x^2 - 6x + 8$ (4) $y = x^2 + 6x + 8$

8. The axis of symmetry of the parabola defined by the equation $y = 3x^2 + 42x + 8$ is
   (1) $x = -7$ (3) $x = -14$
   (2) $x = 7$ (4) $x = 14$

9. Which is the graph of $y = -x^2 + 2x + 3$?

(1)

(3)

(2)

(4)

10. Based on this graph, what are the two solutions to the equation $x^2 - 2x - 1 = 2$?

(1) $x = -3$ and $x = 1$
(2) $x = 3$ and $x = 1$
(3) $x = 3$ and $x = -1$
(4) $x = -3$ and $x = -1$

B. *Show how you arrived at your answers.*

1. The graph of the parabola defined by the equation $y = x^2 + bx + c$ has an axis of symmetry at $x = -3$. Find possible values for b and c.

2. The graph of the parabola defined by the equation $y = x^2 + bx + c$ has x-intercepts at $(1, 0)$ and $(-4, 0)$. What are possible values for b and c?

3. A portion of a parabola is graphed below. It will pass through the three points $(1, 5)$, $(6, 0)$, and vertex $(4, -4)$. What are two other points on this parabola?

4. What are the coordinates of the vertex and the x-intercept(s) of the parabola defined by the equation $y = x^2 - 6x + 9$?

5. Below is the graph of $y = x^2 - 4x - 2$. What is the equation of the axis of symmetry of the graph of $y = (x - 1)^2 - 4(x - 1) - 2$?

6.2 USING THE GRAPHING CALCULATOR TO SOLVE QUADRATIC EQUATIONS

=== KEY IDEAS ===

If a quadratic equation does not require a solution that uses algebra, either because it is a multiple-choice question that does not require showing work, or because it is a free response question that does not say "only an algebraic solution will be accepted," it is possible for the graphing calculator to estimate the answer to any quadratic equation very quickly.

Using the Zeros Feature to Solve Quadratic Equations

There are several ways of using algebra to solve quadratic equations, outlined in Chapter 3. These range from guess and check to factoring to completing the square to the quadratic formula.

If there is an equation already in the form $ax^2 + bx + c = 0$, the solutions can be obtained by graphing the parabola and using the zeros feature of the graphing calculator to find the x-intercepts.

Example 1

Find the two solutions to the equation $2x^2 + 5x - 3 = 0$ on the graphing calculator.

Solution:

For the TI-84:

Plot the graph of $y = 2x^2 + 5x - 3$.

Press [2ND] and [TRACE] to get to the CALCULATE menu. Press [2] for zero. To find the first intercept, move the cursor to the left of it and press [ENTER] to set a lower bound; then move the cursor to the right of the intercept and press [ENTER] to set an upper bound. Move near the intercept and press [ENTER] again. Do this for the other intercept, too.

The first zero is at $x = -3$. The $y = 1E-12$ is a fancy way of saying $1 \cdot 10^{-12}$, which is very close to zero. The second zero is at $x = 0.5$. The x-coordinates of the x-intercepts are also the solutions to the equation $2x^2 + 5x - 3$ so the solutions to this quadratic equation are $x = -3$ or $x = 0.5$.

For the TI-Nspire:

Select [B] from the home screen to get to the graphing Scratchpad. Enter $2x^2 + 5x - 3$ on the entry line after $f1(x)=$ and press [enter].

Graphing Solution Sets for Quadratic Equations

Press [menu], [6], and [1] to select the Zero finding feature.

Select a lower bound and upper bound for each intercept.

The x-intercepts are $(-3, 0)$ and $(0.5, 0)$. The solutions to the original equation $2x^2 + 5x - 3 = 0$ are $x = -3$ and $x = 0.5$.

After the zeros are determined, it is possible to use them to factor the original equation. A quadratic equation with zeros of -3 and $\frac{1}{2}$ can have factors of $(x + 3)$ and $\left(x - \frac{1}{2}\right)$.

But since $(x + 3)\left(x - \frac{1}{2}\right)$ will not get the $2x^2$ as the first term, the factored quadratic must have been $2(x + 3)\left(x - \frac{1}{2}\right)$. Distribute the 2 through the $\left(x - \frac{1}{2}\right)$ term to become $(2x - 1)$.

So $2x^2 + 5x - 3 = (2x - 1)(x + 3)$.

Example 2

Use the calculator to find the two *exact* solutions to the quadratic equation $x^2 - 4x - 7 = 0$.

  (1) $3 \pm \sqrt{11}$ (3) $4 \pm \sqrt{11}$
  (2) $2 \pm \sqrt{11}$ (4) $5 \pm \sqrt{11}$

Solution: Using the Zero feature of the calculator the two x-intercepts of the parabola $y = x^2 - 4x - 7$ are approximately $(5.31, 0)$ and $(-1.32, 0)$. Check each answer choice to see which one is equivalent to these two answers.

Choice 1: $3 + \sqrt{11} = 6.32$, which is not either of the solutions.

Choice 2: $2 + \sqrt{11} = 5.31$, which is one of the answers. $2 - \sqrt{11} = -1.32$, which is the other answer for choice 2.

Solving Quadratic Equations in Different Forms with the Graphing Calculator

The method just described for solving quadratic equations of the form $ax^2 + bx + c = 0$ only works when the equation is in exactly that form. The graphing calculator can also quickly solve equations of the form $ax^2 + bx + c = d$.

Example 3

What are the two solutions to the equation $-x^2 - 4x + 6 = 1$?

Solution: One method for this equation is to subtract 1 from both sides of the equation and use the process for equations of the form $ax^2 + bx + c = 0$. An easier approach is to graph the two sides of the equation separately.

For the TI–84:

To find the intersection, press [2ND], [TRACE], and [5] for the INTERSECT feature. Press enter twice to set which curves to intersect and then move the cursor near the intersection and press [ENTER]. Do this for both intersection points.

Graphing Solution Sets for Quadratic Equations

For the TI-Nspire:

Enter the equation $-x^2 - 4x + 6$ after $f1(x)=$ and 1 after $f2(x)=$

Find the intersection of the parabola with the horizontal line by pressing [menu], [6], and [4].

Just like finding the x-intercepts, select a lower bound and an upper bound for each intersection point. The x-coordinates of the two intersection points are the two solutions to the equation.

Check Your Understanding of Section 6.2

A. Multiple-Choice

1. What are the solutions to $x^2 + 3x - 4 = 0$?
   (1) $x = 1, x = 4$
   (2) $x = 1, x = -4$
   (3) $x = -1, x = 4$
   (4) $x = -1, x = -4$

2. What are the solutions to $x^2 - 3x - 10 = 0$?
   (1) $x = 2, x = 5$
   (2) $x = 2, x = -5$
   (3) $x = -2, x = 5$
   (4) $x = -2, x = -5$

3. Solve for all values of x, rounded to the nearest hundredth, $x^2 - 4x + 1 = 0$.
   (1) 3.65, 0.24
   (2) 3.81, 0.34
   (3) 3.73, 0.27
   (4) 3.96, 0.39

4. Solve for all values of x, rounded to the nearest hundredth, $x^2 + 10x + 23 = 0$.
   (1) −3.41, −6.58
   (2) −3.59, −6.41
   (3) −3.31, −6.64
   (4) −3.62, −6.18

5. Solve for all values of x: $x^2 + 6x + 2 = 0$.
   (1) $-2 + \sqrt{7}, -2 - \sqrt{7}$
   (2) $-3 + \sqrt{7}, -3 - \sqrt{7}$
   (3) $-3 + \sqrt{11}, -3 - \sqrt{11}$
   (4) $-2 + \sqrt{11}, -2 - \sqrt{11}$

Graphing Solution Sets for Quadratic Equations

6. Which graph could be used to find the solutions to the equation $x^2 - 4x = 5$?

(1) [graph showing $y = x^2 - 4x$ and $y = 5$]

(3) [graph showing $y = x^2 - 4x$ and $y = 3$]

(2) [graph showing $y = x^2 - 4x$ and $y = 4$]

(4) [graph showing $y = x^2 - 4x$ and $y = 2$]

7. The x-intercepts of the parabola defined by which equation are the solutions to the equation $x^2 + 5x = 15$?
   (1) $y = x^2 + 5x + 15$
   (2) $y = x^2 + 5x$
   (3) $y = x^2 + 5x - 15$
   (4) $y = x^2 - 5x - 15$

8. The x-coordinates of the intersection of the line and the parabola are the solutions to which equation?

[graph showing $y = x^2 - 3x$ and $y = 8$]

   (1) $x^2 + 3x = 8$
   (2) $x^2 + 8x = 3$
   (3) $x^2 - 3x = 8$
   (4) $x^2 - 8x = 3$

9. Which of these equations does not have the same solutions as the others?
   (1) $x^2 - 8x - 3 = 0$
   (2) $x^2 - 8x + 1 = 2$
   (3) $x^2 - 8x = 3$
   (4) $x^2 - 8x + 1 = 4$

10. This is a portion of the graph of the solution set of $y = x^2 - 14x + 47$. What are the approximate solutions to the equation $x^2 - 14x + 47 = 0$?

   (1) 4 and 10
   (2) 4.5 and 9.5
   (3) 5 and 9
   (4) 5.5 and 8.5

B. *Show how you arrived at your answers.*

1. Below is the graph of the equation $y = x^2 - 10x + 22$. Use it to estimate the solutions to the equation $x^2 - 10x + 22 = 0$.

2. Use the graphing calculator to find the two solutions to the equation $x^2 = x + 1$.

Graphing Solution Sets for Quadratic Equations

3. The x-coordinates of the two intersection points of the line and the parabola are the solutions to what quadratic equation?

4. Use the graphing calculator to find the positive solution to the equation $0 = -16T^2 + 96T + 112$.

5. To solve the equation $x^2 - 5x = 3$, Mary graphs $y = x^2 - 5x$ and $y = 3$. Sofia solves it by graphing $y = x^2 - 5x - 3$. Christopher solves it by graphing $y = x^2$ and $y = 5x + 3$. Who is solving the question correctly?

6.3 SOLVING LINEAR-QUADRATIC SYSTEMS OF EQUATIONS

=== KEY IDEAS ===

A system of equations in which one of the equations is linear (having no exponents greater or equal to two) and the other is quadratic (having at least one exponent equal to two and all others equal to one) is called a *linear-quadratic* system of equations. Two methods for solving such systems are the graphing method and the substitution method.

If the system of equations is

$$y = 2x^2$$
$$y = 2x + 4$$

All the ordered pairs that satisfy both equations at the same time are in the solution set for this system of equations. Unlike a system where both equations are linear and there is at most one ordered pair in the solution set, there are often two ordered pairs in a solution set for a system like this.

Solving a Linear-Quadratic System of Equations with a Table

If you were to make a table of values from $x = -3$ to $x = +3$, it would look like this:

x	$2x^2$	$2x + 4$
-3	$2(-3)^2 = 2(9) = 18$	$2(-3) + 4 = -6 + 4 = -2$
-2	$2(-2)^2 = 2(4) = 8$	$2(-2) + 4 = -4 + 4 = 0$
-1	$2(-1)^2 = 2(1) = 2$	$2(-1) + 4 = -2 + 4 = 2$
0	$2(0)^2 = 2(0) = 0$	$2(0) + 4 = 0 + 4 = 4$
1	$2(1)^2 = 2(1) = 2$	$2(1) = 4 = 2 + 4 = 6$
2	$2(2)^2 = 2(4) = 8$	$2(2) + 4 = 4 + 4 = 8$
3	$2(3)^2 = 2(9) = 18$	$2(3) + 4 = 6 + 4 = 10$

Just from this table, it can be seen that for $x = -1$, both $2x^2$ and $2x + 4$ equals 2. And for $x = 2$, both $2x^2$ and $2x + 4$ equals 8. So, two solutions are $(-1, 2)$ and $(2, 8)$.

Solving a Linear-Quadratic System of Equations with Algebra

If the equations are in the form $y = x^2 + bx + c$ and $y = mx + b$, set the expressions equal to each other and solve for x with algebra.

For the system

$$y = 2x^2$$
$$y = 2x + 4$$

form the equation $2x^2 = 2x + 4$.

Then solve the quadratic equation by first subtracting $2x$ and 4 from both sides so it is in the form where one of the sides of the equation is zero.

$$2x^2 = 2x + 4$$
$$-2x = -2x$$
$$2x^2 - 2x = 4$$
$$-4 = -4$$
$$2x^2 - 2x - 4 = 0$$
$$2(x^2 - x - 2) = 0$$
$$2(x - 2)(x + 1) = 0$$
$$x - 2 = 0 \text{ or } x + 1 = 0$$
$$x = 2 \text{ or } x = -1$$

To get the y values of the solutions, substitute the x values into either equation.

For $x = 2$, $y = 2(2)^2 = 8$ so $(2, 8)$ is one solution.

For $x = -1$, $y = 2(-1)^2 = 2 \cdot 1 = 2$ so $(-1, 2)$ is the other solution. These are the same solutions obtained from the table method used above.

Example 1

Use an algebraic method to find the two solutions to the system of equations:

$$y = x^2 - 5x + 3$$
$$y = 2x - 3$$

Solution: Since both expressions are equal to y, they are equal to each other.

$$x^2 - 5x + 3 = 2x - 3$$

To get this quadratic equation into the form with zero on the right-hand side of the equation, subtract $2x$ from both sides and add 3 to both sides.

$$x^2 - 5x + 3 = 2x - 3$$
$$-2x = -2x$$
$$x^2 - 7x + 3 = -3$$
$$+3 = +3$$
$$x^2 - 7x + 6 = 0$$
$$(x - 6)(x - 1) = 0$$
$$x = 6 \text{ or } x = 1$$

For $x = 6$, $y = 2(6) - 3 = 12 - 3 = 9$
For $x = 1$, $y = 2(1) - 3 = 2 - 3 = -1$

The solutions are $(6, 9)$ and $(1, -1)$.

Solving a Linear-Quadratic System of Equations with a Graph

Using the table at the top of the previous page, the graph can be produced below. The two points where the line and the parabola intersect are $(-1, 2)$ and $(2, 8)$.

Solving a Linear-Quadratic System of Equations with the Graphing Calculator

Just as with a system of two linear equations, by graphing the two equations on the graphing calculator and using the INTERSECT feature, the calculator can estimate the points of intersection that correspond to the solution set to the system of equations.

For the TI-84:

Press [y=] and enter both equations.

Press [GRAPH] then [2ND], [TRACE], and [5]. Select the two curves and move the cursor near one of the intersections and press [ENTER]. Then do the same for the other intersection.

For the TI-Nspire:

From the home screen, press [B] for the Scratchpad Graph. Enter $2x^2$ after $f1(x)=$ and $2x + 4$ after $f2(x)=$

Graphing Solution Sets for Quadratic Equations

Press [menu], [6], and [4] for Intersection.

Move the hand to the left of the intersection and click; then move the hand to the right of the intersection and click again.

The other intersection is not visible with the default window of -10 to 10 on the x-axis and -7.39 to 7.39 on the y-axis. To zoom out, press [menu], [4], and [4] for Zoom—Out. This doubles all the maximum values. Now follow the same steps you did for the first intersection point.

The x-coordinates of the two intersections 2 and -1 are the two solutions to the system of equations.

$$y = 2x^2$$
$$y = 2x + 4$$

Example 2

Use the graphing calculator to find the solutions to the system of equations.

$$y = x^2$$
$$y = 2x + 3$$

Solution: There are two intersections (–1, 1) and (3, 9). So $x = -1$, $y = 1$ is one solution and $x = 3$, $y = 9$ is the other solution.

Check Your Understanding of Section 6.3

A. *Multiple-Choice*

1. Use the graphing calculator to find which ordered pair is a solution to the system.

$$y = x^2$$
$$y = x + 2$$

   (1) (1, 4)   (2) (2, 4)   (3) (3, 4)   (4) (4, 4)

Graphing Solution Sets for Quadratic Equations

2. Use the graphing calculator to find which ordered pair is a solution to the system.

$$y = 2x^2$$
$$y = -x + 3$$

(1) (1, 1)   (2) (1, 2)   (3) (1, 3)   (4) (1, 4)

3. Use the graphing calculator to find which ordered pair is a solution to the system.

$$y = x^2 - 8x + 17$$
$$y = -x + 7$$

(1) (3, 2)   (2) (4, 2)   (3) (5, 2)   (4) (6, 2)

4. Solve this system of equations using algebra.

$$y = x^2$$
$$y = 2x + 3$$

(1) (–1, 1) and (9, 3)
(2) (–1, 1) and (3, 9)
(3) (–1, 1) and (–3, 9)
(4) (1, –1) and (9, 3)

5. Use the graphing calculator to find both solutions to the system.

$$y = 2x^2$$
$$y = 2x + 4$$

(1) (–1, –2) and (–2, 8)
(2) (1, –2) and (2, –8)
(3) (1, 2) and (2, 8)
(4) (–1, 2) and (2, 8)

6. Use the graphing calculator to find both solutions to the system.

$$y = 3x^2$$
$$y = 3x$$

(1) (0, 0) and (–1, 3)
(2) (0, 0) and (1, –3)
(3) (0, 0) and (–1, –3)
(4) (0, 0) and (1, 3)

7. Solve this system of equations algebraically.

$$y = x^2 - 2x + 2$$
$$y = -x + 8$$

(1) (3, 5) and (2, 10)
(2) (3, 5) and (–2, 10)
(3) (–3, 5) and (–2, 10)
(4) (3, –5) and (2, –10)

8. Solve this system using any method you want.

$$y = x^2 + 2x - 2$$
$$y = x + 4$$

(1) (2, 6) and (–3, 1)   (3) (2, –6) and (–3, –1)
(2) (–2, –6) and (3, –1)   (4) (–2, 6) and (–3, 1)

9. Which system of equations could this graph be used to solve?

(1) $y = x^2$
$y = \dfrac{2}{3}x + 1$

(3) $y = x^2$
$y = \dfrac{3}{2}x + 1$

(2) $y = x^2$
$y = \dfrac{3}{2}x - 1$

(4) $y = x^2$
$y = \dfrac{2}{3}x - 1$

10. Graphically find all solutions to the system of equations.

$$y = x^2$$
$$y = 4x - 4$$

(1) (2, 4) only   (3) (2, 4) and (0, 0)
(2) (2, 4) and (–2, 4)   (4) (2, 4) and (2, –4)

B. *Show how you arrived at your answers.*

1. Find the two solutions to the system of equations using the graphing calculator.

$$y = \dfrac{1}{2}x^2$$
$$y = x + 4$$

Graphing Solution Sets for Quadratic Equations

2. Solve this system using an algebraic method. Any other method will not be accepted.

$$y = -x^2 + 4x + 5$$
$$y = x + 5$$

3. Use a graphing calculator to find the two solutions to the system of equations.

$$y = -\frac{1}{2}(x-4)^2 + 6$$
$$y = -\frac{1}{3}x + 6$$

4. Based on the graph below, Kaydence says there is just one solution to the system of equations. Jimena says there are two solutions. Who is correct, and why?

$$y = 3x^2$$
$$y = 6x + 9$$

5. Find a value of b so that the system of equations has exactly (a) 2 solutions and (b) 0 solutions.

$$y = x^2$$
$$y = 2x + b$$

6.4 GRAPHING QUADRATIC EQUATIONS FOR REAL-WORLD APPLICATIONS

=== KEY IDEAS ===

Many real-world scenarios can be modeled with quadratic equations. When a quadratic equation is given as a model for something, the techniques of solving quadratic equations either with algebra or with the graphing calculator can be used to solve the real-world problem.

The Height of a Projectile

When an object is thrown in the air, it rises to a maximum height and then drops back down. When x is the amount of time that has passed since the projectile was thrown and y is the height of the projectile at time x, the graph is a parabola. Its equation will be a quadratic equation with a negative coefficient for the x^2.

Example

A football is thrown in the air. Its height in the air after x seconds is determined by the equation $y = -16x^2 + 80x + 96$. Use the graphing calculator to determine (a) when the football will be 160 feet high and (b) when the football will land on the ground.

Solution:

For the TI–84:

Enter the equation into the graphing calculator.

```
Plot1 Plot2 Plot3
\Y1 = -16X^2+80X+9
6
\Y2=
\Y3=
\Y4=
\Y5=
\Y6=
```

Graphing Solution Sets for Quadratic Equations

If you try to graph by pressing [ZOOM] and [6] for ZStandard, the graph doesn't look much like a parabola. This is because only a small part of the parabola fits in this viewing *window*.

To fix this, the window needs to be adjusted. Press the [WINDOW] key.

```
WINDOW
 Xmin=-10
 Xmax=10
 Xscl=1
 Ymin=-10
 Ymax=10
 Yscl=1
 Xres=1
```

These six values need to be adjusted. Since we are only concerned with what happens after the ball is thrown, the Xmin can be set to 0. Since the ball does not go underground, set the Ymin to 0 also. The Xmax can be set to 6 since in the graph it can be seen that the ball is on the ground at time $x = 6$. Finally, the Ymax needs to be increased since we cannot see where the ball has its maximum height. Set the Ymax to 100. Xscl is the "scale" for how often it puts marks on the *x*-axis. Since there are only 6 units on the *x*-axis, an Xscl of 1 is appropriate. Since there will be 100 units on the *y*-axis, the marks need to be more spread out. A Yscl of 10 is appropriate.

```
WINDOW
 Xmin=0
 Xmax=6
 Xscl=1
 Ymin=0
 Ymax=100
 Yscl=10
 Xres=1
```

This looks better, but the Ymax needs to be increased even more. Try 200.

```
WINDOW
 Xmin=0
 Xmax=6
 Xscl=1
 Ymin=0
 Ymax=200
 Yscl=10
 Xres=1
```

To find when the football will be 160 feet high, graph the line $y = 160$ on the same coordinate plane.

Use the INTERSECT feature to find the two times that the football is 160 feet high, once on the way up and once on the way down.

It is 160 feet high after 1 second and again after 4 seconds.

To find when the football is on the ground, use the ZERO feature.

For the TI-Nspire:

Type $16x^2 + 80x + 96$ after the $f1(x) =$ in the entry line and press [enter].

Graphing Solution Sets for Quadratic Equations

This is just a small portion of the parabola because the viewing window needs to be adjusted. Press [menu], [4], and [1] for Window Settings.

The four numbers for XMin, XMax, YMin, and YMax need to be adjusted. In order to see the vertex of the parabola, the YMax value must be increased. Also it is not necessary to see all the negative values on the x-axis. Make XMin 0, XMax 7, YMin −10, and YMax 100.

The YMax was not high enough. Adjust it to 200.

To find when the football is 160 feet high, graph $y = 160$ by typing 160 after $f2(x) =$.

Now press [menu], [6], and [4] for Intersection and locate the two intersection points.

The x-coordinates of the intersection points are the two answers 1 and 4. The football is 160 feet high after 1 second and after 4 seconds.

To find when the football lands on the ground, press [menu], [6], and [1] for Zero and locate the x-intercept.

Graphing Solution Sets for Quadratic Equations

The x-intercept is $(6, 0)$, which means that the height of the football after 6 seconds is 0 feet, which means that it has landed on the ground after 6 seconds.

This question could also have been solved with algebra.

For the first part,

$$-16t^2 + 80t + 96 = 160$$
$$-160 = -160$$
$$-16t^2 + 80t - 64 = 0$$
$$-16(t^2 - 5t + 4) = 0$$
$$-16(t - 4)(t - 1) = 0$$
$$t - 4 = 0 \text{ or } t - 1 = 0$$
$$t = 4 \text{ or } t = 1$$

The football is 160 feet high after 1 second and after 4 seconds.

For the second part,

$$-16t^2 + 80t + 96 = 0$$
$$-16(t^2 - 5t - 6) = 0$$
$$-16(t - 6)(t + 1) = 0$$
$$t - 6 = 0 \text{ or } t + 1 = 0$$
$$t = 6 \text{ or } t = -1$$

The football is on the ground after 6 seconds. The -1 gets rejected since -1 seconds is before the football was thrown.

Check Your Understanding of Section 6.4

B. *Show how you arrived at your answers.*

1. A projectile's height is modeled by the equation $h = -16t^2 + 96t + 256$. Use the graphing calculator or algebra to determine after how many seconds the projectile will be 336 feet high.

2. The cost in thousands of dollars for x days of work from Fred's construction company is $C = \frac{1}{5}x^2$. The cost for x days of work from Barney's construction company is $C = \frac{2}{5}x + 3$. Use the graphing calculator to determine the number of days that the cost to use Fred's construction company is equal to the cost to use Barney's construction company.

3. A stunt man drops from the top of a 50 foot elevator shaft. His height above ground is determined by the equation $h = -16t^2 + 50$. At the moment the stunt man begins to drop, an elevator goes up so that its height above ground is determined by the equation $h = 34t$. After how many seconds will the stunt man land on top of the elevator?

4. A car falls off a bridge that is 100 feet high. The height of the car off the ground is $h = -16t^2 + 100$. At the moment the car falls, a superhero flies up from the ground. Her height off the ground is $h = 18t$. After how many seconds does the superhero catch the car, and how high in the air will the superhero be when she catches it?

5. In a new video game "Irate Iguanas," two iguanas are shot from slingshots at the same time. Both are traveling to the right at a speed of one foot per second. The height off the ground of the first iguana is determined by the equation $h = -t^2 + 6t + 27$. The height off the ground of the other iguana is $h = \dfrac{3}{2}t + 18$, where t is the amount of time in seconds. When will the two irate iguanas crash into each other?

Chapter Seven

LINEAR INEQUALITIES

7.1 ONE-VARIABLE LINEAR INEQUALITIES

=== Key Ideas ===

A linear equation line $x + 2 = 10$ has just one solution, $x = 8$. This changes when the = is replaced with a > or a <. The inequality $x + 2 > 10$ does not have 8 as a solution, but it does have many other solutions including 9, 10, and 11. Solving one-variable linear inequalities is very similar to solving one-variable linear equations with one important exception, dealing with negative coefficients.

Solving an inequality like $x + 2 > 10$ is a lot like solving the equality $x + 2 = 10$. Just as the addition property of equality and subtraction property of equality allow you to add or subtract to both sides of an equation and keep it true, you can add to, or subtract from, both sides of an inequality and keep it true.

Use algebra to isolate the variable by subtracting 2 from both sides of the inequality.

$$x + 2 > 10$$
$$-2 = -2$$
$$x > 8$$

The solution set includes all the numbers greater than 8. Integers like 9, 10, and 11 are part of the solution set as well as number like 8.1, 8.01, and 8.001. The number 8 is not part of the solution set since it is not true that $8 > 8$.

For an inequality that has a positive coefficient, like $2x > 10$, isolate the variable by dividing both sides of the inequality by the coefficient.

$$\frac{2x}{2} > \frac{10}{2}$$
$$x > 5$$

All numbers greater than 5 satisfy the inequality $2x > 10$.

When both sides of an inequality are divided by a negative number, the direction of the inequality sign must be reversed. For example, it is true that $6 > 4$ but if both sides are divided by -2, the inequality would say $-3 > -2$, which is not true.

213

Linear Inequalities

MATH FACTS

When an inequality has a positive coefficient, you can divide both sides of the inequality by that coefficient without changing the direction of the inequality sign. When an inequality has a negative coefficient, divide both sides by it and reverse the direction of the inequality sign.

For the inequality $-2x > 10$, divide both sides by -2 and reverse the direction of the inequality sign.

$$\frac{-2x}{-2} > \frac{10}{-2}$$
$$x < -5$$

Two-step algebra inequalities are solved the same as two-step algebra equalities, still reversing the direction of the inequality sign when dividing both sides by a negative.

Example 1

What is the solution set to the inequality $-3x + 2 \geq 17$?
   (1) $x > -5$ (2) $x \geq -5$ (3) $x < -5$ (4) $x \leq -5$

Solution:

$$-3x + 2 \geq 17$$
$$-2 = -2$$
$$\frac{-3x}{-3} \geq \frac{15}{-3}$$
$$x \leq -5$$

The answer is choice (4).

Example 2

What is the smallest integer that is a solution for x in the inequality $-5x - 2 < -22$?
   (1) 3
   (2) 4
   (3) 5
   (4) There is no smallest integer solution.

Solution: First, isolate the x to determine the entire solution set.

$$-5x - 2 < -22$$
$$+2 = +2$$
$$\frac{-5x}{-5} < \frac{-20}{-5}$$
$$x > 4$$

All numbers greater than 4 satisfy the original inequality. Of all the integers greater than 4, 5 is the smallest so the answer is 5, choice (3). If the last line had been $x \geq 4$, then the smallest value that would be a solution would be 4.

Example 3

If $3x + ax + 4 > 8$, and if $x = -2$, what is the greatest integer value for a that satisfies this inequality?

   (1) –6   (2) –5   (3) –4   (4) –3

Solution: Substitute $x = -2$ into both x values in the inequality.

$$3(-2) + a(-2) + 4 > 8$$
$$-6 - 2a + 4 > 8$$
$$-2a - 2 > 8$$
$$+2 = +2$$
$$\frac{-2a}{-2} > \frac{10}{-2}$$
$$a < -5$$

Of all the integers less than –5, –6 is the greatest one. The answer is choice (1).

Linear Inequalities

> **Check Your Understanding of Section 7.1**

A. Multiple-Choice

1. What is the solution set for $2x > 10$?
   (1) $x < 5$
   (2) $x \leq 5$
   (3) $x > 5$
   (4) $x \geq 5$

2. What is the solution set for $3x < -18$?
   (1) $x < -6$
   (2) $x > -6$
   (3) $x \geq -6$
   (4) $x \leq -6$

3. What is the solution set for $-4x \geq 20$?
   (1) $x \geq -5$
   (2) $x > -5$
   (3) $x \leq -5$
   (4) $x < -5$

4. What is the solution set for $-5x + 12 \leq -8$?
   (1) $x \leq -4$
   (2) $x \leq 4$
   (3) $x \geq 4$
   (4) $x > 4$

5. What is the smallest integer that satisfies the inequality $-6x < -18$?
   (1) 3
   (2) 4
   (3) 5
   (4) 6

6. What is the smallest integer that satisfies the inequality $-6x \leq -18$?
   (1) 2
   (2) 3
   (3) 4
   (4) 5

7. What is the greatest integer that satisfies the inequality $-3x + 4 > -17$?
   (1) 4
   (2) 5
   (3) 6
   (4) 7

8. What is the greatest integer that satisfies the inequality $-5x - 2 \geq -37$?
   (1) 6
   (2) 7
   (3) 8
   (4) 9

9. Which number satisfies the inequality $-4x + 3 < 23$?
   (1) -7
   (2) -6
   (3) -5
   (4) -4

10. What is the smallest number that satisfies the inequality $-\frac{2}{3}x - 4 \leq 6$?
    (1) –15   (3) –17
    (2) –16   (4) –18

B. *Show how you arrived at your answers.*

1. Graph the solution set to the inequality $3x + 4 \geq 19$ on a number line.

2. Alejandra says that $-2x > 6$ has the solution set $x > -3$. Lorenzo says that it has the solution set $x < -3$. Which student is correct, and why?

3. Genevieve solves the inequality $-4x < 8$ the following way:

    $$-4x < 8$$
    $$+4x = +4x$$
    $$0 < 4x + 8$$
    $$-8 = -8$$
    $$-\frac{8}{4} < \frac{4x}{4}$$
    $$-2 < x$$
    $$x > -2$$

   Is this a valid way of finding the solution set? Explain why or why not.

4. What is the solution set of the inequality $5x - 2 \geq 2x + 10$?

5. Josephine has $60 and wants to go to a restaurant. The bill, including a 20% tip, must be no more than $60. The equation for what the price of the meal can be without the tip is $x + 0.20x \leq 60$. What is the solution set for this inequality?

Linear Inequalities

7.2 GRAPHING TWO-VARIABLE LINEAR INEQUALITIES

=== KEY IDEAS ===

The graph of a linear equation, like $x + y = 10$, is a line. When the = is replaced with a >, <, ≤, or a ≥, the graph becomes a *half-plane*. The process of creating the graph for the solution set to the inequality first requires graphing a line and then deciding which side of the line to shade in.

Graphing Inequalities Involving ≤ or ≥ Signs

The equation $x + y = 10$ had an infinite number of solutions including (2, 8), (3, 7), and (4, 6). When the = is replaced with the ≥ symbol, it becomes the two-variable inequality $x + y \geq 10$.

(2, 8), (3, 7), and (4, 6) are still solutions to this inequality, but also (2, 9), (3, 8), and (4, 7) since $2 + 9 = 11 \geq 10$, $3 + 8 = 11 \geq 10$, and $4 + 7 = 11 \geq 10$. The solution set for the inequality must contain the line and also these other points.

To graph the solution set to the inequality $x + y \geq 10$, first graph the line for the equation $x + y = 10$ using any of the methods from Chapter 5.

Every line divides the plane into two halves. Unless the line passes through (0, 0), one of those halves contains the point (0, 0) and the other one does not.

One of the sides of this line will be shaded. To determine which side to shade, pick a point that is not on the line, generally (0, 0) (as long as the line doesn't pass through (0, 0)) and substitute into the inequality to see if it makes the inequality true.

Testing (0, 0)

$$x + y \geq 10$$
$$0 + 0 \geq 10$$
$$0 \geq 10$$

Since 0 is not greater than or equal to 10, (0, 0) is not part of the solution set so the side of the line that does not contain (0, 0) gets shaded.

Linear Inequalities

Example 1

Graph the solution set to the inequality $y \geq \frac{2}{3}x - 5$.

Solution: First, graph the line $y = \frac{2}{3}x - 5$, which will be the boundary between the shaded and unshaded regions.

Since (0, 0) is not on the line, test to see if (0, 0) satisfies the inequality.

$$y \geq \frac{2}{3}x - 5$$

$$0 \geq \frac{2}{3}(0) - 5$$

$$0 \geq 0 - 5$$

$$0 \geq -5$$

Since 0 is greater than −5, (0, 0) is part of the solution set. Shade the side of the line that contains (0, 0).

Graphing Inequalities Involving > or < Signs

When the two-variable inequality contains a > or < sign, the only difference is that when the line is first graphed it must be a dashed line. For the inequality $x + y > 10$, the ordered pair (2, 8) does not make it true since 2 + 8 is equal to 10, not greater than 10. None of the ordered pairs that satisfy the equation $x + y = 10$ also satisfies the equation $x + y > 10$. But the points (2, 9), (2, 8.1), and (2, 8.01) do make the inequality true so the rule for shading still applies.

To graph $x + y > 10$, first graph $x + y = 10$ (or $y = -x + 10$ if first changed to slope-intercept form) but instead of making the line solid, make it dashed.

Check (0, 0) to determine whether to shade the side containing (0, 0) or the other side of the dashed line.

$$0 + 0 > 10$$
$$0 > 10$$

Since this is not true, shade the side of the dashed line that does not contain (0, 0).

Linear Inequalities

MATH FACTS

If the symbol in a two-variable inequality is a greater than sign (>) or a less than sign (<), the line separating the shaded half plane from the unshaded half plane will be dashed. If the symbol is a greater than or equal to sign (≥) or a less than or equal to sign (≤), the line will be solid.

Example 2

Which of the following could be the graph of $y \leq -2x + 5$?

(1)

(3)

(2)

(4)

Solution: Because of the less than or equal to symbol, the line must be solid, eliminating choices (2) and (4). When (0, 0) is tested into the inequality, it becomes $0 \leq 5$, which is true, so the side of the line containing (0,0) should be shaded, which corresponds to choice (3).

Graphing Two-Variable Inequalities on the Graphing Calculator

Both the TI–84 and the TI–Nspire can graph inequalities. Before entering the inequality into the calculator, it must be in slope-intercept form. For the inequality $2x + 4y > 12$, isolate the y first.

$$2x + 4y > 12$$
$$-2x = -2x$$
$$\frac{4y}{4} > \frac{-2x+12}{4}$$
$$y > -\frac{2}{4}x + 3$$
$$y > -\frac{1}{2}x + 3$$

For the TI–84:

To graph inequalities on the TI–84, press the [APPS] button and then scroll down to an application called Inequalz and press [ENTER] twice. This opens the inequalities application and the Y= menu. When the cursor is moved to the equals sign, the options for the inequality signs appears at the bottom of the screen.

Move the cursor over the = sign after Y1 and press [ALPHA] and then [TRACE] to get the > symbol. Then enter $-\frac{1}{2}x + 3$ and press [GRAPH].

Linear Inequalities

This is the graph of $y > -\left(\dfrac{1}{2}\right)x + 3$ with shading on the proper side of the line and also with the line being a dotted line since there is a > symbol and not a ≥ symbol.

To turn off the Inequalz application, press [APPS] scroll to it, press [ENTER], and then select choice [2] to quit.

Note: If the memory of the TI–84 is completely cleared before the test, the Inequalz application will be deleted also. If just the RAM is cleared, the application will remain on the calculator.

For the TI–Nspire:

Press [b] from the home screen to get to the Scratchpad graph. In the entry line where it says $f1(x)=$, push the [del] key to delete the = sign. A menu of the different inequality signs will appear.

Select [4] and press [enter]. The graph will appear with the proper side of the line shaded in. Notice that on the TI–Nspire it is not easy to see if the line is a dotted line or a solid line. In this case, it is a dotted line since the inequality sign was a > and not a ≥.

224

Check Your Understanding of Section 7.2

A. *Multiple-Choice*

1. Which is the graph of $y < x + 4$?

(1)

(2)

(3)

(4)

Linear Inequalities

2. Which is the graph of $y \geq 2x - 1$?

(1)

(3)

(2)

(4)

3. This is the graph of which inequality?

(1) $2x - 3y > 6$ (3) $2x - 3y \geq 6$
(2) $2x - 3y < 6$ (4) $2x - 3y \leq 6$

4. Which is the graph of the inequality $x + 2y > 10$?

(1)

(2)

(3)

(4)

5. Which is the graph of $y > 4$?

(1)

(2)

(3)

(4)

Linear Inequalities

6. Which is the graph of $x \leq -2$?

(1) (3)

(2) (4)

7. Which is a solution to the inequality $x - y > 3$?
   (1) (10, 8)
   (2) (12, 8)
   (3) (7, 5)
   (4) (4, 1)

8. Below is the graph of $x + y \leq 8$. Which is a point in the solution set?

   (1) (1, 8)   (2) (2, 7)   (3) (4, 5)   (4) (3, 5)

9. All of these ordered pairs are part of the shaded region for the graph of $2x + y < 12$ except

   (1) (2, 7)   (2) (3, 4)   (3) (5, 3)   (4) (5, 0)

10. Which point would not be a good point for checking which side of the line to shade for the inequality $y \leq -x + 4$?

    (1) (0, 5)   (2) (0, 4)   (3) (0, 3)   (4) (0, 2)

B. *Show how you arrived at your answers.*

1. A small elevator has a capacity of 4 people. On the graph below, make points indicating valid combinations of men and women who could be in the elevator.

Linear Inequalities

2. What is the inequality that has this solution set?

3. Ryan is graphing the solution set to the inequality $y \geq 3x$. After drawing the line $y = 3x$, he tests the point $(0, 0)$ and finds that $0 \geq 3(0)$ is true. Does he now have enough information about which side of the line to shade in? Explain.

4. Graph the solution set for the inequality $-y < 2x - 4$.

5. Juice costs $2 a pint. Milk costs $3 a pint. If you have $12, make a graph that shows all the possible combinations of juice and milk you can purchase.

7.3 GRAPHING SYSTEMS OF LINEAR INEQUALITIES

=== KEY IDEAS ===

Systems of two linear equations usually have one ordered pair that satisfies both equations. For systems of two linear inequalities, there are many solutions. These solutions are most easily solved by graphing the two inequalities and finding the intersections of the two shaded regions. The solution will look like a triangle.

Systems of Linear Inequalities

The system of equations

$$x + y = 10$$
$$x - y = 2$$

had the single solution (6,4). In Chapter 4 this was determined with algebra, and in Chapter 5 it was solved by graphing.

The system of inequalities

$$x + y < 10$$
$$x - y \geq 2$$

should be solved by graphing. For the two inequalities, use either two different colors for the shading or two different patterns.

First, graph the inequality $x + y < 10$. Pick a color to shade with if you have different colors, or use a pattern, like horizontal lines.

Then, on the same set of axes, graph $x - y \geq 2$, using either a different color for the shading or a different pattern, like vertical lines.

Linear Inequalities

The two lines form an X that divides the plane into four regions. One of those regions will be shaded with both kinds of shading. This is the solution to the system of linear inequalities. Label it with an S.

Example

Which graph shows the solution set to the system of inequalities?

$$y > -3x + 5$$
$$y \leq x + 2$$

(1)

(2)

(3)

(4)

Solution: The line that has a negative slope must be dashed because the inequality $y > -3x + 5$ has a > symbol. This eliminates choices (1) and (4). Since (0, 0) is not a solution to $y > -3x + 5$, the side of the downward sloping line to be shaded with vertical lines will not contain the point (0, 0). This eliminates choice (1). The answer is choice (2).

Graphing Systems of Inequalities with the Graphing Calculator

To graph systems of inequalities, follow the same steps as you did for graphing single inequalities, but this time enter both inequalities.

Let's consider the system from Example 1:

$$y > -3x + 5$$
$$y \leq x + 2$$

For the TI–84:

With the Inequalz application open, enter both equations, and the calculator will shade one with horizontal lines and the other with vertical lines.

For the TI–Nspire:

Graph both inequalities, and the calculator will use different colors for the shadings and a third color for the overlapping region.

Linear Inequalities

Check Your Understanding of Section 7.3

A. *Multiple-Choice*

1. (7, 1) is in the solution set for which system of inequalities?

(1) $x + y > 5$
   $x - y > 3$
(2) $x + y > 5$
   $x - y < 3$
(3) $x + y < 5$
   $x - y > 3$
(4) $x + y < 5$
   $x - y < 3$

2. Which graph shows the solution to the system of inequalities?

$$y < 2x + 1$$
$$y > \left(\frac{1}{3}\right)x + 4$$

(1)

(2)

(3)

(4)

Linear Inequalities

3. Which system of inequalities does the following graph show the solution for?

(1) $y \geq x - 2$
   $y < -\frac{2}{3}x + 1$

(2) $y \leq x - 2$
   $y > -\frac{2}{3}x + 1$

(3) $y < x - 2$
   $y \geq -\frac{2}{3}x + 1$

(4) $y > x - 2$
   $y \leq -\frac{2}{3}x + 1$

4. Which graph shows the solution to the system of inequalities?

$$y > \left(\frac{1}{2}\right)x + 4$$
$$y > 2x - 1$$

(1)

(2)

(3)

(4)

Linear Inequalities

5. Which graph has the solution set shaded in for the following system of inequalities?

$$y \leq -x + 6$$
$$y \geq \left(\frac{1}{2}\right)x - 1$$

(1)

(2)

(3)

(4)

6. Which graph shows the solution set shaded in for the following system of inequalities?

$$y < x - 2$$
$$y > -\left(\frac{3}{4}\right)x + 3$$

(1)

(2)

(3)

(4)

Linear Inequalities

7. Which is the graph of this system of inequalities?

$$y \geq 0$$
$$x \leq 0$$

(1)

(2)

(3)

(4)

8. What system of inequalities does the graph below show the solution for?

(1) $y > -\left(\frac{1}{3}\right)x + 4$
    $y < -2x + 8$

(2) $y > -\left(\frac{1}{3}\right)x + 4$
    $y > -2x + 8$

(3) $y < -\left(\frac{1}{3}\right)x + 4$
    $y < -2x + 8$

(4) $y < -\left(\frac{1}{3}\right)x + 4$
    $y > -2x + 8$

9. An elevator can hold at most 10 people and at most 1,200 pounds. Children weigh an average of 60 pounds each and adults weigh an average of 130 pounds each. If x is the number of children and y is the number of adults, which system of inequalities can be used to model all the allowable combinations of children and adults that can ride the elevator?

(1) $x + y \leq 10$
    $130x + 60y \leq 1,200$
(2) $x + y \leq 10$
    $60x + 130y \leq 1,200$
(3) $60x + 130y \leq 10$
    $x + y \leq 1,200$
(4) $x - y \geq 10$
    $130x - 60y \geq 1,200$

10. The graph below is the solution set to a system of inequalities where $y \leq 3x - 4$ is one of the equations. What is the other equation?

(1) $y \geq 0$
(2) $y > 0$
(3) $x < 0$
(4) $x \geq 0$

241

Linear Inequalities

B. *Show how you arrived at your answers.*

1. On graph paper, make a graph of the solution to the system of inequalities.

$$y > \left(\frac{2}{3}\right)x - 5$$
$$y \geq -2x + 6$$

2. Find the system of inequalities for which this graph shows the solution.

3. Johan goes to a used bookstore. He wants to buy at most 50 books and spend at most $90. Paperback books cost $0.75 and hard cover books cost $2.50. Create a system of linear inequalities that could be used to determine all the different combinations of paperback and hard cover books he can purchase.

4. Movie tickets cost $8 for children and $12 for adults. A movie theater has 300 seats. The theater wants to collect at least $3,200. What system of equations could be used to determine all the different combinations of children and adult tickets that would fit in the theater and produce the money the theater wants to collect?

5. In economics, the feasible region for production can sometimes be modeled with a system of four inequalities. What would the graph of the solution of the system below look like, based on a problem from economics?

$$x \geq 0$$
$$y \geq 0$$
$$y \leq -x + 4$$
$$y \leq -2x + 5$$

Chapter Eight

EXPONENTIAL EQUATIONS

8.1 SOLVING EXPONENTIAL EQUATIONS

=== KEY IDEAS ===

An *exponential equation* is one where the variable is an exponent. An example of a one-variable exponential equation is $2^x = 8$. Examples of two-variable exponential equations are $y = 3^x$ and $y = 2 \cdot 5^x$.

Finding Solutions to Exponential Equations

In a two-variable equation like $y = 3^x$, substitute values for x, and find the corresponding value for y to get a solution.

For example, if $x = 2$, then $y = 3^2 = 9$.

On the TI graphing calculators, raising a number to a power requires using the [^] button.

When the equation has a number multiplied by the exponential expression, be careful not to multiply the number by the base. For the equation $y = 2 \cdot 5^x$, if $x = 2$, then $y = 2 \cdot 5^2 = 2 \cdot 25 = 50$. It would NOT be correct to multiply the 2 and 5 together to get $y = 10^2 = 100$.

=== MATH FACTS ===

When a number other than 0 is raised to the 0 power, it becomes 1, not 0. So $5^0 = 1$, $10^0 = 1$, and $100^0 = 1$. When a number other than 0 is raised to a negative power, it becomes the reciprocal of that number raised to the positive version of that power. So $3^{-2} = \dfrac{1}{3^2} = \dfrac{1}{9}$ and not -9.

Solving for the Exponent in an Exponential Equation

In a one-variable exponential equation, where the exponent is unknown, isolate the exponential expression and then use guess and check.

Exponential Equations

Example

Solve for x in the equation $2 \cdot 4^x = 32$.

Solution: Isolate the exponential part by dividing both sides of the equation by 2.

$$\frac{2 \cdot 4^x}{2} = \frac{32}{2}$$

$$4^x = 16$$

$$x = 2$$

The last step was done by guess and check.

Check Your Understanding of Section 8.1

A. Multiple-Choice

1. If $x = 2$ and $y = 3^x$, solve for y.
   (1) 8   (2) 9   (3) 10   (4) 11

2. If $x = 3$ and $y = 2 \cdot 3^x$, solve for y.
   (1) 216   (2) 27   (3) 54   (4) 18

3. If $x = 0$ and $y = 4^x$, solve for y.
   (1) $\frac{1}{4}$ (2) 0   (3) 1   (4) 4

4. If $x = 0$ and $y = 5 \cdot 2^x$, solve for y.
   (1) 0   (2) 1   (3) 3   (4) 5

5. If $x = -2$ and $y = 5^x$, solve for y.
   (1) -25 (2) 25   (3) $\frac{1}{25}$ (4) $-\frac{1}{25}$

6. If $y = 64$ and $y = 4^x$, solve for x.
   (1) 16   (2) 8   (3) 5   (4) 3

7. If $y = 512$ and $y = 4 \cdot 2^x$, solve for x.
   (1) 7   (2) 5   (3) 3   (4) 1

8. If $y = \dfrac{1}{125}$ and $y = 5^x$, solve for x.

   (1) $\dfrac{1}{3}$ (2) -3 (3) 3 (4) $-\dfrac{1}{3}$

9. If $x = -3$ and $y = \left(\dfrac{1}{2}\right)^x$, solve for y.

   (1) $-\dfrac{1}{8}$ (2) $\dfrac{1}{8}$ (3) -8 (4) 8

10. If $x = 100$ and $y = \left(1 + \dfrac{1}{x}\right)^x$, solve for y rounded to the nearest hundredth.

    (1) 2.70 (2) 2.72 (3) 2.73 (4) 2.74

B. *Show how you arrived at your answers.*

1. Phoebe put $500 in the bank. The amount of money she has after t years is determined by the equation $A = 500(1.05)^t$. After 4 years, how much money will Phoebe have in the bank?

2. The population of a town after t years can be approximated by the equation $P = 10{,}000(1.023)^t$. (a) According to the formula, what will the population of the town be after ten years? (b) In what year will the population become 14,065?

3. Zoe drinks a cup of coffee that has 100 mg of caffeine. The amount of caffeine in the bloodstream after t hours can be determined by the equation $C = 100 \cdot .5^{\frac{t}{5}}$. How much caffeine will be left in her bloodstream after 20 hours?

4. Food that is 110 degrees is put into a 30 degree freezer. The temperature of the food is related to the number of hours the food is in the freezer by the equation $T = 80 \cdot 0.7^h + 30$. Between which two hours will the food be 32 degrees?

5. Daphne says that 6^x is always greater than 5^x. Julia says that this is not true and that 5^x sometimes is greater than 6^x. Which student is correct? Explain.

Exponential Equations

8.2 GRAPHING SOLUTION SETS TO TWO-VARIABLE EXPONENTIAL EQUATIONS

KEY IDEAS

The solution set to a two-variable exponential equation, like $y = 3 \cdot 2^x$, can be produced with a table or with a graphing calculator. The shape of the graph is not a line or a parabola, but a distinctive shape that looks a bit like a playground slide.

Graphing the Solution Set to an Exponential Equation with a Table

Five points are generally sufficient for graphing the solution set for an exponential equation. x values of 2, 1, 0, –1, and –2 should produce enough points for an accurate graph.

Example 1

Make a table of solutions to $y = 3 \cdot 2^x$, and use them to make a sketch of the graph of the solution set.

Solution: When making the chart, remember that anything raised to the zero power (except zero!) is equal to 1. So $2^0 = 1$. Also, anything raised to a negative power is the reciprocal of the same number raised to the positive version of that power. So $2^{-2} = \left(\frac{1}{2}\right)^2 = \frac{1}{4}$. You can use the [^] button on your calculator to do these calculations if you are not sure.

```
3*2^(-2)
            .75
```

x	y
2	$3 \cdot 2^2 = 3 \cdot 4 = 12$
1	$3 \cdot 2^1 = 3 \cdot 2 = 6$
0	$3 \cdot 2^0 = 3 \cdot 1 = 3$
–1	$3 \cdot 2^{-1} = 3 \cdot \left(\dfrac{1}{2}\right) = \dfrac{3}{2} = 1.5$
–2	$3 \cdot 2^{-2} = 3 \cdot \left(\dfrac{1}{4}\right) = \dfrac{3}{4} = 0.75$

Graph the points that fit on the coordinate plane and join to make the curve.

Making Tables and Graphs on the Graphing Calculator

For the TI–84:

Enter the equation into the Y= menu and then press [2ND] and [GRAPH]; the calculator will produce the chart.

Exponential Equations

Press [ZOOM] and [6] to see the graph on a 20 by 20 coordinate plane.

For the TI–Nspire:

From the home screen press [B], select the graphing Scratchpad and type $3 \cdot 2^x$ after the $f1(x)=$ in the entry line.

Press [enter] to see the graph.

Press [menu], [2], and [5] to show the table.

248

Example 2

Make a table of solutions to the equation $y = 4 \cdot 0.5^x$. Use the table to produce a graph of the solution set.

Solution:

x	y
2	$4 \cdot 0.5^2 = 1$
1	$4 \cdot 0.5^1 = 2$
0	$4 \cdot 0.5^0 = 4$
−1	$4 \cdot 0.5^{-1} = 8$
−2	$4 \cdot 0.5^{-2} = 16$

The graph looks like this:

MATH FACTS

In an exponential equation, the thing being raised to the power is called the *base*. When the base of an exponential equation is between 0 and 1, the graph shows exponential *decay*. When the base is greater than 1, the graph shows exponential *growth*.

Chapter Eight EXPONENTIAL EQUATIONS

Exponential Equations

Check Your Understanding of Section 8.2

A. Multiple-Choice

1. Which ordered pair is in the solution set of $y = 3^x$?
   (1) (0, 0)   (2) (3, 1)   (3) (9, 2)   (4) (4, 81)

2. Which ordered pair is in the solution set of $y = 5 \cdot 2^x$?
   (1) (0, 0)   (2) (2, 25)   (3) (3, 40)   (4) (3, 100)

3. Which is the graph of $y = 2^x$?

(1)

(3)

(2)

(4)

250

4. Which is the graph of $y = \left(\frac{1}{2}\right)^x$?

(1)

(3)

(2)

(4)

5. Below is the graph of which equation?

(1) $y = \left(\frac{1}{3}\right)^x$
(2) $y = 10^x$
(3) $y = 4^x$
(4) $y = 3^x$

251

Exponential Equations

6. Below is the graph of

(1) $y = \left(\frac{1}{3}\right)^x$ (2) $y = \left(\frac{1}{2}\right)^x$ (3) $y = \left(\frac{1}{4}\right)^x$ (4) $y = 3^x$

7. The chart below has ordered pairs for which equation?

x	y
0	6
1	12
2	24
3	48

(1) $y = 12x^2 + 6$ (3) $y = 12^x$
(2) $y = 6x + 6$ (4) $y = 6 \cdot 2^x$

8. The chart below has ordered pairs for which equation?

x	y
0	6
1	3
2	1.5
3	.75

(1) $y = 6 \cdot 2^x$ (3) $y = 6 \cdot \left(\frac{1}{3}\right)^x$

(2) $y = 6 \cdot \left(\frac{1}{2}\right)^x$ (4) $y = -3x + 6$

9. In what interval is the graph of $y = 1.5^x$ increasing?
(1) Always   (3) When $x \geq 0$
(2) Never   (4) When $x \leq 0$

10. Below is the graph of $y = b^x$. What is true about the value of b?

(1) b must be greater than 1.
(2) b must be less than 1 and greater than 0.
(3) b must be less than 0 and greater than -1.
(4) b must be less than -1.

B. *Show how you arrived at your answers.*

1. The graph of $y = b^x$ passes through the point (5, 243). What must the value of b be? $b = 3$

2. After putting $200 into the bank, the amount of money Xavier has after t years is $P = 200 \cdot 1.15^t$. Make a graph showing how the money grows for 5 years. Choose appropriate units for the horizontal t-axis and the vertical P-axis.

3. Below is the graph of $y = 2^x$ and $y = 3^x$ on the same set of axes.

Is it true that $3^x \geq 2^x$ for all values of x? Explain. No when x is negative

4. What is the solution to this system of equations?

$$y = 2^x$$
$$y = 3^x$$

(0, 1)

253

Exponential Equations

5. Below are the graphs of $y = x^2 + 1$ and $y = 2^x$. They both contain the point (0, 1). Is it true that $2^x < x^2 + 1$ for all values of $x > 0$? Explain.

[handwritten note: No for between 0 and 1]

8.3 DISTINGUISHING BETWEEN LINEAR, QUADRATIC, AND EXPONENTIAL EQUATIONS

KEY IDEAS

The graphs of the solution sets of linear, quadratic, and exponential equations are very distinctive so it is possible to tell what type of equation a graph was produced from by just looking at it. When the solution set is given in table form, it is possible to graph the data from the table or to determine which type of equation it came from with fast calculations.

Identifying What Type of Equation It Is By Looking at the Graph of the Solution Set

Linear equations have no exponents greater than or equal to one in them. The graph of a linear equation is always a line.

Quadratic equations have either an x^2 or a y^2 term. The graph of a quadratic equation is a parabola.

Exponential equations have an x as an exponent. The graph of an exponential equation looks like a playground slide.

Example 1

Below is the graph of the solution set of which equation?

(1) $y = x^2 + 1$ (2) $y = 3x + 1$ (3) $y = 0.4^x$ (4) $y = 1.5^x$

Solution: Choice (4). Because of the shape of the graph, the equation must be an exponential equation so choices (1) and (2) can be eliminated. Since the graph is increasing for all values, the base must be greater than 1 so choice (3) can also be eliminated.

Exponential Equations

Example 2

Which could be the graph of the solution set of $y = 0.4^x$?

(1) (3)

(2) (4)

Solution: Choice (1). Choice (2) is the graph of a quadratic equation. Choice (3) is the graph of a linear equation. Choice (4) is the graph of an exponential equation where the base is greater than 1.

Identifying That an Equation Is Exponential By Looking at a Table of Values

When you are given an equation, you can produce a table of solutions to that equation. If you are given a table of solutions to an equation and are given choices of what the equation could be, a table of values for each choice can be made until one matches the question. Graphing the points could help eliminate some choices.

Example 3

Below is a table with solutions to a two-variable equation. Which choice could be the equation?

x	y
2	7
3	25
4	79
5	241

(1) $y = 2^x + 3$ (2) $y = 3^x - 2$ (3) $y = x^2 + 3$ (4) $y = 3x + 1$

The table of values for choice (1) matches the given chart for the first row, but not for the rest.

x	y
2	7
3	12
4	19
5	28

The table of values for choice (2) matches for all the values.

x	y
2	7
3	25
4	79
5	241

Using the Graphing Calculator to Graph Tables of Values

If equations are not given as choices, the quickest way to identify the type of equation a table was based on is by producing a graph of the points and seeing if it has the shape of a line, a parabola, or an exponential equation (like a slide).

Example 4

The following table has solutions to what kind of equation?

x	y
0	250.00
20	435.00
40	756.90
60	1,317.01
80	2,291.59

(1) Linear increasing   (3) Exponential growth
(2) Linear decreasing (4) Exponential decay

Exponential Equations

Solution:

For the TI-84:

Press [STAT] key and then [1] to edit the lists. If data are already in the lists, move the cursor to the L1 in the top row and press [CLEAR]. Do the same for L2.

Enter the five solutions with the x in the L1 column and the y in the L2 column.

Press the [2ND] and then the [Y=] key to get to the STAT PLOT menu. Select option 1. Using the arrow keys, move to On and press [ENTER]. Now press [ZOOM] and then [9].

The calculator will graph the points from the lists and set the window so all the points can be seen.

For the TI-Nspire:

From the home screen select the Add Lists & Spreadsheet icon,

the first column x, and the second column y.

Enter the five x-values into the x column and the five y-values into the y column.

x	y
0	250
20	435
40	756.9
60	1317
80	2291.6

Exponential Equations

Go back to the home screen and select Add Graphs icon.

Press [menu], [3], and [4] for Scatter Plot.

Enter *x* next to the *x* arrow and enter *y* next to the *y* arrow and press [enter]. Then press [menu], [4], and [9] for Zoom—Data.

Of the four choices, this looks most like exponential growth. Choice (3).

Identifying the Type of Equation from a Table Without Graphing

If the x values in the chart are evenly spaced, there is a very fast way to determine if it is an exponential or a linear equation.

If the *difference* between consecutive y values is the same for each pair, the equation is a linear equation.

x	y
0	130.00
15	224.35
30	318.70
45	413.05
60	507.40

Since each x value is 15 more than the previous one, check to see the difference between successive y values:

$$224.35 - 130.00 = 94.35$$
$$318.70 - 224.35 = 94.35$$
$$413.05 - 318.70 = 94.35$$
$$507.40 - 413.05 = 94.35$$

Since all these differences are the same, this was a linear equation.

If the *quotient* (what you get by dividing one y value by the previous y value) of two consecutive y values is the same for each pair, the equation is an exponential equation.

The table from Example 4 was

x	y
0	250.00
20	435.00
40	756.90
60	1,317.01
80	2,291.59

Each x-value is 20 more than the previous x-value. The differences between consecutive y-values are not the same.

$$435.00 - 250.00 = 185$$
$$756.90 - 435.00 = 321.90$$

Exponential Equations

When the quotients are calculated, they become

$$\frac{435.00}{250.00} = 1.74 \qquad \frac{1317.01}{756.90} = 1.74$$

$$\frac{759.90}{435.00} = 1.74 \qquad \frac{2291.59}{1317.01} = 1.74$$

Since these quotients are the same, this is an exponential equation, and since the y values are increasing, it is an example of exponential growth.

Check Your Understanding of Section 8.3

A. Multiple-Choice

1. What type of equation has a graph like the one below?

   (1) Linear (3) Exponential
   (2) Quadratic (4) None of the above

2. What type of equation has a graph like the one below?

   (1) Quadratic (3) Exponential
   (2) Linear (4) None of the above

262

3. What type of equation has a graph like the one below?

(1) Linear (3) Quadratic
(2) Exponential (4) None of the above

4. Which of the graphs below corresponds to a quadratic equation?

(1)

(2)

(3)

(4)

Exponential Equations

5. Which of the graphs below corresponds to an exponential equation?

(1)

(2)

(3)

(4)

6. The ordered pairs in the following chart satisfy which type of equation?

x	y
−2	0
−1	2
0	4
1	6
2	8

(1) Linear
(2) Quadratic
(3) Exponential
(4) None of the above

7. Which of the graphs below corresponds to a linear equation?

(1)

(3)

(2)

(4)

8. The ordered pairs in the following chart satisfy which type of equation?

x	y
-2	$\frac{2}{9}$
-1	$\frac{2}{3}$
0	2
1	6
2	18

(1) Linear
(2) Quadratic
(3) Exponential
(4) None of the above

Exponential Equations

9. The ordered pairs in the following chart satisfy which type of equation?

x	y
-2	-6
-1	6
0	0
1	-6
2	6

(1) Linear
(2) Quadratic
(3) Exponential
(4) None of the above

10. The following scatter can be best modeled with which type of equation?

(1) Linear
(2) Exponential
(3) Quadratic
(4) Sinusoidal

B. Show how you arrived at your answers.

1. The population of a county for several different times is plotted on a graph where the x-axis represents years since the year 1990 and the y-axis represents the population in millions. The graph is below. What type of equation could be used to model this data? Explain.

2. Some data points on a graph are plotted. Myah believes that this is a portion of a quadratic graph. Chloe believes that this is a portion of an exponential graph. Is there enough information to determine who is correct? If not, what more information would be needed?

3. When the number of empty seats on a train is graphed as the y-coordinate and the time since 8:00 A.M. is graphed as the x-coordinate, it makes the graph below. Which type of equation could be used to model this data? What details do you know about the equation besides just the type of equation it is?

4. Below is a chart with ordered pairs that satisfy an equation. The equation is either linear, quadratic, or exponential. Determine which type of equation it is, and fill in the missing number.

x	y
1	50
2	85
3	?
4	245.65
5	417.605

5. A tennis ball is dropped from a height of 30 feet. After each bounce, the highest point of the next bounce is 80% as high as the bounce before. If the height of each bounce is graphed as the y-coordinate and the bounce number is the x-coordinate, will the graph be linear, quadratic, or exponential?

8.4 REAL-WORLD PROBLEMS INVOLVING EXPONENTIAL EQUATIONS

=== KEY IDEAS ===

Exponential equations can be used to model many real-life situations ranging from the way that a cup of hot chocolate cools down to the way that the world population increases. If you are given an exponential equation that is a model for something in the real world, that equation can be used to answer questions about it.

Modeling a Population Increase

In a certain forest, the population of rabbits, in thousands, x years from now can be modeled by the equation $y = 3 \cdot 1.618^x$. With this equation, there are two main questions that can be asked: (1) What will the population of rabbits be after a certain number of years? And (2) In what year will the rabbit population reach a certain number?

The first type of question is very quick to calculate. The second (for this course) requires guess and check or the graphing calculator.

If they ask what the population will be in 5 years, substitute $x = 5$ into the equation $y = 3 \cdot 1.618^x \approx 33.26$. Since this number is in thousands, this represents around 33,000 rabbits.

If they ask how many years it will take for the rabbit population to reach 150,000 rabbits, test numbers until you get close to 150.

This can also be done by intersecting the graph of $y = 3 \cdot 1.618^x$ with the graph of the line $y = 150$.

For the TI-84:

For the TI-Nspire:

Example

The temperature of a slice of pizza that just came out of the oven can be modeled by the equation $y = 280 \cdot .73^x + 70$, where y is the temperature in degrees and x is the number of minutes since the pizza was removed from the oven. What will the temperature of the pizza be after 10 minutes?

Solution: Substitute 10 for x in the equation.

The answer is approximately 82 degrees.

Check Your Understanding of Section 8.4

A. Multiple-Choice

1. The population of a country can be modeled with the equation $P = 250 \cdot 1.07^t$, where P is the population in millions and t is the number of years since 2010. According to this model, rounded to the nearest ten million, what will the population of this country be in 2019?
   (1) 450,000,000
   (2) 460,000,000
   (3) 470,000,000
   (4) 480,000,000

2. The amount of carbon-14 remaining in a substance can be calculated with the formula $Q = P \cdot 0.5^{\frac{t}{5,730}}$, where t is the time in years, P is the original amount of carbon-14, and Q is the amount of carbon-14 remaining. If the bone in a living animal has 200 grams of carbon-14, how many grams of carbon-14 will remain 20,000 years after it dies rounded to the nearest tenth of a gram?
   (1) 17.0 grams
   (2) 17.4 grams
   (3) 17.8 grams
   (4) 18.2 grams

3. Kyle puts $300 into a savings account. The amount of money in the account after t years can be calculated with the formula $A = P \cdot 1.03^t$, where P is the original amount of money and A is the amount of money after t years. How much money will Kyle have after 10 years?
   (1) $397.17
   (2) $399.17
   (3) $401.17
   (4) $403.17

4. A ball is dropped from a window 30 feet above the ground. The height of the nth bounce can be calculated with the formula $h = 30 \cdot .7^n$ where h is the height at the top of the nth bounce. How high will the ball go after the sixth bounce?
   (1) 3.5 feet
   (2) 4.0 feet
   (3) 4.5 feet
   (4) 5.0 feet

5. A cake put into a 400 degree oven from an 80 degree kitchen heats up according to the equation $t = -320 \cdot 0.8^m + 400$ where t is the temperature of the food and m is the number of minutes since it was put into the oven. How hot will the food be after 5 minutes in the oven?
   (1) 290 degrees
   (2) 295 degrees
   (3) 300 degrees
   (4) 305 degrees

6. A cup of tea that is 200 degrees is put into a room that is 80 degrees. The temperature of the tea can be calculated with the formula $t = 120 \cdot 0.9^m + 80$ where m is the number of minutes since the tea was put into the room. What will the temperature of the tea be after 10 minutes?
   (1) 116 degrees
   (2) 118 degrees
   (3) 120 degrees
   (4) 122 degrees

7. The grade Jerry gets on a test is related to the number of hours he studies by the equation $g = -60 \cdot 0.7^h + 100$ where h is the number of hours studied and g is the grade the student gets. According to this equation, what grade will Jerry get if he studies for 6 hours?
   (1) 89
   (2) 91
   (3) 93
   (4) 95

8. Jude weighs 250 pounds and goes on a diet. His weight is determined by the equation $p = 80 \cdot 0.95^t + 170$ where t is the number of weeks since he started the diet and p is his weight in pounds. How much will Jude weigh after 12 weeks?
   (1) 203 pounds
   (2) 208 pounds
   (3) 213 pounds
   (4) 218 pounds

9. A YouTube video of a cat wearing rain boots goes viral. The number of views d days after the video was posted can be calculated with the equation $V = 50 \cdot 1.7^d$. How many times will the video be viewed 20 days after it was posted, rounded to the nearest million?
   (1) 2,000,000
   (2) 3,000,000
   (3) 4,000,000
   (4) 5,000,000

10. The number of shaded triangles in the first picture is 3. The number of shaded triangles in the second picture is 9. If this pattern continues, the number of shaded triangles in the nth picture can be calculated with the formula $T = 3^n$. How many shaded triangles would be in the fifth picture?

    (1) 27   (2) 81   (3) 243   (4) 729

Exponential Equations

B. *Show how you arrived at your answers.*

1. The population of a country can be determined by the equation $P = 300 \cdot 1.13^t$ where P is the population in millions and t is the number of years since 2010. (a) Use this formula to determine the population of the country in 2019. (b) Use this formula to determine when the population of the country will be 800 million.

2. Ice cream is removed from a freezer and put into an 80 degree room. The formula that can be used to determine the temperature of the ice cream m minutes after being removed from the freezer is $t = -48 \cdot 0.9^m + 80$. (a) Use this formula to determine what the temperature of the ice cream will be after 5 minutes. (b) After how many minutes will the ice cream be 72 degrees?

3. Lucas wants to put his $100 into a bank. For the first bank, the formula for how much money he will have after t years is $A = P + t \cdot 0.4P$, where P is the amount of money originally deposited. For the second bank, the formula for how much money he will have after t years is $A = P(1.06)^t$. (a) If he is going to withdraw his money after 10 years, which bank is the better choice? (b) If he is going to withdraw his money after 54 years, which bank is the better choice?

4. The first picture has 4 line segments. The second picture has 16 line segments. If the equation for the number of line segments of the nth picture is $s = 4^n$, how many line segments will be in the fifth picture?

5. Zion has a goal to bench press 250 pounds. The equation that can be used to determine the amount he can bench press after t weeks of exercise is $P = -220 \cdot 0.93^t + 300$. (a) How much will he be able to bench press after 5 weeks? (b) When will he be able to bench press 250 pounds?

Chapter Nine

CREATING AND INTERPRETING EQUATIONS FROM REAL-WORLD SCENARIOS

9.1 CREATING AND INTERPRETING LINEAR EQUATIONS

=== KEY IDEAS ===

Many real-world scenarios can be modeled with two-variable linear equations. The scenarios generally have a *fixed* part, which is the same for every situation, and a *variable* part, which changes depending on the situation. These equations often resemble the slope-intercept form $y = mx + b$.

Identifying the Fixed and Variable Parts of a Real-World Scenario

At a carnival, it costs people $10 to enter and an additional $3 for each ride they go on. The total amount people spend at the carnival depends on how many rides they choose to go on. If they go on no rides, they pay $10. If they go on one ride, they pay $13. If they go on 20 rides, they pay $70. The goal is to create an equation that relates the variable C, for total cost, to the variable R, for number of rides.

One way to create this equation is to first pick some number of rides and calculate how much it would cost to enter the carnival and go on that number of rides. It doesn't matter which number you pick for this method. If you picked 5 rides, you would figure out how much that costs.

For five rides, the person would spend $10 to enter and then $3 · 5 = $15 for rides, for a total of $25. This could be written as $10 + 3 \cdot 5 = 25$. If the number of rides was 20, the calculation would be $10 + 3 \cdot 20 = 70$.

For the variable R rides, it would still cost $10 to get into the carnival. That is the fixed part. But to go on R rides would cost $3R. So the total cost is $10 + 3R$.

As a two-variable equation with C for cost and R for rides, the equation can be written several ways:

$$C = 10 + 3R \text{ or } C = 3R + 10 \text{ or } 3R + 10 = C \text{ or } 10 + 3R = C$$

The $C = 3R + 10$ is the most common form since it resembles $y = mx + b$ form.

Example 1

An empty ice cream cone costs $3. Each scoop of ice cream on the cone costs $1.50. Create an equation that relates the cost of the cone with the ice cream, C, with the number of scoops, S.

Solution: If there are no scoops, the cost is $3. If there is one scoop, the cost is $3 + $1.50 · 1. If there are 5 scoops (careful not to let it tip over!), the cost is $3 + $1.50 · 5.

For S scoops, the cost is $C = 3 + 1.50S$ or, more commonly, $C = 1.50S + 3$.

Example 2

Briggs has $150 in the bank. Each week he takes out $25. Create an equation that relates the total amount he has in the bank, called B, and the number of weeks that have happened, called W.

Solution: After one week he has $150 − $25 = $125 in the bank. After two weeks he has $150 − $25 · 2 = $100 in the bank, and after three weeks he has $150 − $25 · 3 = $75 in the bank. If W is substituted for the number of weeks in any of these equations, the equation becomes $150 − 25W = B$. This can be written as $B = -25W + 150$, though it does not have to.

Example 3

Lydia has $21.00. After she buys one banana, she has $20.25 left. After she buys a second banana, she has $19.50 left. Create an equation that relates the amount of money she has left, M, to the number of bananas she has purchased, B. Use this equation to determine (a) how much money she will have left after buying 10 bananas and (b) how many bananas she can buy before she runs out of money.

Solution: For this question you are not told what the cost of each banana is, but you are given enough information to determine it yourself. The price of each banana is $0.75 since $21.00 − 20.25 = 0.75$. After buying B bananas, she will have $21.00 − 0.75B$ dollars left so the equation is $M = 21 − 0.75B$, which may also be written as $M = -0.75B + 21$.

(a) By substituting 10 for B in the equation, the amount of money she has left after purchasing 10 bananas can be calculated.

$$M = 21 - 0.75 \cdot 10$$
$$M = 21 - 7.50$$
$$M = 13.50$$

After buying 10 bananas, she will have $13.50 remaining.

(b) By substituting 0 for M in the equation, the number of bananas she can purchase can be calculated.

$$0 = 21 - 0.75B$$
$$-21 = -21$$

$$\frac{-21}{-.75} = \frac{-.75B}{-.75}$$

$$28 = B$$

Interpreting Linear Equations

Just as a real-life scenario can sometimes be modeled as a two-variable linear equation, it is also possible to reverse the process. Given a two-variable linear equation that resembles $y = mx + b$ form, it is possible to *interpret* what the constant and the coefficient represent.

For this topic, you will be given a scenario and a two-variable equation. For example: At a pizzeria, the cost of a medium pizza with T toppings is $C = 2.50T + 8.00$. The two numbers 2.50 and 8.00 represent different things about the price of the pizza. One way to approach a question like this is to substitute $T = 0$ into the equation. This would lead to $C = 2.50 \cdot 0 + 8.00 = 8.00$ so a zero-topping pizza would be $8.00. Notice that 8.00 is the constant in the equation.

If you substitute $T = 1$ into the equation, it would become $C = 2.50 \cdot 1 + 8.00 = 2.50 + 8.00 = 10.50$. A pizza with one topping is $10.50, which is $2.50 more than a pizza with zero toppings. Notice that 2.50 is the coefficient of the T.

The 8.00 in the equation $C = 2.50T + 8.00$ must represent the cost of the pizza, not including the toppings, and the 2.50 must represent the cost of each topping.

Example 4

The cost of riding a taxi M miles can be modeled with the equation $C = 0.75M + 2.50$. What does the 0.75 in the equation represent?
 (1) The cost of getting into the taxi
 (2) The number of miles traveled
 (3) The cost of the entire taxi ride
 (4) The cost of traveling each mile

Solution: Choice (4). If $M = 0$, the cost is $2.50, which is the price to get into the taxi. If $M = 1$, the cost is $0.75 + 2.50 = 3.25$, which is $0.75 more than it cost to go zero miles. So 0.75 is the cost for each mile.

Creating and Interpreting Equations from Real-World Scenarios

Example 5

A video store makes a profit for each game sold. The store begins each month by paying rent, however, and has no other expenses. The total profit after paying rent that a video store makes selling G games can be modeled by the equation $P = 5V - 250$. What does the 250 represent in this equation?

(1) The profit for each game
(2) The rent the store has to pay
(3) The number of games sold
(4) The total amount of profit after paying rent

Solution: Choice (2). Substitute 0 for V to get the amount of total profit after selling 0 games. The equation becomes $P = 5 \cdot 0 - 250 = -250$. So if they make no money from selling the games, they pay $250 so 250 must be the cost of the rent.

MATH FACTS

When an equation has the form $y = mx + b$, b generally represents the *fixed* part and the *mx* is the *variable* part. The variable part is the part that changes depending on the value of x.

Check Your Understanding of Section 9.1

A. Multiple-Choice

1. It costs $10 to go to the movies and $3 for each bag of popcorn. Which equation relates the total cost (C) to the number of bags of popcorn purchased (P)?
   (1) $C = 3P + 10$
   (2) $C = 10P + 3$
   (3) $P = 3C + 10$
   (4) $P = 10C + 3$

2. A salad costs $6 for the lettuce and $2 for each topping. Which equation relates the total cost (C) to the number of toppings purchased (T)?
   (1) $T = 6C + 2$
   (2) $T = 2C + 6$
   (3) $C = 6T + 2$
   (4) $C = 2T + 6$

3. A tablet computer costs $400 and $2 for each app. Which equation relates the total cost (C) to the number of apps purchased (A)?
   (1) $A = 2 + 400C$
   (2) $A = 400 + 2C$
   (3) $C = 2 + 400A$
   (4) $C = 400 + 2A$

4. An ice cream store collects $5 for each ice cream. At the end of the day, the store must pay $300 total for all the employees. Which equation relates the total amount of money remaining (M) after paying the employees to the number of ice creams sold (I)?
   (1) $I = 5M - 300$
   (2) $M = 5I - 300$
   (3) $M = 5I + 300$
   (4) $M = 300I - 5$

5. A cable TV plan costs $80 a month plus $10 extra for each premium channel. Which equation relates the monthly bill (B) to the number of premium channels ordered (C)?
   (1) $B = 80C + 10$
   (2) $B = 10C + 80$
   (3) $C = 80B + 10$
   (4) $C = 10B + 80$

6. Muhammad has saved $40. Each week he earns $15 delivering newspapers. Which equation relates the amount Muhammad has saved (S) to the number of weeks working (W)?
   (1) $W = 15S + 40$
   (2) $W = 40S + 15$
   (3) $S = 15W + 40$
   (4) $S = 40W + 15$

7. Lydia wants to buy a DVD player and some DVDs. The equation that relates the total cost for the DVD player and N DVDs is $P = 20N + 200$. What does the number 200 in the equation represent?
   (1) The cost of the DVD player
   (2) The cost of each DVD
   (3) The cost of all N DVDs
   (4) The total cost of the DVD player and all N DVDs

8. Amelia buys an empty sticker album and some sticker sheets. The equation that relates the total cost for the empty sticker album and N sticker sheets is $P = 0.75N + 3.00$. What does the number 0.75 in the equation represent?
   (1) The cost of the empty sticker album
   (2) The cost of each sticker sheet
   (3) The cost of all N sticker sheets
   (4) The total cost of the empty sticker album and all N sticker sheets

9. Riley has a cell phone that requires a certain amount of money to be paid for each gigabyte of data. If the equation that relates the cost of the phone together with N gigabytes of data is $C = 5N + 80$, what is the cost of one gigabyte of data?
   (1) 80
   (2) $5N$
   (3) 5
   (4) $5N + 80$

Creating and Interpreting Equations from Real-World Scenarios

10. Colton is training to run a marathon. The first month he runs 1 mile a day. The second month he runs 3 miles a day. The third month he runs 5 miles a day. Which equation relates the number of months training (M) to the distance he runs each day (D) in the Mth month?
    (1) $M = D + 2$
    (2) $M = 2D - 1$
    (3) $D = M + 2$
    (4) $D = 2M - 1$

B. *Show how you arrived at your answers.*

1. A teacher has a starting salary of $40,000 and each year she gets a $5,000 raise. Create an equation that relates her salary (S) to the number of years (Y) she has been working.

    S = 5000Y + 40000

2. For a taxi ride it costs a certain amount to get into the taxi and an additional amount for each mile driven. If the equation that relates the total cost (C) to travel M miles is $C = 2M + 4$, what do the numbers 2 and 4 represent?

    2 = $ each mile
    4 = amount to get in taxi

3. A 5 foot tree was planted in the year 2000. This chart shows four sets of values where Y is the number of years since 2000 and H is the height of the tree in that year. Create an equation that relates H and Y.

    H = 3Y + 5

Y	H
0	5
1	8
2	11
3	14

4. Owen, who weighs 260 pounds, goes on a diet where he loses the same amount of weight each month. Below is a chart where M represents the number of months Owen has been on the diet and W represents Owen's weight after M months on the diet. Create an equation that relates W and M.

    W = 260 - 2M

M	W
0	260
1	258
2	256
3	254

5. Make up a real-world situation that can be represented with the equation $C = 6N + 20$.

    The price of getting into a carnival is $20 while each ride costs $6

9.2 CREATING AND INTERPRETING EXPONENTIAL EQUATIONS

> **KEY IDEAS**
>
> Many real-life scenarios are more accurately modeled with a two-variable exponential equation rather than a linear equation. These include questions about population growth, money and interest, and cooling liquids. Many exponential equations can be written in the form $y = a \cdot (1 + r)^x$, where a is related to the starting value and r is related to how fast something is growing or decaying.

Percent Increase

Exponential equations are about something that is increasing (or decreasing) by the same percent each time period. Calculating how something changes when it is increased or decreased by a certain percent is the key to creating exponential equations.

Bank A offers 15% interest compounded annually. You put $100 into the bank on January 1, 2015. One year later, January 1, 2016, you will get $100 · 15 = $15 interest. Together with the original $100, you have $115 in the bank. Another year later, January 1, 2017, you will get $115 · 0.15 = $17.25 interest. Together with the $115 you already had, your total is now $132.25.

It took two steps to calculate that you will have $115 on January 1, 2016, multiplying $100 by 0.15 to get $15 and then adding the $15 to the original $100. There is a way to get this same answer in one step by multiplying the original $100 by 1 + 0.15 which is 1.15.

$$\$100 \cdot 1.15 = \$115$$

To get the amount on January 1, 2017, you can multiply the $115 by 1.15.
$115 · 1.15 = $132.25.

To get the amount on January 1, 2018, you can multiply the $132.25 by 1.15.

$$\$132.25 \cdot 1.15 = \$152.09$$

To get from the original $100 to the amount 10 years later, you can calculate $100 · 1.15 · 1.15 · 1.15 · 1.15 · 1.15 · 1.15 · 1.15 · 1.15 · 1.15 · 1.15, which can be done much more quickly as $100(1.15)^{10}$.

Creating and Interpreting Equations from Real-World Scenarios

Example 1

Bank B offers 12% interest compounded annually. You deposit $200 into the bank on January 1, 2015. Which expression could be used to calculate the amount of money in the bank 8 years later, January 1, 2023?
   (1) $200(1.12)^8$
   (2) $\$200 + 8(\$200 \cdot 0.12)$
   (3) $\$200 + (1.12)^8$
   (4) $\$200(8 \cdot \$200 \cdot 0.12)$

Solution: After one year, the amount will be $\$200 \cdot 1.12$. After two years, it will be $\$200 \cdot 1.12 \cdot 1.12 = 200(1.12)^2$, and after three years. it will be $\$200 \cdot (1.12)^3$. So after 8 years it will be $\$200 \cdot (1.12)^8$, which is choice (1).

Percent Decrease

If instead of getting interest of 15% at a bank you had to pay a fee of 15% of the amount you have, each year your savings would decrease. If you started with $200 on January 1, 2015, then on January 1, 2016, you would have to pay $\$200 \cdot 0.15 = \$30$. You would then have $\$200 - \$30 = \$170$ left. This required two calculations, multiplying by 0.15 and then subtracting $30 from $200. Another way to calculate this is to multiply $200 by $(1 - 0.15)$ which is $\$200(0.85) = \$170$. If you take 15% off, it leaves you with 85% left.

After 2 years, you will have $\$200(0.85 \cdot 0.85) = \$200(0.85)^2 = \$144.50$. In general, you will have $\$200(0.85)^T$ after T years.

Example 2

Mila weighs 170 pounds. She goes on a diet, which causes her to lose in a month 2% of whatever her weight was at the beginning of that month. Which expression can be used to determine her weight at the end of 10 months?
   (1) $170(0.98)^{12}$ (3) $170(0.2)^{10}$
   (2) $170(0.98)^{10}$ (4) $170(1.02)^{10}$

Solution: For 2% off, subtract $1 - 0.02$ to get 0.98, which is what needs to be multiplied each month. After 10 months, the expression will be $170(0.98)^{10}$, choice (2).

Two-variable Percent Increase Equations

If in a real-world scenario a starting value is given and the percent increase, or growth rate, is also given, a two-variable equation can be created to model the scenario.

The population in the United States in 2014 is approximately 317 million. Each year the population increases by approximately 1%. Write an equation that relates the population, P, to the number of years that have passed since 2014, T.

To increase something by 1%, multiply it by $1 + 0.01 = 1.01$. So after one year, the population will be $317(1.01)$. After two years, the population will be $317(1.01 \cdot 1.01) = 317(1.01)^2$.

The equation is $P = 317(1.01)^T$.

MATH FACTS

The exponential growth equation is $A = P(1 + r)^T$ where P is the starting value, r is the growth rate, and T is the amount of time periods that have happened.

Example 3

On the first day of school, there are 14 students who know a rumor. Each day the number of students who know the rumor increases by 25%. Create an equation that relates the number of students who know the rumor, N, to the number of days that have passed since the first day of school, T. Approximately how many people will know the rumor after 10 days?

Solution: Since the growth rate is 25%, each new day the number of people who know the rumor is $1 + 0.25 = 1.25$ times the number of people who knew it the day before.

After 3 days, the number of people who know the rumor is $14(1.25 \cdot 1.25 \cdot 1.25) = 14(1.25)^3$. In general, the equation that relates N and T is $N = 14(1.25)^T$.

Substituting 10 for T, it becomes

$$N = 14(1.25)^{10} \approx 130 \text{ people}$$

Two-Variable Percent Decrease Problems

The difference between a percent increase and a percent decrease equation is that instead of multiplying by 1 plus the growth rate each time, you multiply by 1 minus the growth rate.

When a tennis ball is dropped, the height of each bounce is 20% less than the previous bounce. If it is initially dropped from a height of 7 feet, what equation relates the height of the bounce, H, to the number of bounces, B?

To get the number inside the parentheses, calculate 1 minus the growth rate. For this example, this is $1 - 0.20 = 0.80$.

The height of the first bounce is 7(0.8) = 5.6. The height of the second bounce is $7(0.8 \cdot 0.8) = 7(0.8)^2 = 4.48$. In general, the equation is $H = 7(0.8)^B$.

MATH FACTS

The exponential decay equation can be written as $A = P(1 - r)^T$ where P is the starting value, r is the decay rate, and T is the number of time periods that have happened.

Interpreting Exponential Equations

When a real-world scenario can be modeled with an exponential equation of the form $y = a \cdot (1 + r)^x$, the a and the r are related to the starting value and the growth rate. If the scenario has exponential growth, the r will be positive. If the scenario has exponential decay, r will be between -1 and 0 so that $1 + r$ will be between 0 and 1.

Interpreting Exponential Growth Equations

The equation $A = 500(1.07)^T$ can be used to model many things. If the problem says that A is the amount of money a customer will have in the bank after T years, it is possible to interpret what the numbers 500 and 1.07 represent.

In an equation like $y = a \cdot (1 + r)^x$, the a is the starting amount so the 500 must represent the starting amount of money. One less than what gets raised to the power, which is the growth rate. In this case, the growth rate is .07 or 7%.

Example 4

The equation $P = 25(1.31)^T$ models the number of rabbits in a pet store after T months. How many rabbits were there originally, and what is the growth rate of the rabbits?

Solution: First rewrite the equation as $P = 25(1 + 0.31)^T$. The number in front of the parentheses is the starting amount so there were 25 rabbits originally. The number added to the 1 is the growth rate so in this case the growth rate is 0.31 or 31% each month.

Interpreting Exponential Decay Equations

When an exponential equation represents an exponential decay scenario, the expression being raised to the power will be less than one. Two examples of expressions that represent exponential decay are $T = 80(0.75)^M$ and $H = 100(1 - 0.4)^N$. The number in front of the parentheses will represent the starting value. The number that is being subtracted from 1 is the growth rate. When the growth rate is between 0 and 1 it is also known as the decay rate or rate of decay.

Example 5

When you take a pizza out of an oven and put it into a freezer for M minutes, the temperature of the pizza, T, can be determined by the equation $T = 170(0.92)^M$. What was the temperature of the pizza when it came out of the oven? What is the rate of cooling?

Solution: If you substitute 0 for M, T evaluates to 170, so 170 is the starting value. 0.08 is the rate of decay since it is equivalent to $1 - 0.92$.

Check Your Understanding of Section 9.2

A. Multiple-Choice

1. There were 900 birds in a forest. Each year the bird population increases by 12%. Which equation relates the bird population (P) to the number of years that have passed (t)?
   (1) $P = 900(1.12)^t$
   (2) $P = 900(0.12)^t$
   (3) $P = 900(0.88)^t$
   (4) $t = 900(1.12)^P$

2. Clara deposits $300 into a bank. The bank offers 5% interest compounded annually. Which equation relates the amount of money in the bank (A) to the number of years that have passed (t)?
   (1) $A = 300(1.05)^t$
   (2) $A = 300(1.5)^t$
   (3) $A = 300(0.95)^t$
   (4) $t = 300(1.05)^A$

3. A bouncing ball is dropped from 20 feet high. After each bounce, the height of the next bounce is 65% as high as the last bounce. Which equation relates the height of the bounce (H) to the number of bounces that have happened (N)?
   (1) $H = 20(0.35)^N$
   (2) $H = 20(1.65)^N$
   (3) $H = 20(0.65)^N$
   (4) $N = 20(0.65)^H$

Creating and Interpreting Equations from Real-World Scenarios

4. Food that is 80 degrees is put into a freezer. Each minute the temperature of the food decreases by 18%. Which equation relates the temperature of the food (T) to the number of minutes since the food was put into the freezer (M)?
   (1) $T = 80(1 + 0.18)^M$
   (2) $T = 80(1 - 0.18)^M$
   (3) $T = 80(0.18)^M$
   (4) $M = 80(1 + 0.18)^T$

5. The population (P) of a town after t years can be modeled with the equation $P = 20{,}000(1.07)^t$. What does the 20,000 represent?
   (1) The growth rate
   (2) The percent increase each year
   (3) The population after t years
   (4) The starting population of the town

6. Trinity deposits some money into a bank that offers interest compounded annually. The amount of money in the bank (A) after t years can be modeled with the equation $A = 700(1.03)^t$. What does 1.03 represent?
   (1) The percent increase each year
   (2) One plus the growth rate
   (3) The amount of money in the bank after t years
   (4) The starting amount of money

7. After Allie takes some medicine, each hour the number of milligrams of medicine (M) remaining in her body after t minutes can be modeled with the equation $M = 200(0.73)^t$. Which number represents the decay rate?
   (1) 0.73
   (2) 200
   (3) 0.27
   (4) 146

8. A ball is dropped from the window of a building. The height of the bounce (H) is related to the number of bounces (N) by the equation $H = 50(0.4)^N$. Which number represents the height the ball was originally dropped from?
   (1) 50
   (2) 0.4
   (3) 0.6
   (4) 20

9. Mason puts money into a bank that offers interest compounded annually. The formula relating the amount of money in the bank (A) to the number of years it has been in the bank (t) is $A = 800(1.2)^t$. What is the interest rate the bank offers?
   (1) 1.2%
   (2) 2%
   (3) 20%
   (4) 120%

10. An exponential equation $y = a(b)^x$ models exponential decay. What is known about the value of b?
    (1) It is greater than 2.
    (2) It is between −1 and 0.
    (3) It is between 1 and 2.
    (4) It is between 0 and 1.

B. *Show how you arrived at your answers.*

1. The population of a city is 300,000. If the population increases by 4% each year, create an equation that relates the population of the city (P) to the number of years that have passed (t).

2. Blake is trying to quit drinking soda. At the beginning of the year he drinks 64 ounces a day. Each week the amount of soda he drinks each day is 75% of the amount he drank each day the week before. Create an equation that relates the amount of soda he drinks each day (S) to the number of weeks (t) that have passed since he started trying to quit.

3. When Aria started school in kindergarten she had 40 minutes of homework each night. Each year the number of minutes of homework she had was equal to 1.15 times the number of minutes of homework she had the year before. Create an equation that relates the number of minutes of homework she has each night (H) to the number of years that have passed since kindergarten (t).

4. A company's annual profits can be modeled by the equation $P = 200,000(1.36)^t$, where P is the amount of profit and t is the number of years they have been in business. What do the numbers 200,000 and 0.36 represent?

5. Bailey is training to run a marathon. The number of minutes it takes to run the marathon after t weeks of training is $M = 300(0.95)^t$. What do the numbers 300 and 0.05 represent?

Chapter Ten: FUNCTIONS

10.1 DESCRIBING FUNCTIONS

=== **KEY IDEAS** ===

A *function* is like a machine that converts numbers into other numbers. Functions are often named with the letters f, g, and h. If function f converts the number 2 into the number 7, we write f(2) = 7. Functions can be described with an equation, a graph, or a list of ordered pairs.

What Is a Function?

A function can be thought of as a machine with an "in" slot into which you can put numbers on a card and another "out" slot from which numbers on cards come out. If the function's name is f and you put the number 2 on a card and feed it into in slot on the box, another card with the number 7 might come out of the out slot. If 7 comes out when 2 is fed into the function machine, we say $f(2) = 7$. Inside the parentheses tells what number is being fed into the function machine and does not have anything to do with multiplication.

Representing Functions as a List of Ordered Pairs

The only thing we know about the function called f described above is that $f(2) = 7$. We do not know $f(3), f(4)$, or anything else. One way to know more about the function is if a set is given like this:

$$F = \{(1, 4), (2, 7), (3, 10), (4, 13), (5, 16)\}$$

The ordered pair (1, 4) in this list indicates that if you put 1 into the in slot of the function machine, the number 4 will come out of the out slot. Also,

Functions

if you put the number 2 into the in slot, the number 7 will come out of the out slot.

Since there are five ordered pairs in the set, there are five things we now know about this function:

$$f(1) = 4, f(2) = 7, f(3) = 10, f(4) = 13, \text{ and } f(5) = 16$$

Example 1

If the function g is defined as $g = \{(1, 7), (2, 9), (3, 11), (4, 13), (5, 15), (6, 17), (7, 19)\}$, what is the value of $g(7)$?
   (1) 1   (2) 9   (3) 15   (4) 19

Solution: Since the ordered pair that has a 7 as the first coordinate is (7, 19), this means that $g(7) = 19$. Choice (4).

If the function g is defined as in the last example as $g = \{(1, 7), (2, 9), (3, 11), (4, 13), (5, 15), (6, 17), (7,19)\}$, and you are asked for the value of $g(8)$, there is no answer since there is no ordered pair in the set that has 8 as its first coordinate. When this happens we say that the function is *undefined* at 8.

Domain and Range of a Function Described by a Set of Ordered Pairs

The function g described in Example 1 is only defined for the values 1, 2, 3, 4, 5, 6, and 7. If any other number is put into the in slot of this function machine, it will be undefined. These seven numbers are said to be in the *domain* of function g. The domain is usually written as a set of values. The domain of function g is $\{1, 2, 3, 4, 5, 6, 7\}$. The function is only defined for numbers in its domain.

The only values that can come out of the g function's out slot described in Example 1 are 7, 9, 11, 13, 15, 17, and 19. The set of these values $\{7, 9, 11, 13, 15, 17, 19\}$ is called the *range* of the function.

Example 2

What is the domain of the function $h = \{(2, 5), (4, 17), (7, 50)\}$?
   (1) $\{5, 17, 50\}$ (3) $\{5, 4, 7\}$
   (2) $\{2, 17, 50\}$ (4) $\{2, 4, 7\}$

Solution: Choice (4). The set of all the first coordinates is the domain. $h(2) = 5$, $h(4) = 17$, and $h(7) = 50$. For any other input value, the function is undefined; for example, $h(3)$ is undefined.

Example 3

What is the range of the function $f = \{(2, 5), (3, 7), (6, 5)\}$?

Solution: There are only two different y values so the set for the range will just be $\{5, 7\}$. In a set, you do not list "repeats" so the answer is *not* $\{5, 5, 7\}$.

Domain and Range in Real-World Applications

In some real-world situations, certain numbers don't make sense. For example, you can't buy half of a car or eat negative 5 pieces of pizza. When a function is related to a real-world situation, there may be limits on the domain and range.

If a movie theater sells tickets for $10 and there is a function $f(x) = 10x$ for calculating the amount of money collected for x tickets, the domain is not going to be all real-numbers. Since someone cannot buy half a ticket or negative 10 tickets, the domain is just going to be the numbers $\{0, 1, 2, 3, 4, 5, ...\}$.

When a Set of Ordered Pairs Cannot Describe a Function

If in a function f, $f(2) = 7$, this means that any time the number 2 is put into the function machine, the number 7 comes out. There will never be a time that 2 is put in and the number 8 comes out. If it did, we would want to get our function machine repaired!

When a function is described by a list of ordered pairs, no two ordered pairs will have the same first coordinate. If (2, 7) is in the set, there will not be a (2, 8) or a (2, 9) or anything else with a 2 as the first coordinate.

Function	Not a Function
$f = \{(2, 7), (3, 9), (4, 13), (5, 13)\}$	$f = \{(2, 7), (2, 9), (4, 13), (5, 17)\}$
It is OK to have matching y-values	↑ ↑

If a function were described as $f = \{(2, 7), (2, 9), (4, 13), (5, 17)\}$ and they asked the value of $f(2)$, there would be two possible answers, 7 or 9. For this reason, a function cannot have different output values for the same input value.

Functions

Example 4

Which of the following cannot be the description of a function?
(1) {(1, 4), (3, 6), (5, 8), (7, 10)}
(2) {(1, 4), (3, 4), (5, 4), (7, 4)}
(3) {(1, 6), (3, 10), (5, 4), (7, 8)}
(4) {(1, 4), (3, 6), (5, 8), (5, 10)}

Solution: Since choice (4) has two ordered pairs that both have 5 as the first coordinate, it cannot be the description of a function.

Check Your Understanding of Section 10.1

A. Multiple-Choice

1. If a function f is defined as $f = \{(1, 2), (2, 3), (3, 1), (4, 4)\}$, what is $f(2)$?
   (1) 1   (2) 2   (3) 3   (4) 4

2. If a function g is defined as $g = \{(1, 4), (3, 2), (4, 3), (5, 1)\}$, what is $g(2)$?
   (1) 1   (2) 2   (3) 3   (4) Undefined

3. If $f(4) = 7$, which could *not* be the definition of the function?
   (1) $f = \{(1, 3), (4, 7), (5, 6), (7, 4)\}$
   (2) $f = \{(2, 5), (4, 7), (5, 1), (6, 5)\}$
   (3) $f = \{(1, 3), (4, 8), (5, 6), (7, 4)\}$
   (4) $f = \{(1, 3), (4, 7), (5, 6), (4, 7)\}$

4. If $f = \{(1, 4), (2, 8), (3, 7), (4, 1)\}$ and $g = \{(1, 5), (2, 9), (3, 6), (4, 2)\}$, which of the following is true?
   (1) $f(1) > g(1)$ (3) $f(3) > g(3)$
   (2) $f(2) > g(2)$ (4) $f(4) > g(4)$

5. Which of the following *cannot* be the definition of a function?
   (1) $f = \{(1, 5), (2, 7), (2, 8), (4, 9)\}$
   (2) $f = \{(1, 2), (2, 2), (3, 2), (4, 2)\}$
   (3) $f = \{(0, 0), (1, 1), (-1, 1), (2, 4), (-2, 4)\}$
   (4) $f = \{(6, 1)\}$

6. What is the domain of the function defined as $f = \{(1, 4), (3, 7), (4, 8), (5, 8)\}$?
   (1) {4, 7, 8}   (3) {1, 3, 4, 5}
   (2) {1, 3, 4, 5, 7, 8}   (4) {4}

7. What is the range of the function defined as $f = \{(1, 4), (3, 7), (4, 8), (5, 8)\}$?
   (1) {1, 3, 4, 5, 7, 8}
   (2) {1, 3, 4, 5}
   (3) {4}
   (4) {4, 7, 8}

8. If $f = \{(1, 4), (2, 3), (3, 2), (4, 1)\}$ and $g = \{(1, 3), (2, 4), (3, 1), (4, 2)\}$ what is $f(g(1))$?
   (1) 1
   (2) 2
   (3) 3
   (4) 4

9. A function g takes as an input a number representing the number of gallons of gasoline purchased and outputs the price of that many gallons of gasoline. What is the domain of this function?
   (1) All numbers greater than or equal to zero
   (2) All numbers
   (3) All nonnegative integers
   (4) All numbers greater than zero

10. A function t takes as an input the day of the year in New York and outputs the average temperature for that day. What is a reasonable range for this function?
    (1) Numbers between −80 and 200
    (2) Numbers between −10 and 100
    (3) Numbers between 60 and 80
    (4) Numbers between 0 and 60

B. Show how you arrived at your answers.

1. Explain why this is or is not a definition of a function.
   $f = \{(3, 5), (4, 5), (5, 5), (6, 5)\}$

2. If $f = \{(1, 4), (2, 1), (3, 2), (4, 3)\}$, calculate (a) $f(1) + f(2)$; (b) $f(1 + 2)$; (c) $f(f(1))$.

3. If $f = \{(1, 3), (4, 9), (5, 2), (6, 8)\}$ and $f(a) = 8$, what are all possible values for a?

4. Is $f = \{(3, 4), (3, 5), (4, 7), (5, 1)\}$ a function? Explain why or why not?

5. William says the range of a function always has either the same number of numbers as the domain or more numbers than the domain. Mia says the range of a function can either have the same number of numbers as the domain or fewer numbers than the domain. Alice says that the range of a function can have more numbers, fewer numbers, or the same number of numbers as the domain. Which of the three students is correct and why?

Chapter Ten **FUNCTIONS**

Functions

10.2 FUNCTION GRAPHS

=== **KEY IDEAS** ===

A list of ordered pairs is not always the most efficient way to describe the input and output values of a function. A graph of these values is especially useful when there are an infinite number of numbers in the domain of the function. The function graph can be used to evaluate the function at different values and can also be used to determine the domain and range of the function.

Graphing a Function

If a function is represented as a set of ordered pairs, those ordered pairs can be graphed to form the graph of a function.

If function f is described as $f = \{(1, 1), (2, 3), (3, 5), (4, 7), (5, 9)\}$, the graph of the function will contain just five points corresponding to the five ordered pairs in the set. The graph of these five points looks like this.

This graph contains the same information as the set of ordered pairs. Since there is a point at (1, 1), this means that $f(1) = 1$. Since there is a point at (2, 3), this means that $f(2) = 3$.

Sometimes you are given the graph of a function and not the set of ordered pairs. From this graph, it is possible to answer questions about the function. For example, on the next page is the graph of a different function called f.

To find the value of $f(2)$, look for a point on the graph that has an x-coordinate of 2.

Step 1: Locate 2 on the x-axis.
Step 2: Look up and down the vertical line through the 2 on the x-axis to see if there is a point that has an x-coordinate of 2.

Step 3: If there is no point that has this x-coordinate, the function is undefined at that 2.
Step 4: If there is a point that has this x-coordinate, the y-coordinate of that point is the value of $f(2)$.

Since the point on the graph that has an x-coordinate of 2 has a y-coordinate of 1, $f(2) = 1$.

Functions

Example 1

Using the same graph of f from the above explanation, find the values of $f(4), f(5),$ and $f(7)$.

    Solution: $f(4) = 2$ since the point $(4, 2)$ is on the graph. $f(5) = 6$ since the point $(5, 6)$ is on the graph, and $f(7)$ is undefined since there is no point on the graph that has an x-coordinate of 7.

Example 2

Below is the graph of a function called g. Which point could be used to find the value of $g(4)$?

(1) A   (2) B   (3) C   (4) D

    Solution: To find $g(4)$, locate the point that is on the vertical line through $(4, 0)$. This line would pass through point B. The value of $g(4) = 2$, though they are not asking for the value of $g(4)$, just the point that can be used to find $g(4)$, which is choice (2).

Example 3

The graph of h is given below. Use it to approximate the value of $h(4.5)$.

(1) 3   (2) 4   (3) 5   (4) 6

Solution: From the point (4.5, 0) on the *x*-axis, draw a vertical line up toward the line representing the function. When you reach the point on the line, draw a horizontal line to the left until you reach the *y*-axis. It seems to be near the point (4.5, 4), so there is only one choice that is feasible, choice (2).

Determining if a Graph Represents a Function

The set f = {(1, 3), (1, 5), (2, 6)} does not represent a function because two of the ordered pairs have the same first coordinate. $f(1)$ can only equal one thing, but from this set, it seems to be both 3 and 5. When these points from this nonfunction are graphed, the two points with the same *x* value will lie on the same vertical line.

If the graph of a set of ordered pairs has at least two points that lie on the same vertical line, the graph fails the *vertical line test* and the set cannot represent a function.

Functions

Example 4

Which of the following can be the graph of a function?

(1) (2) (3) (4)

Solution: Choice (3). The other three graphs all fail the vertical line test. In the graph for (4) the vertical line at $x = 5$ passes through all the points on the graph.

Finding the Domain and Range from the Graph of a Function

If the graph of a function is a bunch of points, the domain is the set of all the x-coordinates and the range is the set of all the y-coordinates of the points.

In the graph below, the domain is {1, 2, 4, 5}, and the range is {3, 4}.

If the graph of the function is some kind of curve, the domain and range cannot be described by a list since there are an infinite number of values in each set.

In the graph of the function below, there are an infinite number of points on the line segment.

For every number between 1 and 7, including 1 and 7, there is a point on the line segment that has that as an x-coordinate. The domain can be described as $1 \leq x \leq 7$. Likewise, the range can be described as $3 \leq y \leq 5$.

Functions

It is possible for the domain or the range to contain every real number. In this case, write "All real numbers." For example, if the graph of the function is given below, the domain is all real numbers, and the range is $y \geq 2$.

Check Your Understanding of Section 10.2

A. Multiple-Choice

1. Which is the graph of the function $f = \{(1, 2), (2, 5), (3, 1), (4, 1)\}$?

(1)

(2)

(3)

(4)

298

2. Below is the graph for which function?

(1) $f = \{(1, 1), (2, 6), (3, 5), (4, 1), (5, 2)\}$
(2) $f = \{(1, 1), (6, 2), (5, 3), (6, 4), (2, 5)\}$
(3) $f = \{(1, 1), (2, 2), (3, 3), (4, 4), (5, 5)\}$
(4) $f = \{(1, 1), (2, 4), (3, 6), (4, 2), (5, 3)\}$

3. Below is the graph of $y = f(x)$. What is the value of $f(3)$?

(1) 1   (2) 2   (3) 3   (4) 4

4. A portion of the graph of $y = f(x)$ is shown below. What is the value of $f(100)$?

(1) 5   (2) 6   (3) 7   (4) 100

299

Functions

5. Below is the graph of $y = f(x)$. Which point could be used to determine the value of $f(4)$?

(1) A (2) B (3) C (4) D

6. Below is a graph of $y = g(x)$. What is the approximate value of $g(5)$?

(1) 5 (2) 6 (3) 7 (4) 8

7. Below is the graph of $y = f(x)$. What is the domain of f?

(1) {1, 2, 3, 4} (3) {1, 2, 3, 5}
(2) {2, 3, 5, 6} (4) {1, 2, 5, 6}

300

8. Below is the graph of $y = f(x)$. What is the range of f?

(1) {1, 2, 3, 5}
(2) {1, 2, 3, 6}
(3) {2, 3, 5, 6}
(4) {1, 2, 5, 6}

9. Below is the graph of $y = f(x)$. What is the domain and range of f?

(1) Domain $90 < x < 450$, Range $-1 < y < 3$
(2) Domain $90 \leq x \leq 450$, Range, $-1 \leq y \leq 3$
(3) Domain $-1 < x < 3$, Range $90 < y < 450$
(4) Domain $-1 \leq x \leq 3$, Range $90 \leq y \leq 450$

Functions

10. Which is the graph of a function?

(1)

(3)

(2)

(4)

B. *Show how you arrived at your answers.*

1. Alisha says that this is not the graph of a function because there are two points that have the same y-coordinate, like (3, 9) and (−3, 9). Tyson says that it is the graph of a function. Who is right and why?

Tyson because every input has one output

2. What is the domain and range of the function whose graph is below?

3. Using the two graphs below, determine the value of $f(g(3))$.

$y = f(x)$ $\qquad$ $y = g(x)$

4. Below is the graph of $y = f(x)$. What is the value of $f(f(f(f(f(f(f(f(2))))))))$?

Functions

5. In the graph of the function below, list all values that satisfy the equation $f(a) = 4$.

10.3 DEFINING A FUNCTION WITH AN EQUATION

=== **KEY IDEAS** ===

The most convenient way to describe a function is with an equation. Not every function has an equation to describe it, but those that do can be quickly evaluated.

Equations for Functions

Often a function is described with an equation. For example, it might say that for a question f is defined by the equation $f(x) = 2x - 1$. When a function is defined this way, determining an output value for something like $f(5)$ just requires substituting 5 into the equation.

$$f(5) = 2 \cdot 5 - 1 = 10 - 1 = 9$$

If 5 is put into the input slot of this function, the number 9 will come out of the output slot.

Example 1

If the function g is defined as $g(x) = x^2 - 3x$, what is the value of $g(5)$?

Solution: 10 because $g(5) = 5^2 - 3 \cdot 5 = 25 - 15 = 10$.

Example 2

If the function h is defined as $h(x) = x^2 + 2x - 5$, write an expression for $h(x - 2)$.

Solution: Just as $h(3) = 3^2 + 2 \cdot 3 - 5 = 9 + 6 - 5 = 10$, and $h(a) = a^2 + 2a - 5$, when $(x - 2)$ is substituted in for the x in the function, it becomes $h(x - 2) = (x - 2)^2 + 2(x - 2) - 5$. This can be further simplified to $h(x) = x^2 - 2x - 5$.

Defining Functions Based on Other Functions

When a function's description includes another function, it requires two steps to evaluate it. An example of this is $g(x) = f(x + 2) - 1$. In order to determine the value of $g(5)$, substitute 5 in for x in all places. The equation would then become $g(5) = f(5 + 2) - 1 = f(7) - 1$. Unless f is defined somehow, the expression can't be simplified any further.

If it is known, for this example, that $f(x) = x^2 - 10$, then $g(5) = f(7) - 1 = 7^2 - 10 - 1 = 49 - 10 - 1 = 38$ would be the answer.

Example 3

If $f(x) = 2x - 5$ and $g(x) = f(x - 3)$, what is the value of $g(8)$?

Solution: $g(8) = f(8 - 3) = f(5)$ and $f(5) = 2 \cdot 5 - 5 = 10 - 5 = 5$ so $g(8) = 5$.

Graphing a Function from Its Equation

The function $f(x) = 2x - 5$ has the same graph as the solution set of the equation $y = 2x - 5$. This is the graph of $y = f(x)$. All the rules for graphing linear, quadratic, and exponential equations described already in this book apply to graphing linear, quadratic, and exponential functions. Functions can be graphed on the graphing calculator by pressing [Y=] and entering the equation for the function.

Example 4

Make a graph of $y = f(x)$ if $f(x)$ is defined as $f(x) = x^2 - 8x + 12$.

Solution: This is the same process as graphing the two-variable equation $y = x^2 - 8x + 12$. It can be done by hand or with the graphing calculator.

Functions

For the TI-84:
Use the $y =$ menu.

For the TI-Nspire:
Type the equation after $f1(x) =$ in the entry bar of the graphing Scratchpad.

Using the Graphing Calculator to Determine Function Values

If a function is defined by an equation, there are three ways to use a graphing calculator to determine output values of the function for different input values. If the function is defined as $f(x) = x^2 - 8x + 12$, then to determine $f(3)$:

For the TI-84:

Method #1: Graph the equation and press [2ND] and then [TRACE] to get to the CALC menu. On the CALC menu, select [1] for value. The graph will appear and there will be an "X=" in the bottom left corner of the screen. Type 3, and it will calculate $f(3)$ and also show the corresponding point on the graph. In this example, $f(3) = -3$.

Method #2: After entering the equation for the function into the Y= menu, press [2ND] and then [GRAPH] to get to the TABLE menu. Use the up or down arrows to more the cursor to the *x* value you want to evaluate the function at.

Method #3: After entering the equation for the function into the Y= menu, press [2ND] and then [MODE] to quit. Then press [VARS], select Y-VARS by pressing the right arrow, and then select [1] for Function and [1] again for Y_1. A "Y_1" will appear on the screen. To evaluate the function at a value, press [(], [3], [)] after the Y_1 and press [ENTER].

For the TI-Nspire:

Method #1: From the home screen press [B] for the Graph Scratchpad.

Functions

Type $x^2 - 8x + 12$ after the $f1(x)=$ and press [enter].

Press [menu], [5], and [1] for Graph Trace.

To see the value of $f(3)$, press [3] and [enter].

Method #2: After graphing $f(x) = x^2 - 8x + 12$, hold down the [ctrl] and press [t]. This will display a table of values. To see what $f(3)$ is, scroll down to the 3 in the x column.

Method #3: From the home screen press [A] for the Calculator Scratchpad.

Press [menu], [1], and [1] for Define.

Functions

After the word "Define," type $f(x) = x^2 - 8x + 12$ and press [enter].

To evaluate the function f at the value 3, type $f(3)$ and press [enter].

Piecewise Functions

A *piecewise function* is one that uses different equations for different input values. An example of a piecewise function is.

$$f(x) = \begin{cases} x-1, x < 3 \\ 2x+1, x \geq 3 \end{cases}$$

For input values less than 3, it uses the top equation. $f(2)$ will equal $2 - 1 = 1$. For input values greater than or equal to 3, it uses the bottom equation. $f(4) = 2 \cdot 4 + 1 = 9$.

A portion of the chart for this piecewise function looks like this.

x	$f(x)$
–2	–2 – 1 = –3
–1	–1 – 1 = –2
0	0 – 1 = –1
1	1 – 1 = 0
2	2 – 1 = 1
3	2 · 3 + 1 = 7
4	2 · 4 + 1 = 9

Plotting these seven values looks like this:

To fill in the rest of the graph, plot an open circle at the point (3, 2), which is what you get when you substitute 3 into the top equation. Even though this is not a point on the graph, it serves as the boundary between the two different pieces of the graph.

Functions

Graphing Piecewise Functions on the Graphing Calculator

With a TI-Nspire, piecewise functions can be quickly graphed. From the Graph Scratchpad in the entry line press the [template] key next to the [9] and select the piecewise template on the top row, third from the right. It will prompt you to enter the number of pieces in the function.

Enter the equations $x - 1$, $x < 3$, $2x + 1$, and $x \geq 3$ into the template. To get the $<$ and $\geq$ symbols, press [ctrl] and [=] and select the symbol you want from the menu. Then press [enter].

312

Notice that it is not clear which endpoint is an open dot and which is a closed dot on the graphing calculator.

Determining Intervals for When a Function Is Positive or Negative

Unlike numbers that are always either positive, negative, or 0, functions can be positive for some input values, negative for others, and 0 for others. The function $f(x) = 2x - 6$ can be negative, $f(0) = -6$; positive, $f(5) = 4$; or 0, $f(3) = 0$.

The quickest way to determine when a function is positive, negative, or 0 is to graph it and see where the graph is below the x-axis, above the x-axis, or crossing the x-axis.

Example 5

In what interval is the function $f(x) = 3x + 6$ negative?

Solution: The graph of $y = 3x + 6$ has an x-intercept of $(-2, 0)$.

For every x value less than -2, $f(x)$ will be negative. For every x value greater than -2, $f(x)$ will be positive. The solution is that $f(x)$ will be negative when $x < -2$.

Functions

Check Your Understanding of Section 10.3

A. Multiple-Choice

1. If $f(x) = 3x + 7$, what is $f(2)$?
   (1) 10   (2) 11   (3) 12   (4) 13

2. If $g(x) = -x^2 + 7x + 1$, what is $g(2)$?
   (1) 11   (2) 19   (3) 27   (4) 35

3. If $f(x) = 2x + 5$, what is $f(a)$?
   (1) 7   (2) $2a - 5$ (3) $2a + 5$ (4) $a + 5$

4. If $f(x) = 5x - 2$, what is $f(x - 1)$?
   (1) $5x - 2$ (2) $5x - 3$ (3) $5x + 1$ (4) $5x - 7$

5. If $f(x) = x^2 + 5$ and $g(x) = f(x - 2)$, what is $g(6)$?
   (1) 31   (2) 21   (3) 11   (4) 1

6. If $g(x) = 2x - 5$, which is the graph of $y = g(x)$?

314

7. If $f(x) = 2x + 5$, for what value of a does $f(a) = 11$?
   (1) 1   (2) 3   (3) 5   (4) 7

8. For what values of x is the graph of the function $h(x) = 2^x$ increasing?
   (1) For $x > 0$
   (2) For $x < 0$
   (3) All values
   (4) No values

9. What are the zeros of the function $f(x) = x^2 + 10x + 16$?
   (1) –2 and –8
   (2) 2 and 8
   (3) 2 and –8
   (4) –2 and 8

10. If $f(x) = x^2 - 6x + 5$, for what values of x is the function $f(x) < 0$?
    (1) For values between 1 and 5
    (2) For values greater than 5
    (3) For values less than 1
    (4) For all values

B. Show how you arrived at your answers.

1. If $f(x) = x^2 + 8x + 7$, what are (a) $f(3)$ and (b) $f(x + 1)$?

2. If $f(x) = 2x - 3$ and $g(x) = x^2 + 2$, what are (a) $f(g(3))$ and (b) $g(f(3))$?

3. $f(x) = x + 2$, $g(x) = 2x$, and $h(x) = x^2$. Manuel says that the expression $2(x + 2)^2$ can be written as $h(g(f(x)))$. Elizabeth says that this is not correct. Which student is right and why?

4. If $f(x) = x^2$ and $g(x) = f(x - 3)$, what does the graph of $g(x)$ look like?

5. If $g(x) = 2x + 1$ and $g(f(3)) = 15$, what is the value of $f(3)$?

315

10.4 FUNCTION TRANSFORMATIONS

=== **KEY IDEAS** ===

The graph of the function f(x) looks very similar to the graphs of f(x + k), f(x − k), f(x) + k, and f(x) − k. The only difference is that the graphs of the other four are *translated*, in other words shifted, left, right, up, or down. The graphs of the other four can be produced quickly if you know which direction to shift and what the original graph looks like.

Vertical Translations

The graph of $f(x) + k$ will be the same as the graph of $f(x)$ only shifted up by k units. If the graph of $y = f(x)$ is shown below, the graph of $y = f(x) + 3$ will have all points shifted 3 units up.

The graph of $f(x) - k$ will be the same as the graph of $f(x)$ only shifted down by k units. If the graph of $y = f(x)$ is shown below, the graph of $y = f(x) - 3$ will have all points shifted 3 units down.

Horizontal Translations

The graph of the function $f(x + 3)$ is the same as the graph of $f(x)$ when translated 3 units to the left. The graph of the function $f(x - 3)$ is the same as the graph of $f(x)$ when translated 3 units to the right. In general, the graph of the function $f(x + k)$ is the same as the graph of $f(x)$ when translated to the left by k units. The graph of the function $f(x - k)$ is the same as the graph of $f(x)$ when translated to the right by k units.

Example 1

The graph of $y = f(x)$ is on the left and the graph of $y = g(x)$ is on the right. How could $g(x)$ be defined?

(1) $g(x) = f(x) + 4$
(2) $g(x) = f(x) - 4$
(3) $g(x) = f(x + 4)$
(4) $g(x) = f(x - 4)$

Solution: Since the graph on the right is like the graph on the left shifted down by 4 units, the solution is choice (2).

Functions

Example 2

If this is the graph of $y = f(x)$, which could be the graph of $y = g(x)$ if $g(x) = f(x + 2) + 3$?

Solution: This has two transformations the +2 in the parentheses means the transformed graph will be shifted 2 units to the left. The +3 outside the parentheses means the transformed graph will be shifted 3 units up. The vertex of the original graph is (4, −1). The vertex of the transformed graph will be 2 to the left and 3 above (4, −1), which is at (2, 2). Choice (4).

Example 3

The graph of $y = f(x)$ is a parabola with a vertex at (5, 7). If $g(x) = f(x - 4)$, what are the coordinates of the vertex of the parabola that is the graph of $y = g(x)$?

(1) (1, 7)   (2) (9, 7)   (3) (5, 11)   (4) (5, 3)

Solution: Every point on the graph of $y = f(x - 4)$ is four units to the right of the corresponding point on $y = f(x)$. The vertex of the original function is at (5, 7) so the vertex of the translated function will be 4 units to the right, which is (9, 7). Choice (2).

MATH FACTS

The graph of these four transformations will be congruent to the graph of the original function. The only difference is that it is shifted horizontally, vertically, or some combination of both. The four basic transformations are

1. $f(x) + k$ The graph shifts up by k units.
2. $f(x) - k$ The graph shifts down by k units.
3. $f(x + k)$ The graph shifts left by k units.
4. $f(x - k)$ The graph shifts right by k units.

Functions

Check Your Understanding of Section 10.4

A. Multiple-Choice

1. If below is the graph of $y = f(x)$, which is the graph of $y = f(x) - 5$?

(1)

(2)

(3)

(4)

320

2. If below is the graph of $y = f(x)$, which is the graph of $y = f(x) + 5$?

(1)

(2)

(3)

(4)

Functions

3. If below is the graph of $y = f(x)$, which is the graph of $y = f(x - 5)$?

(1)

(2)

(3)

(4)

4. If below is the graph of $y = f(x)$, which is the graph of $y = f(x + 5)$?

Functions

5. Below is the graph of f(x) on the left and g(x) on the right. Which is equivalent to g(x)?

(1) $f(x) + 4$
(2) $f(x) - 4$
(3) $f(x + 4)$
(4) $f(x - 4)$

6. Below is the graph of f(x) on the left and g(x) on the right. Which is equivalent to g(x)?

(1) $f(x) + 4$
(2) $f(x) - 4$
(3) $f(x + 4)$
(4) $f(x - 4)$

7. Below is the graph of f(x) on the left and g(x) on the right. Which is equivalent to g(x)?

(1) f(x) + 4
(2) f(x) − 4
(3) f(x + 4)
(4) f(x − 4)

8. Below is the graph of f(x) on the left and g(x) on the right. Which is equivalent to g(x)?

(1) f(x) + 4
(2) f(x) − 4
(3) f(x + 4)
(4) f(x − 4)

9. If the graph of y = f(x) is a parabola with the vertex at (5, 1), what is the vertex of the graph of the parabola y = f(x − 2)?
(1) (5, 3)
(2) (5, −1)
(3) (7, 1)
(4) (3, 1)

Functions

10. If the graph of $y = g(x) + 2$ is below, which is the graph of $y = g(x)$?

(1)

(2)

(3)

(4)

B. Show how you arrived at your answers.

1. Below is a chart of x values and corresponding $f(x)$ values. (a) If $g(x) = f(x) + 1$, what is the value of $g(3)$? (b) If $h(x) = f(x + 1)$, what is $h(3)$?

x	$f(x)$
1	4
2	3
3	1
4	5
5	2

2. Below is the graph of $y = f(x)$. Make a graph of $y = f(x - 2)$.

3. The graph of $y = f(x)$ has a domain of $2 \leq x \leq 7$. (a) What is the domain of the graph of $y = f(x) + 4$? (b) What is the domain of the graph of $y = f(x + 4)$?

Functions

4. The graph of $y = f(x)$ is below. If $g(x) = f(x + 2)$ and $h(x) = g(x) + 3$, make a graph of $y = h(x)$.

5. If $f(x) = x^2 + 2x - 3$ and $g(x) = x^2 + 6x + 5$, how can $g(x)$ be written as a transformation of $f(x)$? Explain.

Science the graph $g(x)$ is the graph of $f(x)$ shifted to the left 2 units it can be expressed as $g(x) = f(x+2)$

Chapter Eleven

SEQUENCES

11.1 TYPES OF SEQUENCES

=== KEY IDEAS ===

A sequence is a list of numbers. It is sometimes possible to predict the next number on a list by examining the numbers before and detecting a pattern. That pattern could be adding a number to the term before, multiplying a number by the term before, or something even more complicated.

Sequence Notation

An example of a sequence is 3, 7, 11, 15, 19, ...

The sequence is often named with the letter a. The individual elements of the sequence are named by the name of the sequence with a subscript identifying the element's position in the list. For this list, $a_1 = 3$, $a_2 = 7$, $a_3 = 11$, $a_4 = 15$, and $a_5 = 19$. Sometimes instead of a subscript, the term number is put into parentheses, like with functions, like $a(1) = 3$ or $a(2) = 7$.

Arithmetic Sequences

To find the value of a_6, check to see if this is the kind of sequence where each number can be calculated by adding or subtracting the same thing from the previous number. When this happens, the sequence is called an *arithmetic sequence*.

In this example since $3 + 4 = 7$, $7 + 4 = 11$, $11 + 4 = 15$, and $15 + 4 = 19$, this seems to be an arithmetic sequence. To continue the pattern, $a_6 = 19 + 4 = 23$.

The sequence 40, 32, 24, 16 is also an arithmetic sequence. By subtracting 8 from each term, the next term is obtained. When each term is smaller than the previous term, it is a *decreasing* sequence. When each term is larger than the previous term it is an *increasing* sequence.

Sequences

Geometric Sequences

The sequence 3, 6, 12, 24, 48, ... is not an arithmetic sequence. 3 + 3 = 6, but 6 + 3 is not 12. In this sequence, each term is gotten by multiplying the previous term by 2. When this happens, the sequence is called a *geometric sequence*. The next term is 48 · 2 = 96. Since each term is larger than the previous term, this is an increasing geometric sequence. A sequence like 160, 80, 40, 20, ... is a decreasing geometric sequence since each term is 1/2 the previous term.

Check Your Understanding of Section 11.1

A. Multiple-Choice

1. In the sequence $a = 1, 3, 6, 10, 15, \ldots$, what is the value of a_2?
   (1) 1   (2) 3   (3) 6   (4) 10

2. What type of sequence is 3, 7, 11, 15, ...?
   (1) Increasing arithmetic   (3) Increasing geometric
   (2) Decreasing arithmetic   (4) Decreasing geometric

3. What type of sequence is 25, 18, 11, 4, ...?
   (1) Increasing arithmetic   (3) Increasing geometric
   (2) Decreasing arithmetic   (4) Decreasing geometric

4. What type of sequence is 4, 8, 16, 32, ...?
   (1) Increasing arithmetic   (3) Increasing geometric
   (2) Decreasing arithmetic   (4) Decreasing geometric

5. What type of sequence is 1, 215, 405, 135, 45, ...?
   (1) Increasing arithmetic   (3) Increasing geometric
   (2) Decreasing arithmetic   (4) Decreasing geometric

6. What is the next number in the sequence 8, 13, 18, 23?
   (1) 26   (2) 27   (3) 28   (4) 29

7. What is the next number in the sequence 14, 6, −2, −10?
   (1) −17   (2) −18   (3) −19   (4) −20

8. What is the next number in the sequence 20, 10, 5, $\frac{5}{2}$?
   (1) $\frac{5}{4}$ (2) $\frac{5}{8}$ (3) $\frac{5}{16}$ (4) $\frac{5}{32}$

9. A sequence begins with 2, 6. If this is a geometric series, what is the next number?
   (1) 10   (2) 12   (3) 14   (4) 18

10. What type of sequence is $2^1, 2^2, 2^3, 2^4, 2^5, \ldots$?
    (1) Increasing arithmetic   (3) Increasing geometric
    (2) Decreasing arithmetic   (4) Decreasing geometric

11.2 RECURSIVELY DEFINED SEQUENCES

=== KEY IDEAS ===

Describing the terms of a sequence, including the rule of how to get the next term, requires a shorthand that resembles function notation. This *recursive* description identifies the first number in the sequence and also the rule of how to use any term to obtain the next term.

Listing an Arithmetic Sequence from the Recursive Rule

The sequence $a = 3, 7, 11, 15, 19, \ldots$ can be described by first identifying the first element, $a_1 = 3$. This is sometimes called the *base case*. For this sequence, a_2 can be calculated by adding 4 to a_1. Then a_3 can be calculated by adding 4 to a_2, and so on.

The way to write this rule is $a_n = a_{n-1} + 4$ for $n > 1$. The a_n is the term you are trying to calculate, and the a_{n-1} is the previous term. $a_n = a_{n-1} + 4$ is a fancy way of saying that to get any term, add four to the previous term.

Together these two pieces make up the rule:

$$a_1 = 3$$
$$a_n = a_{n-1} + 4 \text{ for } n > 1$$

When $n = 2$, the bottom formula says $a_2 = a_{2-1} + 4 = a_1 + 4 = 3 + 4 = 7$.
When $n = 3$, the bottom formula says $a_3 = a_{3-1} + 4 = a_2 + 4 = 7 + 4 = 11$.
When $n = 4$, the bottom formula says $a_4 = a_{4-1} + 4 = a_3 + 4 = 11 + 4 = 15$.

Sequences

Example 1

The recursive rule for a sequence is

$$a_1 = 5$$
$$a_n = a_{n-1} + 6 \text{ for } n > 1$$

Find the values of a_2, a_3, a_4, and a_5.

Solution: The rule says that to get the next term, add 6 to the previous term. From that rule, the next numbers 11, 17, 23, and 29 can be found. This can also be solved by substituting 2, 3, 4, and 5 into the bottom formula.

For $n = 2$, $a_2 = a_{2-1} + 6 = a_1 + 6 = 5 + 6 = 11$.
For $n = 3$, $a_3 = a_{3-1} + 6 = a_2 + 6 = 11 + 6 = 17$.
For $n = 4$, $a_4 = a_{4-1} + 6 = a_3 + 6 = 17 + 6 = 23$.
For $n = 5$, $a_5 = a_{5-1} + 6 = a_4 + 6 = 23 + 6 = 29$.

Listing a Geometric Sequence from the Recursive Rule

The geometric sequence 3, 6, 12, 24, 48, ... can also be described recursively. Since the first term is 3, $a_1 = 3$. Then since the next term is calculated by multiplying the previous term by 2, the recursive part of the equation is $a_n = 2 \cdot a_{n-1}$ for $n > 1$. The complete rule is

$$a_1 = 3$$
$$a_n = 2 \cdot a_{n-1} \text{ for } n > 1$$

When given the recursive rule for a geometric sequence, it is possible to list all the terms of it.

Example 2

A sequence is defined by the recursive rule

$$a_1 = 2$$
$$a_n = 3 \cdot a_{n-1} \text{ for } n > 1$$

What are the first four terms of the sequence?

Solution: The first term, a_1, is 2 since it is defined in the first line of the rule. The second term can be calculated by substituting $n = 2$ into the bottom line of the rule.

$$a_2 = 3 \cdot a_{2-1} = 3 \cdot a_1 = 3 \cdot 2 = 6$$

The third term can be calculated by substituting $n = 3$ into the bottom line of the rule.

$$a_3 = 3 \cdot a_{3-1} = 3 \cdot a_2 = 3 \cdot 6 = 18$$

And the fourth term, $a_4 = 3 \cdot a_{4-1} = 3 \cdot a_3 = 3 \cdot 18 = 54$.

So, the sequence is 2, 6, 18, 54, ...

Creating Recursive Rules for Arithmetic and Geometric Sequences

---
MATH FACTS

The recursive part of the rule for an arithmetic sequence is $a_n = a_{n-1} + d$ where d is what needs to be added (or subtracted if d is negative) to the previous term to get the next term. The recursive part of the rule for a geometric sequence is $a_n = r \cdot a_{n-1}$ where r is what needs to be multiplied by the previous term to get the next term.

Creating the recursive rule for a sequence requires the same analysis as trying to find the next term of a sequence. Examine the first few terms to determine if it is an arithmetic sequence where each term is obtained from the previous term by adding or subtracting or if it is a geometric sequence where each term is obtained from the previous term by multiplying.

If the sequence is 8, 11, 14, 17, ..., first decide how to get the *next* term. In this case, each term is 3 more than the one before it, so the next term seems to be $17 + 3 = 20$. Each term is 3 more than the previous term.

So, the recursive rule is

$$a_1 = 8$$
$$a_n = a_{n-1} + 3 \text{ for } n > 1$$

The sequence 4, 20, 100, 500, ... is not an arithmetic sequence since 20 is 16 more than 4 and 100 is 80 more than 20. To see if it is geometric, divide the second term by the first term. 20 divided by 4 is 5. Then divide the third term by the second term. 100 divided by 20 is also 5. This is a geometric series where each term is 5 times the previous term.

So, the recursive rule is

$$a_1 = 4$$
$$a_n = 5 \cdot a_{n-1} \text{ for } n > 1$$

Recursive Definitions of Sequences for Picture Patterns

Instead of providing a list of numbers, there might be a sequence of pictures representing a sequence. A recursive formula can be created for these by either turning the pictures into a number list or by examining the pictures to determine the pattern.

Example 3

Create a recursive formula that describes the number of shaded circles in the nth picture of this sequence

1st	2nd	3rd
○	⚬⚬⚬⚬	⚬⚬⚬⚬⚬⚬⚬

Solution: The numbers follow a pattern 1, 4, 7, ... Since the first picture has 1 circle, $a_1 = 1$. Since each picture has three more circles than the previous picture, the recursive part is $a_n = a_{n-1} + 3$ for $n > 1$.

Recursive Definitions of Sequences for Real-World Scenarios

A sequence of numbers could be related to a real-world scenario. The process of finding the equation is the same as for a list.

Suppose there is a plant that grows according to an arithmetic progression. It is 4 inches tall at the end of the first week, 7 inches tall at the end of the second week, and 10 inches tall at the end of the third week. Finding a recursive rule to describe the height of the plant after n weeks is the same as finding the rule for the sequence 4, 7, 10, ...

$$a(1) = 4$$
$$a(n) = a(n-1) + 3 \text{ for } n > 1$$

Check Your Understanding of Section 11.2

A. Multiple-Choice

1. Find the value of a_2 in the sequence defined by
$$a_1 = 4$$
$$a_n = 3 + a_{n-1} \text{ for } n > 1$$
   (1) 3 (2) 5 (3) 7 (4) 12

2. Find the value of a_2 in the sequence defined by
$$a_1 = 10$$
$$a_n = a_{n-1} - 6 \text{ for } n > 1$$
   (1) 6 (2) 4 (3) 10 (4) 16

3. Find the value of a_2 in the sequence defined by
$$a_1 = 4$$
$$a_n = 3 \cdot a_n - 1 \text{ for } n > 1$$
   (1) 7 (2) 10 (3) 12 (4) 16

4. Find the value of a_2 in the sequence defined by
$$a_1 = 6$$
$$a_n = \frac{1}{3} a_{n-1} \text{ for } n > 1$$
   (1) 1 (2) 2 (3) 3 (4) 18

5. The sequence 5, 11, 17, 23, ... can be generated by which definition?
   (1) $a_1 = 5$ $a_n = 6a_{n-1}$ for $n > 1$
   (2) $a_1 = 5$ $a_n = \frac{11}{5} a_{n-1}$ for $n > 1$
   (3) $a_1 = 5$ $a_n = -6 + a_{n-1}$ for $n > 1$
   (4) $a_1 = 5$ $a_n = 6 + a_{n-1}$ for $n > 1$

6. The sequence 3, 15, 75, 375, ... can be generated by which definition?
   (1) $a_1 = 3$ $a_n = 5a_{n-1}$ for $n > 1$
   (2) $a_1 = 3$ $a_n = 12 + a_{n-1}$ for $n > 1$
   (3) $a_1 = 3$ $a_n = \frac{1}{5} a_{n-1}$ for $n > 1$
   (4) $a_1 = 3$ $a_n = -12 + a_{n-1}$ for $n > 1$

Sequences

7. Which definition generates a sequence of numbers to describe the total number of rectangles for each picture?

(1) $a_1 = 2$ $a_n = 3 + a_{n-1}$ for $n > 1$
(2) $a_1 = 2$ $a_n = \frac{5}{2} a_{n-1}$ for $n > 1$
(3) $a_1 = 2$ $a_n = a_{n-1} - 3$ for $n > 1$
(4) $a_1 = 2$ $a_n = 5 a_{n-1}$ for $n > 1$

8. The side length of the square in the first picture is one unit. What sequence describes the area of each of the pictures in the pattern?

(1) $a_1 = 1$ $a_n = 1 + a_{n-1}$ for $n > 1$
(2) $a_1 = 1$ $a_n = \frac{1}{2} a_{n-1}$ for $n > 1$
(3) $a_1 = 1$ $a_n = 2 + a_{n-1}$ for $n > 1$
(4) $a_1 = 1$ $a_n = 2 \cdot a_{n-1}$ for $n > 1$

336

9. A person jogs 3 miles on the first day of training and 5 more miles each day after that. Which definition can generate the sequence of the number of miles run each day?

(1) $a_1 = 3 \quad a_n = \frac{8}{3} a_{n-1}$ for $n > 1$
(2) $a_1 = 3 \quad a_n = 5 + a_{n-1}$ for $n > 1$
(3) $a_1 = 3 \quad a_n = 5 a_{n-1}$ for $n > 1$
(4) $a_1 = 3 \quad a_n = \frac{8}{3} \cdot a_{n-1}$ for $n > 1$

10. A ball is dropped from a window 50 feet above the ground. Each bounce is $\frac{4}{5}$ the height of the previous bounce. Which definition would generate the height of the bounces?

(1) $a_1 = 50 \quad a_n = -10 + a_{n-1}$ for $n > 1$
(2) $a_1 = 50 \quad a_n = 10 + a_{n-1}$ for $n > 1$
(3) $a_1 = 50 \quad a_n = \frac{4}{5} a_{n-1}$ for $n > 1$
(4) $a_1 = 50 \quad a_n = \frac{5}{4} a_{n-1}$ for $n > 1$

B. *Show how you arrived at your answers.*

1. A sequence can be generated by the definition

$$a_1 = 5$$
$$a_n = 4 + a_{n-1} \text{ for } n > 1$$

What are the first four terms of this sequence?

2. The recursive part of a sequence definition is $a_n = 3 a_{n-1}$. If $a_{17} = 58$, what is a_{18} and a_{19}?

3. Madelyn puts $50 in a bank that offers simple interest. This chart lists the amount in the bank at the beginning of each year. What recursive definition would generate these numbers?

Year	Amount
1	$50
2	$55
3	$60
4	$65
5	$70

Sequences

4. Archimedes estimated the circumference of a circle by measuring regular polygons with increasing numbers of sides.

   [Figures showing polygons with Sides = 6, Sides = 12, Sides = 24, Sides = 48, Sides = 96]

   What recursive definition could be used to generate the number of sides for each picture?

5. What are the first five terms of the sequence defined by this recursive definition?

$$a_1 = 3$$
$$a_2 = 4$$
$$a_n = 2a_{n-1} + a_{n-2} \text{ for } n > 2$$

11.3 CLOSED FORM DEFINED SEQUENCES

=== **KEY IDEAS** ===

A recursive rule is useful for finding the next term of a sequence. If, however, you need to find the 100th term of a sequence, the recursive rule is not very convenient as it would require finding all 100 terms. A *closed form* definition of a sequence is a formula that relates the position of the number in the sequence to the number in that position. For arithmetic and geometric sequences, there is a short way to create the closed form definition.

Listing a Sequence from the Closed Form Definition

Like a function, the closed form definition takes the position of the number on the list as an input, and it outputs the number that goes into that position.

If the closed form definition of a sequence is $a_n = 3n + 7$, then the sequence can be created by substituting 1, 2, 3, and so on for n. This formula does not require looking at the previous term.

$$a_1 = 3 \cdot 1 + 7 = 3 + 7 = 10$$
$$a_2 = 3 \cdot 2 + 7 = 6 + 7 = 13$$
$$a_3 = 3 \cdot 3 + 7 = 9 + 7 = 16$$

So the sequence begins 10, 13, 16, ...

Example 1

The sequence 10, 13, 16, 19, ... has a closed from definition of $a_n = 3n + 7$. What is the 100th term of the sequence?

Solution: To get a_{100}, substitute $n = 100$ into the formula to get $a_{100} = 3 \cdot 100 + 7 = 300 + 7 = 307$. To get this answer with the recursive definition would take a very long time!

Finding the Closed Form Definition of an Arithmetic Sequence

The sequence 8, 11, 14, 17, ... is an arithmetic sequence. Since $11 = 8 + 3 \cdot 1$, $14 = 8 + 3 \cdot 2$, and $17 = 8 + 3 \cdot 3$, the closed form definition for this sequence is

$$a_n = 8 + 3(n - 1)$$

MATH FACTS

The closed form formula for describing an arithmetic sequence is $a_n = a_1 + d(n - 1)$ where a_1 is the first term in the sequence and d is the difference between any two consecutive terms.

Example 2

Find the closed form definition to calculate a_n for the sequence 5, 12, 19, 26, ... and determine the 50th number in the sequence.

Solution: Since $a_1 = 5$ and $d = 7$, the formula is $a_n = 5 + 7(n - 1)$ which can be simplified to $a_n = 7n - 2$. Substitute 50 into either of these formulas.

$$a_5 = 5 + 7(50 - 1) = 5 + 7 \cdot 49 = 5 + 343 = 348$$

Sequences

After the closed form definition of an arithmetic sequence is found, it can be used to determine when in a sequence a particular number will appear.

Example 3

What position in the sequence 5, 12, 19, 26, ... will the number 651 be in?

Solution: From example 2, the closed form definition of the sequence is $a_n = 5 + 7(n - 1)$. To find the position of 651, replace a_n with 651 and solve for n.

$$656 = 5 + 7(n - 1)$$
$$-5 = -5$$
$$\frac{651}{7} = \frac{7(n-1)}{7}$$
$$93 = n - 1$$
$$+1 = +1$$
$$94 = n$$

Finding the Closed Form Definition of a Geometric Sequence

2, 10, 50, 250, 1250, ... is a geometric series since each term is equal to the previous term multiplied by 5. The first three terms of this sequence can also be written as:

$$a_1 = 2 = 2g5^0$$
$$a_2 = 10 = 2g5^1$$
$$a_3 = 50 = 2g5^2$$

In general, the nth term will be $a_n = 2g5^n$

MATH FACTS

The closed from formula for describing a geometric sequence is $a_n = a_1 gr^{n-1}$ where a_1 is the first term and r is what you have to multiply by the previous term to get to the next term.

Example 4

What is the tenth term of the geometric sequence that begins 3, 12, 48, 192, ...?

Solution: Since the first term is 3 and each term is 4 times the previous term, $a_1 = 3$ and $r = 4$. The closed form formula for the terms of this sequence, then, is $a_n = 3g4^{n-1}$. For $n = 10$, this becomes $a_{10} = 3g4^{10-1} = 3g4^9 = 3g262144 = 786432$.

Closed Form Definitions of Sequences for Real-World Scenarios

If a real-world scenario has numbers that form an arithmetic sequence, the closed form definition can be used to describe the real-world scenario.

Suppose there is a swimming pool being filled by a hose according to an arithmetic progression. It has 4 inches of water after the first hour, then 7 inches of water at the end of the second hour, and then 10 inches of water at the end of the third hour. Finding a closed form rule to describe the height of the water in the pool after n hours is the same as finding the rule for the sequence 4, 7, 10, 13, ….

Since $a_1 = 4$ and $d = 3$, the formula is $a_n = 4 + 3(n - 1)$, which can be simplified to $a_n = 3n + 1$.

Closed Form Definitions of Sequences for Picture Patterns

If a picture pattern models an arithmetic sequence, the closed form equation will describe that sequence too.

Example 5

Create a closed formula that describes the number of shaded circles in the nth picture of this sequence. Use the formula to determine how many shaded circles would be in the 20th picture.

1st	2nd	3rd
○	○○○○ (cluster)	○○○○○○○ (larger cluster)

Solution: $a_1 = 1$ and $d = 3$, so the closed form equation is $a_n = 1 + 3(n - 1)$, which can also be simplified to $a_n = 3n - 2$. The value of a_{20} is

$$a_{20} = 1 + 3(20 - 1) = 1 + 3 \cdot 19 = 1 + 57 = 58$$

Sequences

Check Your Understanding of Section 11.3

A. Multiple-Choice

1. What are the first three terms of the sequence generated by the definition $a_n = 2 + 3(n-1)$?
   (1) 3, 5, 7   (2) 2, 5, 8   (3) 2, 8, 14   (4) 0, 5, 10

2. What is the fifth term of the sequence generated by the definition $a_n = 10 - 4(n-1)$?
   (1) 2   (2) –2   (3) –6   (4) –10

3. What is the third term of the sequence generated by the definition $a_n = 3 \cdot 4^{n-1}$?
   (1) 3   (2) 4   (3) 12   (4) 48

4. What is the fifth term of the sequence generated by the definition $a_n = 6 \cdot \left(\dfrac{2}{3}\right)^{n-1}$?
   (1) 96   (2) 48   (3) $\dfrac{32}{27}$ (4) $\dfrac{16}{9}$

5. What definition would produce the sequence 4, 13, 22, 31, …?
   (1) $a_n = 4 + 9n$ (3) $a_n = 9 + 4n$
   (2) $a_n = 4 + 9(n-1)$ (4) $a_n = 9 + 4(n-1)$

6. What definition would produce the sequence 4, 12, 36, 108, …?
   (1) $a_n = 4 \cdot 3^{n-1}$ (3) $a_n = 3 \cdot 4^{n-1}$
   (2) $a_n = 4 \cdot 3^n$ (4) $a_n = 3 \cdot 4^n$

7. Which expression could be used to find the 20th term of the sequence 5, 9, 13, 17, 21, …?
   (1) $5 + 4(20)$ (3) $4 + 5(20)$
   (2) $5 + 4(19)$ (4) $4 + 5(19)$

8. Which expression could be used to find the 20th term of the sequence 5, 15, 45, 135, …?
   (1) $3 \cdot 5^{19}$ (3) $5 \cdot 3^{19}$
   (2) $3 \cdot 5^{20}$ (4) $5 \cdot 3^{20}$

9. A person has saved $20. Each week he adds $7 to the savings. Which definition can be used to generate a sequence of the amount of money saved at the beginning of each week?
 (1) $a_n = 7 + 20n$
 (2) $a_n = 7 + 20(n-1)$
 (3) $a_n = 20 + 7n$
 (4) $a_n = 20 + 7(n-1)$

10. What definition will generate a sequence of the number of hexagons in each picture?

 (1) $a_n = 1 + 6(n-1)$
 (2) $a_n = 1 + 6n$
 (3) $a_n = 1 \cdot 6^{n-1}$
 (4) $a_n = 6 + 1(n-1)$

B. Show how you arrived at your answers.

1. Liam says that the sequence 7, 12, 17, 22, ... can be written as $a_n = 7 + 5(n-1)$. Cecilia says it can be written as $a_1 = 7$, $a_n = 5 + a_{n-1}$. Who is right? Explain.

2. What is the 100th term of the sequence 26, 19, 12, 5, ...?

3. A folded sheet of paper is $\frac{1}{633,600}$ miles thick. If a sheet of paper is folded the thickness doubles. (a) Create a definition that will generate a sequence where the first number is $\frac{1}{633,600}$ and the next number is always double the previous number. (b) Use the definition to determine the height of the folded paper after 38 folds. Will this reach the Moon (240,000 miles away)?

4. The first terms of sequence a are 1,000, 1,200, 1,400, 1,600, ... The first terms of sequence b are 1, 2, 4, 8, ... Which is greater, a_{20} or b_{20}?

5. To find the 100th term of the sequence 5, 8, 11, 14, ..., would it be faster to use the closed form definition or the recursive definition of the sequence? Explain.

Chapter Twelve

REGRESSION CURVES

12.1 LINE OF BEST FIT

KEY IDEAS

Two points can always be joined with a line. If there are more than two points, however, there may not be one line that passes through all of them. If we need a linear equation for a line that comes close to all the points, there is a feature of the graphing calculator that will calculate the slope and *y*-intercept of this line.

Perfectly straight lines are not common in nature. They also are rare when real-world data is graphed. Below is a chart of some scientific measurements and a graph of the data.

Temperature	Number of people at the beach
30	5
40	21
50	45
60	65
70	77
80	102
90	129

Regression Curves

A line drawn through any of the two points will come close but will not pass through any of the other points.

Though there is not one line that passes through all the points, there is a *line of best fit* that will come close to all the points. The equation for the line of best fit will be useful in calculating approximate coordinates for other points on the line.

For the TI-84:

The first step in the process of determining the line of best fit for a data set is to enter the data into a list.

Press [STAT] and then [1] to edit the list. If the lists already have numbers in them, use the arrows to move the cursor to L1 and L2 and press [CLEAR] (not [DEL]!). Enter the temperatures into L1 and the number of people into L2.

Press [Y=] and use the arrow keys to move the cursor to the Plot1 on the top left. Press [ENTER] to highlight it.

Then press [ZOOM] and [9] to create the scatter plot of the data.

Since the points line up nicely, a line of best fit is appropriate. Press [STAT] and then move to [CALC] and [4]. Then press [ENTER].

For this example, the line of best fit is approximately $y = 2.02x - 57.86$ or, if using T for temperature and P for people, $P = 2.02T - 57.86$.

This is a graph of the scatter plot and the line of best fit. It does not go through any of the points but comes very close to all of them.

For the TI-Nspire:

From the home screen select the Add Lists & Spreadsheet icon.

Regression Curves

In the first row, label the A column "temp" and the B column "people." Enter the temperature values in cells A1 through A7. Enter the corresponding people values in cells B1 through B7.

Go back to the home screen, and select the Add Data & Statistics icon.

To create the scatter plot, set the horizontal axis to temp and the vertical axis to people. Do this by clicking on the Click to add variable box and selecting the variables temp and people from the list.

To see the line of best fit and its equation, select [menu], [4], and [6] for Regression. Then select [1] for Show Linear (mx+b) and press [enter].

The equation for the line of best fit is $y = 2.02x - 57.86$ or $P = 2.02T - 57.86$.

Using the Line of Best Fit to Answer Questions About the Real-World Scenario

After the equation is determined, it can be used to answer questions. In the temperature/people example above, the equation became $P = 2.02T - 57.86$.

One question that can be answered with this equation is: Approximately how many people will be at the beach when the temperature is 100 degrees?

To solve, substitute 100 for T and solve for P.

$$P = 2.02(100) - 57.86 = 202 - 57.86 = 144.14$$

Approximately 144 people will be there when it is 100 degrees.

Another question that can be answered with this equation is: At approximately what temperature will there be 100 people on the beach?

To solve, substitute 100 for P and solve for T. This will require algebra.

Regression Curves

$$100 = 2.02T - 57.86$$
$$+57.86 = +57.86$$
$$\frac{157.86}{2.02} = \frac{2.02T}{2.02}$$
$$78.15 = T$$

At a temperature of approximately 78 degrees, there will be 100 people at the beach.

Check Your Understanding of Section 12.1

A. Multiple-Choice

1. Calculate the equation for the line of best fit for the following set of data in $y = mx + b$ form. Round m and b to the nearest tenth.

x	y
1	3
2	5
3	4
4	6
5	8

    (1) $y = 1.1x + 1.9$
    (2) $y = 1.4x + 1.7$
    (3) $y = 1.7x + 1.4$
    (4) $y = 1.9x + 1.1$

2. Calculate the equation for the line of best fit for the following set of data in $y = mx + b$ form. Round m and b to the nearest tenth.

x	y
1	7
2	6
3	6
4	5
5	4

    (1) $y = 7.7x - 0.7$
    (2) $y = -0.5x + 9.2$
    (3) $y = 9.2x - 0.5$
    (4) $y = -0.7x + 7.7$

3. Calculate the equation for the line of best fit for the following set of data in $y = mx + b$ form. Round m and b to the nearest tenth.

x	y
1	1
2	3
3	8
4	7
5	9

(1) $y = 2x - 0.4$
(2) $y = -0.4x + 2$
(3) $y = 2.5x - 0.3$
(4) $y = -0.3x + 2.5$

4. Calculate the equation for the line of best fit for the following set of data in $y = mx + b$ form. Round m and b to the nearest tenth.

x	y
1	4
2	3
3	5
4	4
5	5

(1) $y = 2.8x + 0.4$
(2) $y = 0.4x + 2.8$
(3) $y = 0.3x + 3.3$
(4) $y = 3.3x + 0.3$

5. Calculate the equation for the line of best fit for the following set of data in $y = mx + b$ form. Round m and b to the nearest tenth.

x	y
2	16
4	12
6	2
8	6
10	2

(1) $y = 18.2x - 1.3$
(2) $y = -1.3x + 18.2$
(3) $y = 17.8x - 1.7$
(4) $y = -1.7x + 17.8$

Regression Curves

6. Calculate the equation for the line of best fit for the following set of data in $y = mx + b$ form. Round m and b to the nearest tenth.

x	y
10	33
20	20
30	10
40	14
50	6

(1) $y = 34.6x - 0.6$
(2) $y = -0.6x + 34.6$
(3) $y = 31.8x - 0.7$
(4) $y = -0.7x + 31.8$

7. Calculate the equation for the line of best fit for the following set of data in $y = mx + b$ form. Round m and b to the nearest tenth.

x	y
10	250
25	310
37	450
46	560
59	820

(1) $y = 11.1x + 70.3$
(2) $y = 70.3x + 11.1$
(3) $y = 11.5x + 69.8$
(4) $y = 69.8x + 11.5$

8. What is the equation for the line of best fit for the points on this scatter plot?

(1) $y = -0.5x + 7$
(2) $y = 0.7x - 0.5$
(3) $y = -0.4x + 0.9$
(4) $y = 0.9x - 0.4$

9. Of these four choices, which line appears to be the best fit for this scatter plot?

Regression Curves

10. The equation $y = x + 3$ is a line of best fit for which scatter plot?

(1)

(2)

(3)

(4)

B. *Show how you arrived at your answer.*

1. A bird-watching group tracks the number of birds they see and the temperature for six different occasions. The data are collected on the table below. (a) Make a scatter plot of the data. (b) Calculate the line of best fit that relates birds (B) to temperature (T). Round values to the nearest hundredth. (c) Use your equation of the line of best fit to predict how many birds they would see if the temperature was 55 degrees.

Temperature	Birds seen
40	30
50	41
60	72
70	91
80	94
90	89

2. A car dealership keeps track of how many cars they sell at different prices. The data are collected on the table below. (a) Make a scatter plot of the data. (b) Calculate the line of best fit that relates cars sold (C) to price (P). Round values to the nearest hundredth. (c) Use your equation of the line of best fit to predict how many cars they would sell at a cost of $31,000.

Price (in thousands of dollars)	Cars sold
25	150
30	127
35	118
40	91
50	65

3. How is it possible to find the equation for the line of best fit for the scatter plot below without the use of a calculator?

Regression Curves

4. The two scatter plots below have all points the same except the one on the left has the point (3, 1) while the one on the right has the point (3, 10). Will they have the same lines of best fit? If not, how will the line of best fit for the scatter plot on the left be different from the line of best fit for the scatter plot on the right?

12.2 THE CORRELATION COEFFICIENT

KEY IDEAS

The points on some scatter plots line up better than the points on other scatter plots. A measure of how well the points in a scatter plot resemble a straight line is called the *correlation coefficient* and is denoted by the letter r. The correlation coefficient is a number between -1 and $+1$. Correlation coefficients very close to -1 or $+1$ indicate that the points very nearly line up. Correlation coefficients close to 0 indicate that the points do not line up very well.

Calculating the Correlation Coefficient

The temperature / people example from Section 12.1 had a scatter plot that looked like this:

For the TI-84:

To find the correlation coefficient, press [2ND] and then [0] to view the CATALOG. Use the arrows to scroll down to DiagnosticOn and then press [ENTER] and [ENTER] again. This setting will remain until the calculator's memory is reset. If the test proctor has you clear the memory of the calculator, you will have to do this process again.

Now when you do the steps from the previous section for determining the line of best fit, the *r* value will be displayed with it.

For the TI-Nspire:

From the home screen, select the Add Lists & Spreadsheet icon. Enter the data as you would to determine the line of best fit.

Regression Curves

Press [menu], [4], and [1] for Stat Calculations.

Press [3] for Linear Regression (mx+b) and press [enter]. In the X List field, enter "temp." In the Y List field, enter "people." Click on [OK].

The *r* value of 0.996244 will be in row 6.

The correlation coefficient for this data set is 0.9962442173, which is very close to 1 because the points on the scatter plot very nearly line up.

Example 1

What is the correlation coefficient of the line of best fit for the scatter plot below?

(1) 0.27   (2) 0.92   (3) 0.99   (4) 1

Solution: The r value is approximately 0.92, choice (2).

Regression Curves

Comparing Correlation Coefficients by Looking at the Scatter Plots

The data set from the temperature/people scenario had a correlation coefficient very close to 1. If the points on the scatter plot did not line up so nicely, the correlation coefficient would have been lower. Here are two data sets that have lower correlation coefficients.

L1	L2	L3
30	5	
40	19	
50	13	
60	37	
70	64	
80	53	
90	82	

L3(1)=

L1	L2	L3
30	5	
40	69	
50	122	
60	193	
70	137	
80	229	
90	171	

L3(1)=

The first set has a correlation coefficient of 0.94, and the second has a correlation coefficient of 0.84.

Example 2

Which scatter plot appears to have the lowest correlation coefficient?

(1)

(2)

(3)

(4)

Solution: Since the points on choice (4) do not line up very well, it has the lowest correlation coefficient. The correlation coefficient for choice 4 is about 0.15.

Negative Correlation Coefficients

If the line of best fit has a negative slope, the correlation coefficient will be negative. The more the points line up, the closer the correlation coefficient will be to −1, whereas the less they line up, the closer the correlation coefficient will be to 0.

The scatter plot below has a correlation coefficient of −1.

The scatter plot below has a correlation coefficient of −0.86.

In the first example, the correlation coefficient is −1 because all the points land exactly on the line of best fit. In the second example, the correlation coefficient is −0.86, which is closer to 0 because the points are farther away from the line of best fit.

Regression Curves

Check Your Understanding of Section 12.2

A. Multiple-Choice

1. What is the correlation coefficient (r), rounded to the nearest hundredth, for the line of best fit for the data on the table below?

x	y
3	10
6	13
9	27
12	38
15	40

    (1) 0.97   (2) 0.96   (3) 0.95   (4) 0.94

2. What is the correlation coefficient (r) for the line of best fit for the data represented on the scatter plot below?

    (1) −0.45   (2) −0.52   (3) −0.59   (4) −0.67

362

3. For which scatter plot is the correlation coefficient closest to 1?

(1)

(3)

(2)

(4)

4. Which scatter plot has a correlation coefficient closest to −1?

(1)

(3)

(2)

(4)

363

Regression Curves

5. Which scatter plot has a correlation coefficient (r) closest to zero?

(1)

(3)

(2)

(4)

6. Of the four choices, which is closest to the correlation coefficient for this scatter plot?

(1) 0.8   (2) −0.8   (3) 1   (4) −1

7. Of the four choices, which is closest to the correlation coefficient for this scatter plot?

(1) 1   (2) –1   (3) 0.7   (4) –0.7

8. What is the correlation coefficient (r) for the line of best fit for the data represented on the scatter plot below?

(1) 0.6856   (2) 0.6860   (3) 0.6864   (4) 0.6868

9. What could be the correlation coefficient for the line of best fit for a scatter plot that has only two points?
(1) 0
(2) 1
(3) 0.5
(4) There is not enough information to answer this question.

Regression Curves

10. The scatter plot below has a line of best fit with equation $y = 0.5x + 4$ and a correlation coefficient of 0.76. Which point, if added to the scatter plot, would most increase the value of the correlation coefficient?

(1) (6, 1)   (2) (6, 5)   (3) (6, 6)   (4) (6, 7)

B. *Show how you arrived at your answers.*

1. Which scatter plot has the higher correlation coefficient? Explain your reasoning.

Scatter plot on the left because it has the higher correlation coefficient

2. Ronald says that a correlation coefficient cannot be negative since completely random placed points have a correlation coefficient of 0, and the points cannot get any more randomly placed. Jett says that it is possible to have a negative correlation coefficient. Who is correct? Explain.

Jeff is correct because if the line of best fit is negative so is the correlation coef.

3. The seven points on the scatter plot below all lie on the curve $y = \left(\dfrac{1}{16}\right)x^2$.

   What can be determined about the correlation coefficient of the line of best fit for these seven points without using a calculator?

4. Create a five-point scatter plot that has a line of best fit with the equation $y = x + 3$ and a correlation coefficient (r) of $+1$.

12.3 PARABOLAS AND EXPONENTIAL CURVES OF BEST FIT

KEY IDEAS

Not every scatter plot resembles a straight line. When data are plotted from real-world scenarios, sometimes the plot resembles a parabola or an exponential curve. When this happens, the line of best fit is no longer appropriate. Instead the graphing calculator can also be used to find the parabola of best fit or the exponential curve of best fit.

The Parabola of Best Fit

When the scatter plot resembles a parabola, it is possible to find the equation of the parabola of best fit using a process similar to finding the line of best fit. Below is a data set with the scatter plot next to it.

Regression Curves

x	y
1	2
2	5
3	11
4	15
5	12
6	8
7	4
8	1

There is a line of best fit for this scatter plot, but it will have a very low correlation coefficient because the points look more like the shape of a parabola than a line. When this happens, it is more appropriate to find the parabola of best fit.

For the TI-84:

Press [STAT] and [1] and enter the x values into L1 and the y values into L2. Press [STAT] and go to the CALC menu. Press [5] for QuadReg.

```
QuadReg
y=ax²+bx+c
a=-.9642857143
b=8.392857143
c=-5.928571429
R²=.8893752487
```

For the TI-Nspire:

From the home screen, select Add Lists & Spreadsheet icon. Enter the data for x into column A and the data for y into column B. Press [menu], [4], [1], and [6] for Quardratic Regression.

The equation of the parabola of best fit is approximately
$$y = -0.96x^2 + 8.39x - 5.93$$

Here is the graph of the scatter plot and the parabola plot on the same coordinate plane.

Regression Curves

The Exponential Curve of Best Fit

If the scatter plot does not resemble a line or a parabola, it may resemble the playground slide shape of an exponential curve.

Here is a data set with its scatter plot.

x	y
1	1.3
2	1.9
3	2.1
4	2.1
5	3.7
6	4
7	6.7

The graphing calculator can quickly find the exponential curve of best fit.

For the TI-84:

Press [STAT], [1] to enter the x values into L1 and the y values into L2. Then press [STAT], [right arrow],[0] for ExpReg, and [ENTER].

ExpReg
$y=a*b^x$
$a=1.001475564$
$b=1.282856853$
$r^2=.9322025577$
$r=.9655063737$

For the TI-Nspire:

From the home screen, select the Add Lists & Spreadsheet icon. Enter the data for x into column A and the data for y into column B. Press [menu], [4], [1], and [A] for Exponential Regression.

The exponential curve of best fit is approximately $y = 1.0 \cdot 1.28^x$.
Here is the scatter plot with the exponential curve of best fit.

Choosing Which Model Is Most Appropriate

Some questions could indicate which model to use—linear, quadratic (for parabolas), or exponential. Others could ask for you to choose the most appropriate model. By looking at, or by making, the scatter plot, decide if it

Regression Curves

looks more like a line, a parabola, or an exponential curve. Then choose either linear regression, quadratic regression, or exponential regression.

Example

The following data were collected from an experiment. What model would be most appropriate for determining an equation for this data?

Years	Value
0	$100
10	$217
20	$471
30	$1,022
40	$2,217
50	$4,812
60	$10,441

(1) Linear with a positive correlation coefficient
(2) Linear with a negative correlation coefficient
(3) Exponential
(4) Quadratic

Solution: Make the scatter plot for the data. Since it looks more like an exponential curve than a parabola or a line, the answer is choice (3). This could also be done by noticing that the x values are equally spaced and that each number in the y column is 2.17 times greater than the number above it.

Check Your Understanding of Section 12.3

A. Multiple-Choice

1. Find the equation of the line of best fit for the data on the table below.

x	y
1	2
2	2
3	3
4	4
5	6
6	8
7	10
8	14

(1) $y = 1.27x - 1.73$
(2) $y = -1.73x + 1.27$
(3) $y = 1.68x - 1.43$
(4) $y = -1.43x + 1.68$

2. Find the equation of the parabola of best fit for the data on the table below.

x	y
1	2
2	2
3	3
4	4
5	6
6	8
7	10
8	14

(1) $y = 0.31x^2 - 67x + 3.53$
(2) $y = 0.24x^2 - 0.52x + 2.23$
(3) $y = 0.58x^2 - 0.41x + 4.12$
(4) $y = 0.73x^2 - 0.72x + 6.87$

Regression Curves

3. Find the equation of the exponential curve of best fit for the data on the table below.

x	y
1	2
2	2
3	3
4	4
5	6
6	8
7	10
8	14

(1) $y = 1.28 \cdot 1.35^x$
(2) $y = 1.35 \cdot 1.28^x$
(3) $y = 1.47 \cdot 1.39^x$
(4) $y = 1.39 \cdot 1.47^x$

4. For which scatter plot would an exponential curve of best fit be most appropriate?

(1)

(2)

(3)

(4)

374

5. For which scatter plot would a parabola of best fit be most appropriate?

(1)

(2)

(3)

(4)

6. For which scatter plot would a line of best fit be most appropriate?

(1)

(2)

(3)

(4)

Chapter Twelve REGRESSION CURVES

Regression Curves

7. Which equation would be the most accurate model for this scatter plot?

   (1) $y = 0.33x + 6.20$
   (2) $y = 1.17x^2 - 9.03x + 20.02$
   (3) $y = 5.35 \cdot 1.04^x$
   (4) $y = -0.33x + 6.20$

8. The most appropriate model for this scatter plot is

   (1) Linear with a negative correlation coefficient
   (2) Linear with a positive correlation coefficient
   (3) Exponential decay
   (4) Exponential growth

9. The height of a ball at different times is graphed on this scatter plot. Which type of equation is most appropriate for modeling this?

(1) Quadratic with a negative a value
(2) Quadratic with a positive a value
(3) Exponential growth
(4) Exponential decay

10. Which type of equation would be most appropriate for describing the data in this table?

x	y
10	350
20	420
30	490
40	560
50	630
60	700

(1) Decreasing linear function
(2) Increasing linear function
(3) Decreasing exponential function
(4) Increasing exponential function

Regression Curves

B. Show how you arrived at your answers.

1. The population of a country at different times was recorded on the following chart where T represents the number of years since 1980 and P represents the population in millions. (a) Find the most appropriate type of equation to model this data. Round all numbers to the nearest hundredth. (b) Use your equation (with the rounded values) to estimate what the population was in 2000 ($T = 20$).

T (years since 1980)	P (population in millions)
1.5	1.59
7.04	2.12
12.32	3.08
16.09	5.90
18.70	8.46
24.17	14.63
27.5	23
30.56	30.32

2. Food is put into a 0 degree freezer. The temperature of the food is recorded at different times after it is put into the freezer on the following chart where M represents the number of minutes since the food was put into the freezer and D represents the temperature of the food in degrees. (a) Find the most appropriate type of equation to model this data. Round all numbers to the nearest hundredth. (b) Use your equation (with the rounded values) to estimate what the temperature of the food will be after 40 minutes.

M (minutes since putting in freezer)	D (temperature in degrees)
3	58
9	31
15	16
20	10
27	5
31	3

378

3. When a computer company sets the price of a smart phone too low, they don't make a lot of profit. When they set the price too high, few people buy the phone so the company also does not make a lot of profit. Below is a chart showing the amount of profit the company made for six different prices. (a) Decide which model is most appropriate and use it to find an equation to relate the amount of profit they make (F) to the price of the phone (R). Round numbers to the nearest thousandth. (b) Use this equation to determine the price they should charge to make the most amount of profit.

R (price of phone in dollars)	F (profit in millions)
50	2
75	4
100	7
125	8
150	6
175	3

4. A person uses a linear equation to model the data on this scatter plot.

(a) Was a linear model the most appropriate choice? Why or why not?
(b) Find the line of best fit and the correlation coefficient for the data set.

Regression Curves

5. Tyler wants to find a curve that best fits the data on this scatter plot and chart.

Since it looks like half a parabola and also like an exponential curve, he is not sure whether a quadratic or an exponential model is best. How can he use the calculator to determine which is a more accurate type of equation?

An exponential equation is better fit.

When finding the equation r^2 value is slightly closer to +1 then the r^2 value

Chapter Thirteen

STATISTICS

13.1 MEASURES OF CENTRAL TENDENCY

=== KEY IDEAS ===

A *measure of central tendency* is a number that summarizes a list of numbers. Since one number cannot describe everything someone might want to know about a set of data, there are many different measures of central tendency.

Mean

The *mean* of a data set is also known as the arithmetic mean or the average. To calculate the mean, add up all the numbers in the data set and divide that result by the number of numbers in the set. The symbol for mean is $\bar{x}$.

This is a set of scores on 11 quizzes 70, 80, 73, 91, 75, 83, 85, 96, 85, 79, 85. To calculate the mean, add the 11 numbers and divide the result by 11.

$$\bar{x} = (70 + 80 + 73 + 91 + 75 + 83 + 85 + 96 + 85 + 79 + 85) \div 11 = 902 \div 11 = 82$$

The mean is 82, and it is a single number that gives information about the eleven numbers.

Example 1

On the first four quizzes, Jose gets 85, 88, 90, and 84. What must he get on the fifth quiz to have a mean of 88 for the five quizzes?

Solution: If the unknown quiz grade is called x, the sum of the five quizzes would be $85 + 88 + 90 + 84 + x$. When divided by 5, it needs to equal 88. This becomes the equation

$$\frac{85+88+90+84+x}{5} = 88$$

$$\frac{347+x}{5} = 88$$

$$5 \cdot \frac{347+x}{5} = 5 \cdot 88$$

$$347 + x = 440$$
$$-347 = -347$$
$$x = 93$$

Statistics

Mode

The *mode* of a data set is the most "popular" number. Of the eleven numbers, there are three 85s but only one of each other number. The mode of this data set, therefore, is 85.

Median

The median is the "middle" number after the numbers have been arranged from least to greatest. For the data set of the eleven quizzes, the numbers arranged from least to greatest are 70, 73, 75, 79, 80, 83, 85, 85, 85, 91, 96. In this case, 83 is the median because there are five numbers greater than 83 and five numbers less than 83.

One way to locate the median is to cross off the first and the last number of the list to make it

$$\cancel{70}, 73, 75, 79, 80, 83, 85, 85, 85, 91, \cancel{96}$$

Then cross off the second and the second to last number of this smaller list to get

$$\cancel{70}, \cancel{73}, 75, 79, 80, 83, 85, 85, 85, \cancel{91}, \cancel{96}$$

Keep doing this until you have either one or two numbers left in the middle.

$$\cancel{70}, \cancel{73}, \cancel{75}, \cancel{79}, \cancel{80}, 83, \cancel{85}, \cancel{85}, \cancel{85}, \cancel{91}, \cancel{96}$$

The median for this data set is 83.

If the list has an even number of numbers, add the two middle numbers and divide by 2. If the data set was six numbers, which when arranged in order were 70, 73, 75, 79, 80, and 83, after crossing off the first and last number twice, the list would look like this

$$\cancel{70}, \cancel{73}, 75, 79, \cancel{80}, \cancel{83}$$

There are two numbers left, 75 and 79. Add them and divide by 2 to get the median.

$$(75 + 79) \div 2 = 154 \div 2 = 77$$

Quartiles

The median is also called the *second quartile* since it is the number that is greater than 50% of the numbers in the set. The *first quartile* is the number that is greater than 25% of the numbers on the list. The *third quartile* is the number that is greater than 75% of the numbers on the list.

To calculate the first and third quartile, find the median of the list. Make a list of all the numbers less than the median and another list of all the numbers greater than the median. The medians of these two lists are the first quartile and the third quartile.

With the 11 quiz scores arranged in order from least to greatest, the median was 83.

$$70, 73, 75, 79, 80, 83, 85, 85, 85, 91, 96$$

The five numbers less than 83 are 70, 73, 75, 79, 80.
Since the median of these five numbers is 75, 75 is the first quartile.
The five numbers greater than 83 are 85, 85, 85, 91, 96. Since the median of these five numbers is 85, 85 is the third quartile.

Example 2

Find the mean, first quartile, and third quartile of the fifteen numbers

$$16, 19, 20, 13, 13, 11, 10, 17, 10, 14, 13, 20, 10, 19, 16$$

Solution: Arrange the numbers from least to greatest

$$10, 10, 10, 11, 13, 13, 13, 14, 16, 16, 17, 19, 19, 20, 20$$

The median of the set is 14.

There are seven numbers less than the median. They are 10, 10, 10, 11, 13, 13, 13. Since the median of those seven numbers is 11, the first quartile is 11.

There are seven numbers greater than the median. They are 16, 16, 17, 19, 19, 20, 20. Since the median of those seven numbers is 19, the third quartile is 19.

Interquartile Range

The difference between the third quartile and the first quartile is called the *interquartile range*. For some data sets this number could be small and for others it could be large. In the example with the eleven quizzes, the difference between the third quartile and the first quartile is $85 - 75 = 10$.

Example 3

What is the interquartile range of the data from Example 2?

Solution: Since the third quartile was 19 and the first quartile was 11, the interquartile range is $19 - 11 = 8$.

Statistics

Using the Graphing Calculator to Determine Maximum, Minimum, Median, First Quartile, and Third Quartile

For the TI-84:

The graphing calculator can calculate the five measure of central tendency. First enter all the numbers into L1 by pressing [STAT] and [1] for Edit.

To find the minimum, first quartile, median, third quartile, and maximum of the seven numbers 10, 4, 8, 12, 6, 16, 14, enter them into L1. Then press [STAT] and [1] for 1-Var Stats and press [ENTER].

The screen will display

The $\bar{x} = 10$ is for the mean. The $n = 7$ means that there were seven elements in the list. For the minimum, first quartile, median, third quartile, and maximum, press the down arrow five times.

minX is for the minimum, Q1 is for the first quartile, Med is for the median, Q3 is for the third quartile, and maxX is for the maximum element.

For the TI-Nspire:

From the home screen, select the Add Lists & Spreadsheet icon. Name column A x and fill in cells A1 through A7 with the numbers 10, 4, 8, 12, 6, 16, 14.

Press [4], [1], [and 1] for One-Variable Statistics.

Press [OK] since there is just one list. Set the X1 List to x since that was what the column with the data was named in the spreadsheet.

Statistics

In cells C2 through C13, the one-variable statistics will be displayed. The median is the $\bar{x}$. n is for the number of numbers. MinX is the smallest number. Q1X is the first quartile. MedianX is the median. Q3X is the third quartile. MaxX is the largest number.

Two-Way Frequency Tables

A *two-way frequency table* is a good way to organize data collected from a survey. Instead of just displaying the results of the survey, the two-way frequency table is more specific, displaying how certain categories of people responded to the survey.

Interpreting a Two-Way Frequency Table

Below is a two-way frequency table based on a survey given to 40 men and 60 women about whether or not math is their favorite subject.

	YES	NO	TOTAL
MEN	32	8	40
WOMEN	54	6	60
TOTAL	86	14	100

From this two-way frequency table, there are nine pieces of information:

100 people took the survey.
Of the 100 people, 40 were men and 60 were women.
Of the 100 people, 86 said "yes" and 14 said "no."
32 men said "yes."
8 men said "no."
54 women said "yes."
6 women said "no."

Creating a Two-Way Frequency Table

If a two-question survey is given to a group of people, the results can be collected on a two-way frequency table. Below are 12 surveys. Create a two-way frequency table based on these 12 surveys.

Circle one answer for each question. Grade: (9), 10, 11 Course: (Algebra), Geometry	Circle one answer for each question. Grade: (9), 10, 11 Course: (Algebra), Geometry	Circle one answer for each question. Grade: 9, (10), 11 Course: Algebra, (Geometry)
Circle one answer for each question. Grade: 9, (10), 11 Course: Algebra, (Geometry)	Circle one answer for each question. Grade: (9), 10, 11 Course: (Algebra), Geometry	Circle one answer for each question. Grade: 9, 10, (11) Course: (Algebra), Geometry
Circle one answer for each question. Grade: 9, (10), 11 Course: (Algebra), Geometry	Circle one answer for each question. Grade: 9, 10, (11) Course: Algebra, (Geometry)	Circle one answer for each question. Grade: (9), 10, 11 Course: (Algebra), Geometry
Circle one answer for each question. Grade: 9, (10), 11 Course: Algebra, (Geometry)	Circle one answer for each question. Grade: (9), 10, 11 Course: Algebra, (Geometry)	Circle one answer for each question. Grade: 9, 10, (11) Course: Algebra, (Geometry)

Statistics

There needs to be a Total row and column and also a row and column for the labels, so for this data set, make five rows and four columns.

	Algebra	Geometry	Total
9th			
10th			
11th			
Total			

Tally the numbers from each combination 9, 10, 11 with Algebra, Geometry and fill in those numbers first.

	Algebra	Geometry	Total
9th	4	1	
10th	1	3	
11th	1	2	
Total			

Make sure these numbers add up to 12: 4 + 1 + 1 + 3 + 1 + 2 = 12.

Add each row and put the result into the total row. Add each column and put the result into the total column.

	Algebra	Geometry	Total
9th	4	1	5
10th	1	3	4
11th	1	2	3
Total	6	6	

Add up the numbers in the total row and in the total column. They should get the same answer. In this example, 6 + 6 = 12 and 5 + 4 + 3 = 12. Put this nuhmber in the bottom right cell.

	Algebra	Geometry	Total
9th	4	1	5
10th	1	3	4
11th	1	2	3
Total	6	6	12

Creating a Two-Way Relative Frequency Table

A relative two-way frequency table is much like a regular two-way frequency table except that all the numbers are recorded as percentages of the total. To convert the table from the last section into a relative two-way frequency table, divide each number by the total, which is 12.

	Algebra	Geometry	Total
9th	$\frac{4}{12} = 0.33 = 33\%$	$\frac{1}{12} = 0.08 = 8\%$	$\frac{5}{12} = 0.42 = 42\%$
10th	$\frac{1}{12} = 0.08 = 8\%$	$\frac{3}{12} = 0.25 = 25\%$	$\frac{4}{12} = 0.33 = 33\%$
11th	$\frac{1}{12} = 0.08 = 8\%$	$\frac{2}{12} = 0.17 = 17\%$	$\frac{3}{12} = .25 = 25\%$
Total	$\frac{6}{12} = 0.50 = 50\%$	$\frac{6}{12} = 0.50 = 50\%$	$\frac{12}{12} = 1.00 = 100\%$

Check Your Understanding of Section 13.1

A. Multiple-Choice

1. Find the mean of this set of numbers {4, 5, 8, 8, 8, 10, 10, 13, 15, 17, 23}.
   (1) 8 (2) 9 (3) 10 (4) 11

2. Find the median of this set of numbers {4, 5, 8, 8, 8, 10, 10, 13, 15, 17, 23}.
   (1) 8 (2) 10 (3) 11 (4) 15

3. Find the mode of this set of numbers {4, 5, 8, 8, 8, 10, 10, 13, 15, 17, 23}.
   (1) 23 (2) 11 (3) 10 (4) 8

4. Find the first quartile of this set of numbers {4, 5, 8, 8, 8, 10, 10, 13, 15, 17, 23}.
   (1) 8 (2) 10 (3) 11 (4) 15

Statistics

5. Find the third quartile of this set of numbers {4, 5, 8, 8, 8, 10, 10, 13, 15, 17, 23}.
   (1) 10 (2) 15 (3) 17 (4) 23

6. Find the interquartile range of this set of numbers {4, 5, 8, 8, 8, 10, 10, 13, 15, 17, 23}.
   (1) 7 (2) 8 (3) 9 (4) 10

7. For the first four days of a five day vacation, the mean temperature was 80 degrees. What must the temperature be on the fifth day in order for the mean temperature to be 82 degrees?
   (1) 88 (2) 89 (3) 90 (4) 91

8. What is the median of the set of numbers {2, 4, 8, 3, 1, 4, 9, 5}?
   (1) 2 (2) 3 (3) 4 (4) 4.5

9. In a set of seven numbers, the largest number is increased by 10. Which measure of central tendency must increase because of this?
   (1) Mean
   (2) Mode
   (3) First quartile
   (4) Interquartile range

10. For which data set is the median greater than the mean?
    (1) {4, 7, 10, 13, 16}
    (2) {8, 9, 10, 18, 19}
    (3) {8, 9, 10, 11, 12}
    (4) {1, 2, 10, 11, 12}

B. Show how you arrived at your answer.

1. Seven numbers out of a set of nine numbers are 16, 17, 19, 19, 21, 21, and 25. If the mode of the eight numbers is 19 and the mean is 20, what are the other two numbers?

2. Two basketball teams each have a mean height of 6 feet and a median height of 6 feet, but the interquartile range of the first team is much greater than the interquartile range of the second team. How can this be? Explain.

3. Zahra says that on five tests she has a mode of 81 but a mean and median of 90. How is this possible? Explain.

4. What fraction of these 11 numbers is between the first quartile and the third quartile?
   3, 17, 4, 8, 4, 9, 9, 15, 14, 19, 8.

5. Lila has taken five algebra tests. Her median score is 90, and her mode score is 90, but her mean score is greater than 90. What is the highest her mean score can be?

390

13.2 GRAPHICALLY REPRESENTING DATA

KEY IDEAS

When data are listed as just a series of numbers, many aspects of the data set are not clear. With a graphical representation of the data, certain information can be seen, even without doing any calculations. Three ways to represent data graphically are dot plots, histograms, and box plots. Each provides more information than a list of numbers would.

Dot Plots

In a *dot plot*, each piece of data is represented by a circle. When two data points represent the same value, they are stacked in a vertical line. With a dot plot, the mode can be easily determined. It is even possible to estimate the interquartile range from a dot plot.

If a data set for the heights, in inches, of 11 professional men's basketball players are

$$76, 67, 76, 77, 72, 77, 80, 70, 84, 77, 80$$

a dot plot is created by putting the numbers from 67 to 84 on a horizontal line and then creating a dot for each number. When a number is repeated, the dot goes on top of the dot, or dots, already representing that number.

From this dot plot, it can be seen that the mode is 77. Also it can be seen that the data are bunched together so the interquartile range will be small for this data set.

Statistics

Histograms

The same data set can be represented graphically with a histogram. In a histogram the data is represented by rectangles. The height of the rectangle is the number of times that number appears in the data set. Unlike a bar graph, in a histogram there are no spaces between bars representing consecutive values. In the picture below, there is no space between the bar representing the two 76s and the bar representing the three 77s.

In a histogram, a range of numbers can be grouped together with a larger bin size. Here is the same data on a histogram with bin size of 5. The first bar is for numbers between 66 and 71, including 71 but not 66. The second bar is for numbers between 71 and 76, including 76 but not 71. The third bar is for numbers between 76 and 81, including 81 but not 76. The fourth bar is for numbers between 81 and 85, including 85 but not 81.

Box Plots

A *box plot* is a way to represent graphically the minimum value, the maximum value, the first quartile, the median, the third quartile, and the maximum value on the same diagram.

To create a box plot, first calculate these five values. For the data set with the 11 heights, the five values are

67	70	72	76	76	77	77	77	80	80	84
min		Q1			med			Q3		max

On a number line, plot these five values.

Make short vertical lines through each of these five points.

Complete the diagram by making a rectangle with the second and fourth segments as two of its sides and also make a horizontal line segment connecting the first point to the second point and another connecting the fourth point to the fifth point.

From a box plot (sometimes called a *box and whisker plot*) the interquartile range is the width of the rectangle in the middle. When the numbers in the data set are close together, this rectangle will be very narrow.

Statistics

Check Your Understanding of Section 13.2

A. Multiple-Choice

1. What is the mode of the data in this dot plot?

(1) 5   (2) 6   (3) 7   (4) 8

2. What is the median of the data in this dot plot?

(1) 3   (2) 4   (3) 5   (4) 6

3. What is the mean of the data in this dot plot, rounded to the nearest hundredth?

(1) 5.13   (2) 5.27   (3) 5.45   (4) 5.78

4. What is the mode of the data in this histogram?

(1) 23   (2) 24   (3) 15   (4) 16

5. What is the median of the data in this histogram?

(1) 34   (2) 35   (3) 36   (4) 37

6. What is the median of the data in this box plot?

(1) 17   (2) 15   (3) 8   (4) 25

395

Statistics

7. What is the interquartile range of the data in this box plot?

(1) 7   (2) 9   (3) 18   (4) 30

8. Which box plot is based on data from this histogram?

(1)

(2)

(3)

(4)

396

9. What is true about this data set: 1, 2, 10, 11, 11?
   (1) The median is greater than the mean.
   (2) The median is equal to the mean.
   (3) The median is less than the mean.
   (4) The median is greater than the mode.

10. For which data set is the interquartile range equal to 0?
    (1) 2, 3, 6, 6, 6, 7, 8
    (2) 2, 6, 6, 6, 6, 6, 8
    (3) 1, 2, 3, 6, 7, 8, 9
    (4) 2, 6, 6, 6, 6, 7, 8

B. *Show how you arrived at your answer.*

1. Create a dot plot for the following set:

   5, 5, 6, 7, 7, 7, 8, 8, 10, 10, 12, 13, 14, 14, 14, 14

2. Create a box plot for the following set:

   20, 12, 4, 8, 4, 8, 12, 17, 2, 12, 9, 3, 2, 6, 15

3. Two basketball teams each have a mean height of 72 inches and median height of 72 inches. The box plots for the two teams are below. How is it that two teams with the same mean and median can have such different box plots?

4. The box plot below is based on 11 numbers. Find two different sets of data that would produce this box plot.

Statistics

5. Make a histogram for this data set using 6 bars, each with a width of 5, using 0 to 5 including 5 but not 0, 5 to 10 including 10 but not 5, 10 to 15 including 15 but not 10, 15 to 20 including 15 but not 20, 20 to 25 including 25 but not 20, and 25 to 30 including 30 but not 25.

2, 2, 4, 6, 7, 7, 8, 8, 9, 11, 12, 12, 13, 16, 17, 17, 18, 18, 18, 19, 21, 21, 22, 22, 22, 23, 24, 28, 29, 29

Chapter Fourteen

TEST-TAKING STRATEGIES

Knowing the material is only part of the battle in acing the Algebra Regents exam. Things like not managing your time, making careless errors, and struggling with the calculator can cost valuable points. In this chapter some test-taking strategies are explained to help you perform your best on test day.

14.1 TIME MANAGEMENT

Don't Rush

The Algebra Regents exam is three hours long. Even though you are permitted to leave after one and a half hours, to get the best grade possible, stay until the end of the exam. Just as it wouldn't be wise to come to the test an hour late, it is almost the same thing if you leave a test an hour early.

Do the Test Twice

The best way to protect against careless errors is to do the entire test twice and compare the answers you got the first time to the answers you got the second time. For any answers that don't agree, do a "tie breaker" third time. Redoing the test and comparing answers is much more effective than simply "looking over" your work. People tend to skim by careless errors when looking over their work. Redoing the questions, you are less likely to make the same careless error.

Bring a Watch

In some classrooms the clock is broken. Without knowing how much time is left, you might rush and make careless errors. Though the proctor should write the time on the board and update it every so often, it is best if you have your own watch.

The TI-84 graphing calculator has a built in clock. Press the [MODE] to see it. If the time is not right, go to SET CLOCK and set it correctly. The TI-Nspire does not have a built-in clock.

Test-Taking Strategies

14.2 KNOW HOW TO GET PARTIAL CREDIT

Know the Structure of the Exam

The Algebra Regents exam has 37 questions. The first 24 of those questions are multiple choice worth two points each. There is no partial credit if you make a mistake on one of those questions. Even the smallest careless error, like missing a negative sign, will result in no credit for that question.

Parts Two, Three, and Four are free-response questions with no multiple choice. Besides giving a numerical answer, you may be asked to explain your reasoning.

Part Two has eight free-response questions worth two points each. The smallest careless error will cause you to lose one point, which is half the value of the question.

Part Three has four free-response questions worth four points each. These questions generally have multiple parts.

Part Four has one free-response question worth six points. This question will have multiple parts.

Explaining Your Reasoning

When a free-response question asks to "Justify your answer," "Explain your answer," or "Explain how you determined your answer," the grader is expecting a few clearly written sentences. For these, you don't want to write too little since the grader needs to see that you understand why you did the different steps you did to solve the equation. You also don't want to write too much because if anything you write is not accurate, points can be deducted.

Here is an example followed by two solutions. The first would not get full credit, but the second would.

Example

Use algebra to solve for x in the equation $\frac{2}{3}x + 1 = 11$. Justify your steps.

Solution 1 (part credit):

| $\begin{aligned} \frac{2}{3}x + 1 &= 11 \\ -1 &= -1 \\ \frac{2}{3}x &= 10 \\ x &= 15 \end{aligned}$ | I used algebra to get the x by itself. The answer was $x = 15$. |

Solution 2 (full credit):

$\frac{2}{3}x + 1 = 11$ $-1 = -1$ $\frac{2}{3}x = 10$ $\frac{3}{2} \cdot \frac{2}{3}x = \frac{3}{2} \cdot 10$ $1x = 15$ $x = 15$	I used the subtraction property of equality to eliminate the +1 from the left-hand side. Then to make it so the x had a 1 in front of it, I used the multiplication property of equality and multiplied both sides of the equation by the reciprocal of $\frac{2}{3}$, which is $\frac{3}{2}$. Then since $1 \cdot x = x$, the left-hand side of the equation just became x and the right-hand side became 15.

Computational Errors vs. Conceptual Errors

In the Part Three and Part Four questions, the graders are instructed to take off one point for a "computational error" but half credit for a "conceptual error." This is the difference between these two types of errors.

If a four point question was $x - 1 = 2$ and a student did it like this,

$$x - 1 = 2$$
$$+1 = +1$$
$$x = 4$$

the student would lose one point out of 4 because there was one computational error since $2 + 1 = 3$ and not 4.

Had the student done it like this,

$$x - 1 = 2$$
$$-1 = -1$$
$$x = 1$$

the student would lose half credit, or 2 points, since this error was conceptual. The student thought that to eliminate the -1, he should subtract 1 from both sides of the equation.

Either error might just be careless, but the conceptual error is the one that gets the harsher deduction.

14.3 KNOW YOUR CALCULATOR

Which Calculator Should You Use?

The two calculators used for this book are the TI-84 and the TI-Nspire. Both are very powerful. The TI-84 is somewhat easier to use for the functions needed for this test. The TI-Nspire has more features for courses in the future.

Test-Taking Strategies

The choice is up to you. This author prefers the TI-84 for the Algebra Regents.

Graphing calculators come with manuals that are as thick as the book you are holding. There are also plenty of video tutorials online for learning how to use advanced features of the calculator. To become an expert user, watch the online tutorials or read the manual.

Clearing the Memory

You may be asked at the beginning of the test to clear the memory of your calculator. When practicing for the test, you should clear the memory too so you are practicing under test-taking conditions.

This is how you clear the memory.

For the TI-84:

Press [2ND] and then [+] to get to the MEMORY menu. Then press [7] for Reset.

```
MEMORY
1:About
2:Mem Mgmt/Del…
3:Clear Entries
4:ClrAllLists
5:Archive
6:UnArchive
7↓Reset…
```

Use the arrows to go to [ALL] for All Memory. Then press [1].

```
RAM  ARCHIVE  ALL
1:All Memory…
```

Press [2] for Reset.

```
RESET MEMORY
1:No
2:Reset
Resetting ALL
will delete all
data, programs &
Apps from RAM &
Archive.
```

The calculator will be reset as if brand new condition. The one setting that you may need to change is to turn the diagnostics on if you need to calculate the correlation coefficient. (See Chapter 12 for this.)

For the TI-Nspire:

The TI-Nspire must be set to Press-To-Test mode when taking the Algebra Regents. Turn the calculator off by pressing [ctrl] and [home]. Press and hold [esc] and then press [home].

While in Press-to-Test mode, certain features will be deactivated. A small green light will blink on the calculator so a proctor can verify the calculator is in Press-to-Test mode.

To exit Press-to-Test mode, use a USB cable to connect the calculator to another TI-Nspire. Then from the home screen on the calculator in Press-to-Test mode, press [doc], [9] and select Exit Press-to-Test.

Use Parentheses

The calculator always uses the order of operations where multiplication and division happen before addition and subtraction. Sometimes, though, you may want the calculator to do the operations in a different order.

Suppose at the end of a quadratic equation, you have to round $x = \frac{-1+\sqrt{5}}{2}$ to the nearest hundredth. If you enter (−) (1) (+) (2ND) (x^2) (5) (/) (2), it displays

which is not the correct answer.

One reason is that for the TI-84 there needs to be a closing parentheses (or on the TI-Nspire, press [right arrow] to move out from under the radical sign) after the 5 in the square root symbol. Without it, it calculated $-1 + \sqrt{\frac{5}{2}}$.

Test-Taking Strategies

More needs to be done, though, since

still is not correct. This is the solution to $-1 + \sqrt{\dfrac{5}{2}}$.

To get this correct, there also needs to be parentheses around the entire numerator, $-1 + \sqrt{5}$.

This is the correct answer.

On the TI-Nspire, fractions like this can also be done with [templates].

Using the ANS Feature

The last number calculated with the calculator is stored in something called the ANS variable. This ANS variable will appear if you start an expression with a +, −, ×, or ÷. When an answer has a lot of digits in it, this saves time and is also more accurate.

404

If for some step in a problem you need to calculate the decimal equivalent of $\frac{1}{7}$, it will look like this on the TI-84:

```
1/7
    .1428571429
```

For the TI-Nspire, if you try the same thing, it leaves the answer as $\frac{1}{7}$. To get the decimal approximation, press [ctrl] and [enter] instead of just [enter].

Now if you have to multiply this by 3, just press [×], and the calculator will display "Ans*"; press [3] and [enter].

```
1/7
    .1428571429
Ans*3
    .4285714286
```

Test-Taking Strategies

The ANS variable can also help you do calculations in stages. To calculate $x = \dfrac{-1+\sqrt{5}}{2}$ without using so many parentheses as before, it can be done by first calculating $-1 + \sqrt{5}$ and then pressing [÷] and [2] and Ans will appear automatically.

```
-1+√(5
         1.236067977
Ans/2
          .6180339887
```

Scratchpad	
-1+√5	1.23607
Ans/2	

Scratchpad	
-1+√5	1.23607
1.2360679774998 / 2	0.618034

The ANS variable can also be accessed by pressing [2ND] and [−] at the bottom right of the calculator. If after calculating the decimal equivalent of $\dfrac{1}{7}$ you wanted to subtract $\dfrac{1}{7}$ from 5, for the TI-84 press [5], [−], [2ND], [ANS], and [ENTER].

For the TI-Nspire press [5], [−], [ctrl], [ans], and [enter].

```
1/7
          .1428571429
```

```
1/7
          .1428571429
5-Ans
          4.857142857
```

Scratchpad	
$\dfrac{1}{7}$	0.142857
5−Ans	

Scratchpad	
$\dfrac{1}{7}$	0.142857
5−0.14285714285714	4.85714

14.4 USE THE REFERENCE SHEET

In the back of the Algebra Regents booklet is a reference sheet with 17 conversion facts, like inches to centimeters and quarts to pints, and also 17 formulas. Many of these conversion facts and formulas will not be needed for an individual test, but the quadratic formula and the arithmetic sequence formula are the two that will come in the handiest.

High School Math Reference Sheet

1 inch = 2.54 centimeters	1 kilometer = 0.62 mile	1 cup = 8 fluid ounces
1 meter = 39.37 inches	1 pound = 16 ounces	1 pint = 2 cups
1 mile = 5280 feet	1 pound = 0.454 kilogram	1 quart = 2 pints
1 mile = 1760 yards	1 kilogram = 2.2 pounds	1 gallon = 4 quarts
1 mile = 1.609 kilometers	1 ton = 2000 pounds	1 gallon = 3.785 liters
		1 liter = 0.264 gallon
		1 liter = 1000 cubic centimeters

Triangle	$A = \frac{1}{2}bh$
Parallelogram	$A = bh$
Circle	$A = \pi r^2$
Circle	$C = \pi d$ or $C = 2\pi r$
General Prisms	$V = Bh$
Cylinder	$V = \pi r^2 h$
Sphere	$V = \frac{4}{3}\pi r^3$
Cone	$V = \frac{1}{3}\pi r^2 h$
Pyramid	$V = \frac{1}{3}Bh$

Pythagorean Theorem	$a^2 + b^2 = c^2$
Quadratic Formula	$x = \frac{-b \pm \sqrt{b^2 - 4ac}}{2a}$
Arithmetic Sequence	$a_n = a_1 + (n-1)d$
Geometric Sequence	$a_n = a_1 r^{n-1}$
Geometric Series	$S_n = \frac{a_1 - a_1 r^n}{1 - r}$ where $r \neq 1$
Radians	1 radian = $\frac{180}{\pi}$ degrees
Degrees	1 degree = $\frac{\pi}{180}$ radians
Exponential Growth/Decay	$A = A_0 e^{k(t - t_0)} + B_0$

407

14.5 HOW MANY POINTS DO YOU NEED TO PASS?

The Algebra Regents exam is scored out of a possible 86 points. Unlike most tests given in the year by your teacher, the score is not then turned into a percent out of 86. Instead each test has a conversion sheet that varies from year to year. For the June 2014 test, the conversion sheet looked like this.

Raw Score	Scale Score	Raw Score	Scale Score	Raw Score	Scale Score
86	100	57	75	28	64
85	99	56	74	27	63
84	97	55	74	26	62
83	96	54	74	25	61
82	95	53	73	24	60
81	94	52	73	23	59
80	92	51	73	22	58
79	91	50	72	21	56
78	90	49	72	20	55
77	89	48	72	19	54
76	88	47	72	18	52
75	87	46	71	17	50
74	86	45	71	16	49
73	85	44	71	15	47
72	84	43	70	14	45
71	83	42	70	13	42
70	82	41	70	12	40
69	82	40	70	11	38
68	81	39	69	10	35
67	80	38	69	9	32
66	79	37	69	8	30
65	79	36	68	7	26
64	78	35	68	6	23
63	78	34	67	5	20
62	77	33	67	4	16
61	77	32	66	3	12
60	76	31	66	2	9
59	76	30	65	1	4
58	75	29	64	0	0

On this test, 30 points became a 65, 57 points became a 75, and 73 points became an 85. This means that for this examination a student who got 30 out of 86, which is just 35% of the possible points, would get a 65 on this exam. 57 out of 86 is 66%, but this scaled to a 75. 73 out of 86, however, is actually 85% and became an 85. So in the past there has been a curve on the exam for lower scores, though the scaling is not released until after the exam.

Answers and Solution Hints to Practice Exercises

CHAPTER 1

Section 1.1
A

1. (4)   3. (1)   5. (3)   7. (4)   9. (2)
2. (3)   4. (2)   6. (1)   8. (1)   10. (1)

B

1. Malachi would have $6x - 2 = 16$ and Wesley would have $3x - 1 = 8$.
2. Addition property of equality, Distributive property of multiplication over addition, Subtraction property of equality, Division property of equality.
3. The first person is wrong. The proper use of the distributive property is $5(3 + 7) = 5 \cdot 3 + 5 \cdot 7$, not $5 \cdot 3 + 7$.
4. There is no commutative property of subtraction. For example, $5 - 2 = 3$ is not equivalent to $2 - 5 = -3$.
5. 30 is correct. There is no distributive property of multiplication over multiplication so it is not true that $a(b \cdot c) = a \cdot b \cdot a \cdot c$.

Section 1.2
A

1. (3)   3. (2)   5. (1)   7. (3)   9. (3)
2. (4)   4. (2)   6. (3)   8. (4)   10. (2)

B

1. It will become the equation $5x = 60$, which can then be solved by dividing both sides of the equals sign by 5.
2. If you multiply both sides by $\frac{5}{3}$ or if you divide both sides by $\frac{3}{5}$, you get the same correct answer.
3. By subtracting -2 from both sides, it becomes $x = 5 - (-2) = 5 + 2 = 7$.
4. The addition property of equality. Add 3 to both sides of the equation.
5. There is nothing that when multiplied by 0 becomes 10. If you try to use the division property of equality and divide both sides of the equation by 0, you get $\frac{10}{0}$, which is not defined.

Section 1.3
A

1. (3)   3. (3)   5. (4)   7. (2)   9. (1)
2. (1)   4. (1)   6. (1)   8. (1)   10. (4)

Answers and Solution Hints to Practice Exercises

B
1. $x = 5$. Distributive property, addition property of equality, and then multiplication property or by multiplication property then addition property of equality.
2. $x = 7$. Addition property of equality then division property of equality.
3. $x = 1$. Distributive property, addition property of equality, and then division property of equality or by division property of equality then addition property of equality.
4. They are both right. Either way gets the answer of $x = 5$.
5. $p = 20$. The sneakers sell for $20.

Section 1.4
A
1. (2) 3. (3) 5. (1) 7. (3) 9. (4)
2. (3) 4. (1) 6. (1) 8. (3) 10. (2)

B
1. $q = \dfrac{p+f}{r}$ 3. $b = \dfrac{2a}{h}$ 5. $n = \dfrac{a_n - a_1}{d} + 1$
2. $w = \dfrac{v}{lh}$ 4. $r = \dfrac{s}{2\pi h}$

Chapter 2

Section 2.1
A
1. (2) 3. (1) 5. (4) 7. (3) 9. (4)
2. (3) 4. (2) 6. (3) 8. (3) 10. (3)

B
1. It is a trinomial with degree 2, also called a quadratic trinomial.
2. It is a binomial with degree 1, also called a linear binomial.
3. Cross off either the $2x^2$, the $-7x$, or the $+9$.
4. Anything of the form ax^4, for example $5x^4$. The variable name can be something other than x.
5. Mark was correct that it was a binomial. It is true that after simplifying it would no longer be a binomial, but before simplifying it is one.

Section 2.2
A
1. (3) 3. (4) 5. (3) 7. (2) 9. (4)
2. (1) 4. (4) 6. (2) 8. (3) 10. (1)

B
1. She is not correct. The answer is $20x^5$ because $x^2 \cdot x^3 = x^5$, not x^6.

Answers and Solution Hints to Practice Exercises

2. When multiplying monomials, multiply the coefficients and add the exponents. So for dividing, do the opposite, divide the coefficients and subtract the exponents to get $4x^4$.
3. Sawyer is right, $x^0 = 1$ provided that $x \neq 0$.
4. 10. There are two ways to do this. One is to multiply the two monomials and get $10x^0$, which is $10 \cdot 1 = 10$. Another way is to rewrite as
$$5x^2 \cdot \frac{2}{x^2} = 10x^0 = 10.$$
5. It will be a fifth degree monomial. For example $3x^2 \cdot 4x^3 = 12x^5$.

Section 2.3
A
1. (2)   3. (3)   5. (4)   7. (2)   9. (3)
2. (4)   4. (3)   6. (1)   8. (4)   10. (4)

B
1. Yes, this is allowed. The expression $x^2(3 + 5)$ can be simplified to $3x^2 + 5x^2$ and also to $8x^2$. Since each expression is equivalent to $x^2(3 + 5)$, $3x^2 + 5x^2$ must equal $8x^2$.
2. Josie is correct. Since $5y^2x$ is equivalent to $5xy^2$ by the commutative property of multiplication, these are like terms and can be combined to become $8xy^2$.
3. The variable a must be equivalent to $2xy^2z$ so that it can be combined with the $5xy^2z$.
4. It becomes $x^3 + 3x^2y + 3xy^2 + y^3$
5. $6 - i - i^2$ or $-i_2 - i + 6$

Section 2.4
A
1. (2)   3. (3)   5. (4)   7. (2)   9. (1)
2. (4)   4. (2)   6. (4)   8. (3)   10. (2)

B
1. Alexander distributes the 5 through the terms in the parentheses to get $10x + 15x = 25x$, which is the same answer as Tucker.
2. Jack is correct. Because of the commutative property of multiplication, it is the same thing as $5x(3x^2 + 2x + 1)$.
3. The a represents $2x$ since $2x \cdot 5x^2 = 10x^3$, $2x \cdot (-3x) = -6x^2$, and $2x \cdot 7 = 14x$.
4. First it becomes $8x^2 + 12x - 4x - 6$. Then after combining like terms becomes $8x^2 + 8x - 6$.
5. First it becomes $6x^2 - 10x - 18x + 30$. Then after combining like terms becomes $6x^2 - 28x + 30$.

Answers and Solution Hints to Practice Exercises

Section 2.5
A

1. (1)	3. (2)	5. (4)	7. (1)	9. (1)
2. (1)	4. (3)	6. (2)	8. (3)	10. (3)

B
1. He did not do it correctly. He should have distributed the negative through the parentheses to get $2x + 3 - 4x - 2$, which would lead to the correct answer of $-2x + 1$.
2. $3x + 5 - 3(2x - 5) = 3x + 5 - 6x + 15 = -3x + 20$
3. The rules for combining polynomials are the same as for combining numbers so this process will still get the right answer.
4. $2(-2x + 5) - 5(3x - 7) = -4x + 10 - 15x + 35 = -19x + 45$
5. $(x^2 + 5x + 2) + (x^2 - 3x + 4) - (x^2 - 2x + 5) =$
   $x^2 + 5x + 2 + x^2 - 3x + 4 - x^2 + 2x - 5 = x^2 + 4x + 1$

Section 2.6
A

1. (1)	3. (3)	5. (4)	7. (2)	9. (1)
2. (3)	4. (1)	6. (4)	8. (3)	10. (1)

B
1. The equation can be $x + 20 = 2x + 3$ with a solution of $x = 17$.
2. One reason for choosing to subtract $2x$ from both sides is to avoid negatives. Also fine to write that the methods are the same degree of difficulty so you would not choose either way over the other.
3. $x = 4$
4. $30x$ is the distance the first train is from its starting point. The second train begins 300 miles from the first train's starting point and then every hour gets 20 miles closer so in x hours it will be $300 - 20x$ miles from the first train's starting point. When both expressions are equal, the trains are the same distance from the first train's starting point so they are passing. The solution to the equation is $x = 6$.
5. The equation is $x + (2x - 5) + (3x + 10) = 125$, which becomes $6x + 5 = 125$ or $x = 20$. So the store sold 20 items on Monday, 35 items on Tuesday, and 70 items on Wednesday.

Section 2.7
A

1. (3)	3. (3)	5. (4)	7. (1)	9. (3)
2. (1)	4. (4)	6. (2)	8. (2)	10. (2)

B
1. $(x + 4)(x - 1) + 2(x + 4) = x^2 + 3x - 4 + 2x + 8 = x^2 + 5x + 4$
2. $4 + 12i + 9i^2 + 1 + 2i = 5 + 14i + 9i^2$
3. She is not correct. There is not a distributive property for exponents. Instead $(x + 4)(x + 4) = x^2 + 4x + 4x + 16 = x^2 + 8x + 16$.

Answers and Solution Hints to Practice Exercises

4. Since this is not a binomial multiplied by a binomial, FOIL can't be used. Instead, use the distributive property.
$(x + 3)x^2 + (x + 3)7x + (x + 3)10 = x^3 + 3x^2 + 7x^2 + 21x + 10x + 30 = x^3 + 10x^2 + 31x + 30$
5. $(50 - 2)(50 + 2)$ has the $(a - b)(a + b)$ pattern so the solution is $50^2 - 2^2 = 2500 - 4 = 2496$.

Section 2.8
Solutions
A
1. (4)   3. (1)   5. (2)   7. (1)   9. (4)
2. (2)   4. (2)   6. (1)   8. (1)   10. (2)

B
1. Ariana is correct. $(x + 5)^2 = x^2 + 10x + 25$, not $x^2 + 25$.
2. First factor out the greatest common factor of x. $x(x^2 + 4x - 21)$. Then factor the quadratic trinomial in the parentheses $x(x + 7)(x - 3)$.
3. First factor out the greatest common factor of 5. $5(x^2 + 4x + 4)$. Then factor the quadratic trinomial in the parentheses $5(x + 2)(x + 2)$ or $5(x + 2)^2$.
4. He is not correct. $(x - 6)(x - 1)$ is $x^2 - 7x + 6$, not $x^2 - 5x + 6$.
5. The other factor is $(x + 5)$. To get the 15 as the last term there needs to be a $+5$ in the other factor. Also to get $2x^2$ in the first term there needs to be a $1x$ in the other factor. To check that this is correct, multiply $(2x + 3)(x + 5) = 2x^2 + 10x + 3x + 15 = 2x^2 + 13x + 15$.

Section 2.9
A
1. (3)   3. (3)   5. (2)   7. (3)   9. (1)
2. (1)   4. (1)   6. (3)   8. (2)   10. (2)

B
1. $x^4 - 81 = (x^2)^2 - 81 = (x^2 - 9)(x^2 + 9)$ Then $x^2 - 9$ can be factored into $(x - 3)(x + 3)$. The solution is $(x - 3)(x + 3)(x^2 + 9)$.
Complete answer is $(x - 3)(x + 3)(x^2 + 9)$.
2. $x^4 + 4x^2 - 5 = (x^2)^2 + 4(x^2) - 5 = (x^2 + 5)(x^2 - 1) = (x^2 + 5)(x - 1)(x + 1)$
3. $x^4 - 13x^2 + 36 = (x^2)^2 - 13x^2 + 36 = (x^2 - 4)(x^2 - 9) = (x - 2)(x + 2)(x - 3)(x + 3)$
4. This is like $x^2 + 2x - 3 = (x + 3)(x - 1)$ but instead of an x there is a $(3x - 5)$. The solution is $(3x - 5 + 3)(3x - 5 - 1)$ which can also be simplified further to $(3x - 2)(3x - 6)$.
5. $x^6 - 8 = (x^2)^3 - 2^3$ so a is x^2 and b is 2. Solution is $(x^2 - 2)(x^4 + 2x^2 + 4)$.

Answers and Solution Hints to Practice Exercises

CHAPTER 3

Section 3.1
A

1. (4)	3. (3)	5. (3)	7. (4)	9. (3)
2. (1)	4. (2)	6. (4)	8. (3)	10. (1)

B
1. The solutions are $\sqrt{15}$ and $-\sqrt{15}$. Rounded to the nearest hundredth, 3.87 and –3.87.
2. $(x + 5)^2 = 64$ becomes $x + 5 = \pm 8$ with solutions $x = 3$, $x = -13$.
3. $(x + 2)^2 = 49$ becomes $x + 2 = 7$ so $x = 5$.
4. $(x + 4)^2$ does not equal $x^2 + 16$. The correct answer is $x = -4 \pm \sqrt{97}$.
5. Taking the cube root of both sides of the equation, it becomes $x + 1 = 2$ (not ± 2 since $(-2)^3 = -8$), or $x = 1$.

Section 3.2
A

1. (1)	3. (2)	5. (2)	7. (2)	9. (2)
2. (4)	4. (3)	6. (4)	8. (1)	10. (4)

Section 3.3
A

1. (1)	3. (1)	5. (2)	7. (2)	9. (1)
2. (2)	4. (4)	6. (3)	8. (1)	10. (4)

B
1. $b = 16, p = 8$ and $b = -16, p = -8$
2. $x^2 - 16x = 7$, $x^2 - 16x + 64 = 7 + 64 = 71$, $(x - 8)^2 = 71$, $x - 8 = \pm\sqrt{71}$, $x = 8 \pm \sqrt{71}$
3. $\frac{22}{7}$ is just an approximation for π. $\frac{22}{7}$ is rational since it is a fraction with an integer numerator and denominator.
4. $x^2 - 5x + (\frac{5}{2})^2 = 14 + (\frac{5}{2})^2$, $(x - \frac{5}{2})^2 = 14 + \frac{25}{4}$, $x - \frac{5}{2} = \pm\sqrt{\frac{81}{4}}$, $x = \frac{5}{2} \pm \frac{9}{2}$, $x = -2, 7$.
5. $x^2 + 2ax + a^2 = b + a^2$, $(x + a)^2 = b + a^2$, $x + a = \pm\sqrt{b + a^2}$, $x = -a \pm \sqrt{b + a^2}$.

Section 3.4
A

1. (1)	3. (4)	5. (2)	7. (1)	9. (2)
2. (1)	4. (2)	6. (3)	8. (3)	10. (3)

Answers and Solution Hints to Practice Exercises

B
1. No, she is not correct. By factoring the left-hand side, it becomes $x(x-3) = 0$ so either $x = 0$ or $x - 3 = 0$, which means there are two solutions 0 and 3.
2. $(x-1)(x-2)(x-3) = 0$ so $x-1 = 0$ or $x-2 = 0$ or $x-3 = 0$ with three solutions 1, 2, and 3.
3. Either $a = 0$ or $b = 0$ or both. Since $a \neq 0$, b must equal 0.
4. It can be factored into $(x^2 - 9)(x^2 - 4) = 0$ and then into $(x-3)(x+3)(x-2)(x+2) = 0$ so $x-3 = 0$ or $x+3 = 0$ or $x-2 = 0$ or $x+2 = 0$ so the four solutions are 3, –3, 2, and –2.
5. When two things have a product of 40, it does not follow that one of them must be 40, the way it is when the product is 0. The most common way to do this question is to start by subtracting 40 from both sides and factoring.

$$x^2 + 6x - 40 = 0$$
$$(x+10)(x-4) = 0$$
$$x = 10 \text{ or } x = 4$$

It actually can be done from $x(x+6) = 40$ through guess and check. Find two numbers where one is 6 larger than the other so that the product is 40. The numbers are 4 and 10, so $x = 4$ is one solution.

Section 3.5
A
1. (2)   3. (2)   5. (1)   7. (4)   9. (3)
2. (3)   4. (3)   6. (1)   8. (2)   10. (2)

B
1. $(x-2)(x+2)(x-4) = 0$
2. $(x+4)(x+7) = 0$ or $x^2 + 11x + 28 = 0$
3. No. The root is a number, whereas the factor is an expression, like $x+2$. So changing the sign would make it $-(x+2) = -x - 2$, which is not the same as the number –2.
4. $2x + 3 = 0$, $x = -\frac{3}{2}$ and $2x + 5 = 0$, $x = -\frac{5}{2}$. When there is a coefficient in front of the x, the roots are not just the opposite of the constant term.
5. The roots are $2 + \sqrt{7}$ and $2 - \sqrt{7}$ so the factors can be written as $(x - (2 + \sqrt{7}))$ and $(x - (2 - \sqrt{7}))$.

Section 3.6
A
1. (1)   3. (2)   5. (1)   7. (3)   9. (1)
2. (2)   4. (4)   6. (4)   8. (1)   10. (2)

B
1. $x = 5 \pm \sqrt{3}$, $x = 6.73, 3.27$

Answers and Solution Hints to Practice Exercises

2. $2a = 2$ so $a = 1$. $-b = -5$ so $b = 5$. $b^2 - 4ac = 17$ so $25 - 4c = 17$ and $c = 2$.

3. $\dfrac{-4 \pm \sqrt{(16-4c)}}{2}$, which can be optionally simplified to

   $\dfrac{-4 \pm \sqrt{4 \cdot (4-c)}}{2} = -2 \pm \sqrt{4-c}$.

4. The solutions are both irrational. $\sqrt{17}$ is irrational so $-3 \pm \sqrt{17}$ are both irrational, and $\dfrac{-3 \pm \sqrt{17}}{2}$ are both irrational.

5. First, get into the form $ax^2 + bx + c = 0$ by subtracting $4x$ and 6 from both sides to get $x^2 - 4x - 3 = 0$. Then use the quadratic formula to get $x = 2 \pm \sqrt{7}$.

Section 3.7
A
1. (4)   3. (3)   5. (4)   7. (4)   9. (1)
2. (1)   4. (2)   6. (2)   8. (3)   10. (4)

B
1. $-16t^2 + 128t + 144 = 384$ becomes $-16t^2 + 128t - 240 = 0$. By factoring or by using the quadratic formula, the solution is $t = 3$ and $t = 5$.
2. The equation is $(4 + 2x)(6 + 2x) = 80$. It becomes $4x^2 + 20x + 24 = 80$ and then $4x^2 + 20x - 56 = 0$ with a solution of $x = 2$.
3. The equation is $(0.99 + 0.10x)(200{,}000 - 10{,}000x) = 223{,}500$. It becomes $-1{,}000x^2 + 10{,}100x - 25{,}500 = 0$ with a solution of $x = 5$.
4. The equation is $x(x - 35) = 884$ becomes $x^2 - 35x - 884 = 0$ with a solution of $x = 52$.
5. Since the perimeter is 44, $2x + 2w = 44$ or $2w = 44 - 2x$ or $w = 22 - x$. The area of the rectangle is $120 = lw = x(22 - x)$. This becomes the equation $120 = x(22 - x)$ or $120 = 22x - x^2$ or $x^2 - 22x + 120 = 0$. This can be factored as $(x - 12)(x - 10) = 0$ to get the solution $x = 12$ or $x = 10$. (Or it can be solved with the quadratic formula.)

CHAPTER 4

Section 4.1
A
1. (2)   3. (4)   5. (1)   7. (1)   9. (4)
2. (1)   4. (1)   6. (4)   8. (3)   10. (2)

B
1. $(4, 1)$ and $(1, 3)$
2. $7 + 1 = 8$, $7 - 1 = 6$ so the solution is $(7, 1)$.

Answers and Solution Hints to Practice Exercises

3. Any pair that satisfies the first equation cannot also satisfy the second equation since if the left-hand side evaluates to 25, then it cannot equal 26.
4. $c = 25$ since $4 + 3 \cdot 7 = 25$
5. There are many, but $(6, -2)$ is the quickest. Since the second equation is obtained by multiplying both sides of the original equation by 3, it has the same solution set, including $(6, -2)$.

Section 4.2
A
1. (4)   3. (1)   5. (3)   7. (4)   9. (2)
2. (3)   4. (4)   6. (3)   8. (1)   10. (1)

B
1. There are many answers to this. One is the equation $5x + 10y = 25$ obtained by adding the two equations together. Any equation that has $(1, 2)$ as its solution is acceptable.
2. By adding $2x$ to both sides of the top equation, it becomes $y = 2x + 4$. Then the y in the second equation can be replaced with $2x + 4$ to solve for the answer $(3, 10)$.
3. By subtracting the two equations, it becomes $0 = 5x - 25$, which becomes $5x = 25$ or $x = 5$. By substituting the y in the second equation with $2x - 9$, it becomes $2x - 9 = -3x + 16$, which also becomes $5x = 25$. The solution is $(5, 1)$.
4. The system is $x + y = 17$, $x - y = 11$. Add them together to get $2x = 28$, $x = 14$, $y = 3$.
5. They are both correct. Adding makes the equation $4y = -24$, or $y = -6$. Subtracting makes $10x = 40$, or $x = 4$. Either way the solution $(4, -6)$ can be obtained.

Section 4.3
A
1. (3)   3. (2)   5. (3)   7. (1)   9. (1)
2. (2)   4. (4)   6. (4)   8. (4)   10. (4)

B
1. Eliminating the x requires less work because it can be done by just changing the first equation, multiplying both sides of the first equation by -3; however, to eliminate the y requires changing both equations because 7 is not a multiple of 5.
2. Multiply both sides of the first equation by -2. It becomes $-2x - 2y = -80$. Add to the second equation to eliminate the x, $2y = 56$ so $y = 28$ and $x = 12$.

Answers and Solution Hints to Practice Exercises

3. It is not accurate. Most ordered pairs do not satisfy either equation, for instance (0, 0). But every ordered pair that satisfies the first equation, like (0, 7) will also satisfy the second equation and so will satisfy the system of equations. There still are an infinite number of ordered pairs, but not all ordered pairs satisfy the system.
4. There are no solutions. The top equation is equivalent to $10x - 4y = 10$ so any ordered pair that satisfies this equation will not satisfy the equation $10x - 4y = 11$. When using elimination, it becomes $0 = -1$, which is never true.
5. Victoria will continue by adding the two equations together to eliminate the x. Porter will continue by subtracting the changed second equation from the first equation to eliminate the x. Either way, the answer (6, 2) will be correct.

Section 4.4
A
1. (4)   3. (1)   5. (2)   7. (1)   9. (1)
2. (1)   4. (3)   6. (2)   8. (3)   10. (2)

B
1. x is the number of red roses, and y is the number of pink roses. The equations are $x + y = 24$, $3x + 2y = 68$. The solution is $x = 20$, $y = 4$ so there are 20 red roses and 4 pink roses.
2. (a) $60; (b) $40; (c) $50; (d) 4 cones and 6 shakes. This could be done with the system of equations $x + y = 10$, $4x + 6y = 52$, or by noticing that if it were 5 and 5, it would cost $50 so to make it cost $2 more, switch one of the cones with a shake to become 4 cones and 6 shakes.
3. x is the number of ounces of marshmallows, and y is the number of ounces of cookies. The cost equation is $2x + 3y = 29$. The calorie equation is $500x + 400y = 4,800$. The solution is 4 ounces of marshmallows and 7 ounces of cookies.
4. x is the cost of a burger, and y is the cost of an order of French fries. $3x + 2y = 24$, $5x + 1y = 33$, solution is $x = 6$ and $y = 3$ so a burger costs $6 and an order of French fries costs $3.
5. x is the number of bicycles, and y is the number of tricycles. $x + y = 58$, $2x + 3y = 134$. The solution is 40 bicycles and 18 tricycles.

Chapter 5

Section 5.1

A
1. (4)　　3. (1)　　5. (2)　　7. (1)　　9. (3)
2. (1)　　4. (3)　　6. (2)　　8. (4)　　10. (2)

B

1. The table could look like

x	y
0	8
1	6
2	4

2. For the x-intercept, $4x - 6(0) = 24$ has a solution of $x = 6$. For the y-intercept $4(0) - 6y = 24$ has a solution of $y = -4$. The x-intercept is $(6, 0)$ and the y-intercept is $(0, -4)$

3. It is $y = 3$ since it contains points $(0, 3)$, $(1, 3)$, and all points that have 3 as the y-coordinate.

Answers and Solution Hints to Practice Exercises

4. Using a chart or the intercept method, graph the two lines. The intersection is located at (5, 1).

5. Substituting either ordered pair into the equation will lead to $k = 8$.

Section 5.2
A

1. (2)	3. (1)	5. (3)	7. (2)	9. (4)
2. (4)	4. (1)	6. (1)	8. (3)	10. (3)

B

1. The product of the slopes of two perpendicular lines is equal to –1. It is not necessary to calculate the slopes for this question, but it is permitted. The slope of line 1 is $-\frac{3}{5}$ so the slope of line 2 is $\frac{5}{3}$ and $-\frac{3}{5} \cdot \frac{5}{3} = -1$.

2. Side AB has the greatest slope. BC has a slope of 0, AB has a slope of $\frac{5}{2}$, and AC has a slope of $\frac{5}{4}$.

3. From (–5, 2) go three to the right and two up to get to (–2, 4), then to (1, 6), then to (4, 8).

4. To get from (–7, 2) to (–3, –1) you have to go 4 to the right and 3 down. From (–3, –1) go 4 right and 3 down to get to (1, –4). Again go 4 to the right and 3 down to get to (5, –7). So the answer is 5.

5. The x-intercept is (3, 0), and the y-intercept is (0, –2). Using the slope formula, the slope of the line is $\frac{2}{3}$.

Section 5.3
A

1. (1)	3. (2)	5. (1)	7. (4)	9. (3)
2. (2)	4. (2)	6. (4)	8. (3)	10. (3)

B

1. Slope is 3, and the y-intercept is at $(0, -9)$.

2. Edwin is correct. After dividing both sides of the equation by 2 to get it into slope-intercept form, it becomes $y = \frac{3}{2}x - 4$.

3. $(0, 3)$. To get from $(4, 5)$ to $(2, 4)$ go down one and left two. Then go down one and left two from $(2, 4)$ and get to $(0, 3)$, which is the y-intercept.

4. The solution is $(3, 2)$.

5. They are all of the form $y = mx + 2$ where m is positive.

Section 5.4
B
1.

Answers and Solution Hints to Practice Exercises

2.

[Scratchpad screen showing graph of $f1(x) = \frac{-3}{4} \cdot x + 2$]

3. The x-coordinate of the intersection is 4, so $x = 4$ is the solution to the equation.

[Scratchpad screen showing intersection at $(4, 5)$ of $f2(x) = 5$ and $f1(x) = 2 \cdot x - 3$]

4.

[Scratchpad screen showing intersection at $(-5, 6)$ of $f1(x) = \frac{-4}{5} x + 2$ and $f2(x) = 2 \cdot x + 16$]

5.

[Scratchpad screen showing intersection at $(8, 3)$ of $f1(x) = \frac{1}{4} x + 1$ and $f2(x) = -x + 11$]

Answers and Solution Hints to Practice Exercises

Section 5.5
A
1. (2)   3. (3)   5. (4)   7. (2)   9. (2)
2. (3)   4. (1)   6. (3)   8. (4)   10. (2)

B
1. The equation is $y = \frac{2}{3}x + 2$, which passes through (5, 5.3) since $\frac{2}{3}(5) + 2 = 5.3$.

2. The equation is $y = -\frac{1}{4}x + 8$, which passes through (18, 3.5) since the equation $3.5 = -\frac{1}{4}a + 8$ becomes $-4.5 = -\frac{1}{4}a$ becomes $18 = a$.

3. The equation is $y = -\frac{4}{5}x + 10$. When $x = 0$, $y = 10$ so the y-intercept is (0, 10). When $y = 0$, $0 = -\frac{4}{5}x + 10$ becomes $-10 = -\frac{4}{5}x$ becomes $x = \frac{50}{4} = 12.5$ so the x-intercept is (12.5, 0).

4. One way to do this is to use the slope formula. The slope is known to be $\frac{1}{3}$ so $\frac{1}{3} = \frac{a-7}{12-6} = \frac{a-7}{6}$. If $\frac{1}{3} = \frac{a-7}{6}$, cross multiply to get $6 = 3(a-7) = 3a - 21$ or $27 = 3a$ or $a = 9$. Another way to do this would be to substitute 12 for x and a for y into the equation to get $a = \frac{1}{3}(12) + 5 = 4 + 5 = 9$.

5. The equation of the line is $y = \frac{1}{5}x + 3$. The line passes through (6, 4.2), which is 0.8 away from (6, 5). The line passes through (7, 4.4), which is 0.6 away from (7, 5). The line passes through (9, 4.8) which is 0.8 away from (9, 4). The sum of the three vertical segments is $0.8 + 0.6 + 0.8 = 2.2$.

Section 5.6
A
1. (4)   3. (2)   5. (4)   7. (3)   9. (4)
2. (3)   4. (1)   6. (4)   8. (1)   10. (2)

B
1. (a) $H = 4Y + 6$; (b) 46 feet tall
2. (a) $P = \frac{3}{2}W + 3$; (b) 3; (c) 14 weeks
3. (a) $V = 300D + 100$; (b) 4,300; (c) 9,400

Answers and Solution Hints to Practice Exercises

4. (a) $P = 10T + 230$; (b) 2,050
5. (a) $E = -4M + 200$; (b) 140; (c) 8:20; (d) 8:50

CHAPTER 6

Section 6.1
A
1. (4)   3. (4)   5. (3)   7. (2)   9. (2)
2. (2)   4. (4)   6. (1)   8. (1)   10. (3)

B
1. $b = 6$ and c can be anything.
2. The equation can be $y = (x - 1)(x + 4) = x^2 + 3x - 4$ so one answer is $b = 3$ and $c = -4$.
3. Because the axis of symmetry passes through the vertex, reflect the point $(6, 0)$ over the line $x = 4$ to get $(2, 0)$. Also reflect $(1, 5)$ over the line $x = 4$ to get $(7, 5)$.
4. The vertex is $(3, 0)$ and the one x-intercept is also $(3, 0)$.
5. $x = 3$. One way to do this is to expand $y = (x - 1)^2 - 4(x - 1) - 2 = x^2 - 2x + 1 - 4x + 4 - 2 = x^2 - 6x + 3$ and then use

$$x = -\frac{b}{2a} = -\frac{-6}{2 \cdot 1} = 3$$

Section 6.2
A
1. (2)   3. (3)   5. (2)   7. (3)   9. (2)
2. (3)   4. (2)   6. (1)   8. (3)   10. (4)

B
1. Approximately 3.5 and 6.5. More accurate answers are 3.3 and 6.7.
2. It becomes the equation $x^2 - x - 1 = 0$ with solutions 1.62 and -0.62.
3. The horizontal line is $y = 5$. The parabola has x-intercepts $(1, 0)$ and $(4, 0)$, so its equation is $y = (x - 1)(x - 4) = x^2 - 5x + 4$. The intersection points solve the equation $x^2 - 5x + 4 = 5$.
4. Graph $y = -16x^2 + 96x + 112$ and find x–intercept at $x = 7$.
5. All three are correct since the equations are all equivalent. One can be turned into the other using the addition property of equality.

Section 6.3
A
1. (2)   3. (3)   5. (4)   7. (2)   9. (3)
2. (2)   4. (2)   6. (4)   8. (1)   10. (1)

B
1. (–2, 2) and (4, 8)
2. (0, 5) and (3, 8)
3. (6, 4) and (2.67, 5.11)
4. Jimena is correct. The second intersection occurs off the screen. The two solutions are (–1, 3) and (3, 27).
5. (a) Any number greater than –1; (b) any number less than –1.

Section 6.4
B
1. At $t = 1$ and $t = 5$. $336 = -16t^2 + 96t + 256$, which becomes $16t^2 - 96t + 80 = 0$. Then $16(t - 1)(t - 5) = 0$.
2. The graphs intersect at (5, 5), which means that for 5 days it costs $5,000 to use either construction company.
3. The graphs intersect at the point (1, 34) so after 1 second the stunt man and the elevator are both 34 feet high and the stunt man has landed on top of the elevator.
4. The solution is (2, 36) so after 2 seconds the superhero catches the car 36 feet in the air.
5. The solution to the system of equations is (6, 27) so after 6 seconds the iguanas will crash into each other.

CHAPTER 7

Section 7.1
A
1. (3)   3. (3)   5. (2)   7. (3)   9. (4)
2. (1)   4. (3)   6. (2)   8. (2)   10. (1)

B
1. The solution set is $x \geq 5$. On a number line, graph a ray with initial point at 5 and going to the right.

2. Lorenzo is correct. When dividing both sides of an inequality by a negative, the direction of the inequality must be switched to keep it true.
3. It is valid. Rather than divide by a negative and switch the direction of the inequality sign, adding or subtracting the same thing from both sides of an inequality does not require changing the direction of the sign.
4. Subtract $2x$ from both sides and add 2 to both sides to transform inequality into $3x \geq 12$ with solution set $x \geq 4$.
5. $1.20x \leq 60$ or $x \leq 50$. So the most the meal can cost before the tip is $50.

Answers and Solution Hints to Practice Exercises

Section 7.2
A
1. (4)   3. (2)   5. (2)   7. (2)   9. (3)
2. (1)   4. (1)   6. (2)   8. (4)   10. (2)

B
1. It is like the graph of $x + y <= 4$

2. $y < \frac{1}{2}x + 3$

3. He does not have enough information. (0, 0) is on the line but he needs to check a point that is not on the line.

4. Divide both sides by –1 and switch the direction of the inequality to get $y > -2x + 4$

5. It is like the equation $2x + 3y \leq 12$. Since you can buy fractions of a pint, this is shaded unlike question B1.

Section 7.3
A
1. (1) 3. (1) 5. (1) 7. (2) 9. (2)
2. (2) 4. (3) 6. (2) 8. (3) 10. (4)

B
1.

2. $y \leq x + 1, y \leq \frac{1}{2}x + 2$
3. $x + y \leq 50, .75x + 2.50y \leq 90$
4. $x + y \leq 300, 8x + 12y \geq 3{,}200$
5.

Chapter 8

Section 8.1
A
1. (2) 3. (3) 5. (3) 7. (1) 9. (4)
2. (3) 4. (4) 6. (4) 8. (2) 10. (1)

B
1. $607.75
2. A) 12,553 after 10 years. B) Using guess-and-check, 14,065 after 15 years.
3. 6.25 mg

427

Answers and Solution Hints to Practice Exercises

4. At $h = 10$, $T = 32.25$. At $h = 11$, $T = 31.58$. So between 10 and 11 hours.
5. Julia is correct. When x is negative, 5^x is greater than 6^x. For example, $5^{-2} = \frac{1}{25}$, which is greater than $6^{-2} = \frac{1}{36}$.

Section 8.2
A
1. (4)   3. (3)   5. (4)   7. (4)   9. (1)
2. (3)   4. (1)   6. (1)   8. (2)   10. (2)

B
1. $b = 3$ since $3^5 = 243$
2.

3. No. When x is negative, $2^x > 3^x$. For example $2^{-2} = \frac{1}{4}$ while $3^{-2} = \frac{1}{9}$.
4. $(0,1)$. Graph both and they intersect at $(0, 1)$.
5. No. For numbers between 0 and 1, $x^2 + 1 < 2^x$. For example, $.5^2 + 1 = 1.25$ but $2^{.5} = 1.41$. The curves cross again at approximately $x = 4.3$, so for any $x > 4.3$, $2x > x^2 + 1$ also.

Section 8.3
A
1. (1)   3. (2)   5. (2)   7. (1)   9. (4)
2. (1)   4. (3)   6. (1)   8. (3)   10. (3)

B
1. It resembles an exponential equation since it is somewhat flat at the beginning and then gets much steeper very quickly.
2. There is not enough information. Helpful would be some points with negative x-coordinates to see if they are close to the x-axis like an exponential curve or if they form a picture with symmetry like a parabola for a quadratic equation.

3. It is an exponential equation. Because the curve is always decreasing, it is exponential decay. The number being raised to the exponent will be between 0 and 1.
4. It is an exponential equation. Since $\frac{85}{50} = 1.7$, each number in the y column is 1.7 multiplied by the number above it. The missing number is $85 \cdot 1.7 = 144.5$.
5. A chart would have the y-values going from 30 to 24 to 19.2 to 15.36, which has the shape of curve with exponential decay.

Section 8.4
A
1. (2)   3. (4)   5. (2)   7. (3)   9. (1)
2. (3)   4. (1)   6. (4)   8. (3)   10. (3)

B
1. A) 901 million. B) By guess and check or with the graphing calculator, when $t = 8$, which is year 2018.
2. A) It will be approximately 52 degrees. B) By guess and check or with the graphing calculator, approximately 17 minutes.
3. A) For 10 years bank A will have $500 while bank B will have $179.08. B) For 54 years bank A will have $2,260 while bank B will have $2,325.50.
4. 1024 line segments
5. A) 147 pounds. B) Using guess and check, after 20 weeks he will be able to bench press about 250 pounds.

CHAPTER 9

Section 9.1
A
1. (1)   3. (4)   5. (2)   7. (1)   9. (3)
2. (4)   4. (2)   6. (3)   8. (2)   10. (4)

B
1. $S = 5,000Y + 40,000$
2. The 2 means that it is $2 for each mile. The 4 means that it is $4 to get into the taxi.
3. $H = 3Y + 5$ or $H = 5 + 3Y$
4. $W = 260 - 2M$
5. Many correct answers. One answer is the price of getting into a carnival is $20 while each ride costs $6. C is the cost of getting into the carnival and going on N rides.

Answers and Solution Hints to Practice Exercises

Section 9.2
A
1. (1) 3. (4) 5. (4) 7. (3) 9. (3)
2. (1) 4. (2) 6. (1) 8. (1) 10. (4)

B
1. $P = 300{,}000(1.04)^t$
2. $S = 64(.75)^t$
3. $H = 40(1.15)^t$
4. 200,000 represents the $200,000 profits in the company's first year. 0.36 is the growth rate, so 1.36 is the growth factor or 1 plus the growth rate.
5. 300 means that when the training begins it takes her 300 minutes to run the marathon. 0.05 is the decay rate, so 0.95 is the growth (or decay) factor or 1 minus the decay rate.

Chapter 10

Section 10.1
A
1. (3) 3. (3) 5. (1) 7. (4) 9. (1)
2. (4) 4. (3) 6. (3) 8. (2) 10. (2)

B
1. Yes, this is a function. For each input value, there is only one output value. There are no repeats in the x-coordinates.
2. (a) $4 + 1 = 5$; (b) $f(3) = 2$; (c) $f(4) = 3$
3. The only value is $a = 6$.
4. No, it is not. $f(3)$ cannot be both 4 and 5.
5. Mia is correct. If there are no repeated y-values, the number will be the same. If there are repeated y-values, there will be fewer numbers in the range. There is no way to have more numbers in the range since that would require having repeated x values, which is not permitted for a function.

Section 10.2
A
1. (3) 3. (4) 5. (4) 7. (4) 9. (2)
2. (1) 4. (1) 6. (2) 8. (1) 10. (3)

B
1. Tyson is correct. It does pass the vertical line test. Two x values may correspond to the same y value.
2. Domain is $0 \leq x \leq 6$; range is $0 \leq y \leq 3$.
3. $f(g(3)) = f(5) = 2$
4. 2

Answers and Solution Hints to Practice Exercises

5. $a = 2$ and $a = -2$ since the points $(2, 4)$ and $(-2, 4)$ are both on the graph.

Section 10.3
A
1. (4)	**3.** (3)	**5.** (2)	**7.** (2)	**9.** (1)
2. (1)	**4.** (4)	**6.** (4)	**8.** (3)	**10.** (1)

B
1. (a) 40; (b) $(x + 1)^2 + 8(x + 1) + 7 = x^2 + 10x + 16$
2. (a) $f(11) = 19$; (b) $g(3) = 11$.
3. Elizabeth is right. $h(g(f(x))) = [2(x + 2)]^2$. The correct expression is $g(h(f(x)))$.
4.

5. Since $g(7) = 15$, $2f(3) + 1 = 15$, so $f(3)$ must be 7.

Section 10.4
A
1. (4)	**3.** (2)	**5.** (4)	**7.** (3)	**9.** (3)
2. (1)	**4.** (3)	**6.** (1)	**8.** (2)	**10.** (2)

B
1. (a) $g(3) = f(3) + 1 = 1 + 1 = 2$; (b) $h(3) = f(3 + 1) = f(4) = 5$.
2.

3. (a) The domain is $2 \leq x \leq 7$ because shifting up does not affect the domain. (b) The domain is $-2 \leq x \leq 3$ since the graph is shifted four units to the left.

431

Answers and Solution Hints to Practice Exercises

4.

5. Since the graph of $g(x)$ is the graph of $f(x)$ shifted to the left two units, it can be expressed as $g(x) = f(x + 2)$.

CHAPTER 11

Section 11.1
A
1. (2) 3. (2) 5. (4) 7. (2) 9. (4)
2. (1) 4. (3) 6. (3) 8. (1) 10. (3)

Section 11.2
A
1. (3) 3. (3) 5. (4) 7. (1) 9. (2)
2. (2) 4. (2) 6. (1) 8. (4) 10. (3)

B
1. 5, 9, 13, 17
2. $a_{18} = 3 \cdot 58 = 174$, $a_{19} = 3 \cdot 174 = 522$
3. $a_1 = 50$, $a_n = 5 + a_{n-1}$ for $n > 1$
4. $a_1 = 6$, $a_n = 2a_{n-1}$ for $n > 1$
5. 3, 4, $2 \cdot 4 + 3 = 11$, $2 \cdot 11 + 4 = 26$, $2 \cdot 26 + 11 = 63$

Section 11.3
A
1. (2) 3. (4) 5. (2) 7. (2) 9. (4)
2. (3) 4. (3) 6. (1) 8. (3) 10. (1)

B
1. They are both right. Liam did the closed form, and Cecilia did the recursive form.
2. $a_{100} = 26 - 7(99) = -667$
3. (a) $a_n = \dfrac{1}{633,600} \cdot 2^n$ (b) $a_{38} = \dfrac{1}{633,600} \cdot 1 \cdot 2^{38} = 433,835$ miles. Yes, it will reach the moon.

4. $b_n = 1 \cdot 2^{n-1}$ so $b_{20} = 1 \cdot 2^{19} = 524{,}288$. $a_n = 1{,}000 + 200(n-1)$ so $a_{20} = 1{,}000 + 200 \cdot 19 = 4{,}800$. So, $b_{20} > a_{20}$.
5. The closed form will be faster. The recursive form would require calculating all 100 values of the sequence.

Chapter 12

Section 12.1

A

1. (1)   3. (1)   5. (4)   7. (3)   9. (3)
2. (4)   4. (3)   6. (2)   8. (4)   10. (1)

B

1.
   (a)

   (b) $B = 1.35T - 18.34$

   (c) If $T = 55$, $B = 1.35 \cdot 55 - 18.34 = 55.91$ or approximately 56 birds.

2.
(a)

(b) $C = -3.39P + 232.41$

(c) $C = -3.39 \cdot 31 + 232.41 = 127.32$ or approximately 127 cars sold.
3. Because the five points all lie on a line, the line of best fit is just the equation of the line through any of the two points. The equation is $y = \frac{2}{3}x + 4$.
4. The one with (3, 1) will have a slope closer to zero since the line of best fit for the graph with that point will need to be closer to that point.

Section 12.2
A
1. (1)	**3.** (2)	**5.** (2)	**7.** (4)	**9.** (2)
2. (4)	**4.** (4)	**6.** (1)	**8.** (3)	**10.** (4)

B
1. The scatter plot on the left has the higher correlation coefficient since the points more closely resemble a straight line.
2. Jett is correct. If the line of best fit has a negative slope, the correlation coefficient will be negative.

Answers and Solution Hints to Practice Exercises

3. The correlation coefficient will be positive since the line of best fit will have a positive slope. Also it will be less than 1 since the points do not all lie on the same line.
4. Pick any five x-coordinates and calculate the y-coordinates with the equation $y = x + 3$. That way all five points will be on the line of best fit so the correlation coefficient will be +1.

Section 12.3
A

| 1. (3) | 3. (1) | 5. (1) | 7. (2) | 9. (1) |
| 2. (2) | 4. (3) | 6. (2) | 8. (3) | 10. (2) |

B

1.

(a) The exponential model is most appropriate. The equation is $P = 1.07 \cdot 1.11^T$. (b) When $T = 20$, $P = 8.63$ or 8.63 million people.

2.

(a) The exponential decay model is most appropriate. The equation is $D = 78.99 \cdot 0.90^M$. (b) When $M = 40$, $D = 1.17$ degrees.

Answers and Solution Hints to Practice Exercises

3.

(a) The quadratic model is most appropriate. The equation is $F = -0.001x^2 + 0.303x - 10.471$. (b) The vertex of the parabola has an x-coordinate of $-\dfrac{0.303}{2 \cdot (-0.001)} = 151.5$ so \$151.50 is the price that will generate the most profit.

4. (a) No. A quadratic model is more appropriate because the points get closer to zero and then further from zero. On the other hand, in a linear model the points either keep getting higher or lower. (b) The line of best fit was $y = 4$. The correlation coefficient was 0.

5. An exponential equation is a better fit. When finding the equation, the r^2 value for the exponential equation is slightly closer to +1 than the r^2 value for the quadratic equation.

Chapter 13

Section 13.1
A

1. (4)	3. (4)	5. (2)	7. (3)	9. (1)
2. (2)	4. (1)	6. (1)	8. (4)	10. (4)

436

B
1. One number is 19 so there will be three 19s making it the mode. The sum of all eight numbers is 160 so the other missing number is 23.
2. The second team must have many players that are close in height to six feet while the first team has some people much shorter than six feet and others much taller than six feet.
3. There can be two 81s, and then a 90 as the first three grades. To get a mean of 90, the sum of all five grades must be 450. The 81, 81, and 90 have a sum of 252 so the other two grades must have a sum of 198. They can't be 99 and 99 since that would change the mode, but they can be 98 and 100.
4. Five out of eleven. The first quartile is 4 and the third quartile is 15 so the numbers 8, 8, 9, 9, and 14 are between them.
5. 94. Her grades can be 90, 90, 90, 100, 100 and still meet the conditions of both median and mode being 90.

Section 13.2
A
1. (2) 3. (2) 5. (2) 7. (3) 9. (1)
2. (3) 4. (4) 6. (1) 8. (4) 10. (2)

B
1.

2. min = 2, Q1 = 4, med = 8, Q3 = 12, max = 20

3. The players on the first team all have heights very close to 72 inches, whereas on the other team there are players that are much shorter and players that are much taller than 72 inches.

Answers and Solution Hints to Practice Exercises

4. Any data set that when arranged from smallest to largest has 10, x, 13, x, x, 18, x, x, 25, x, 30 will produce this box plot.

5.

Glossary of Algebra I Terms

A

Addition property of equality A property of algebra that states that when equal values are added to both sides of a true equation, the equation continues to be true. To solve the equation $x - 2 = 5$, add 2 to both sides of the equation by using the addition property of equality.

Arithmetic sequence A number sequence in which the difference between two consecutive terms is a constant. The sequence 2, 5, 8, 11, 14, ... is an arithmetic sequence because the difference between consecutive terms is always 3.

Axis of symmetry An imaginary vertical line that passes through the vertex of a parabola. The equation for the axis of symmetry of a parabola is defined by $y = ax^2 + bx + c$ is $x = -\frac{b}{2a}$.

B

Base The number being raised to a power in an exponential expression. In the expression $2 \cdot 3^x$, the 3 is the base.

Binomial A polynomial with only two terms. $3x + 5$ is a binomial.

Box plot A graphical way to summarize data. The five numbers represented by the minimum, first quartile, median, third quartile, and maximum are graphed on a number line. A line segment connects the minimum to the first quartile. A rectangle is drawn around the first quartile and third quartile with a vertical line at the median. A line segment connects the third quartile to the maximum.

C

Closed form defined sequence A formula that defines the nth term of a sequence. The formula $n = 3 + 2(n - 1)$ is a closed form definition of a sequence. To get the 50th term of the sequence, substitute 50 for n in the definition.

Coefficient A number multiplied by a variable expression. In the expression $5x + 2$, 5 is the coefficient of x.

Common difference In an arithmetic sequence, the difference between consecutive terms. The common difference in the sequence 2, 5, 8, 11, 14, ... is 3.

Common ratio In a geometric sequence, the ratio between consecutive terms. The common ratio in the sequence 2, 6, 18, 54, 162, ... is 3 since 162/54 = 54/18 = 18/6 = 6/2 = 3.

Commutative property of addition or multiplication The law from arithmetic that states that the order in which two numbers are added or multiplied does not matter. Because of the commutative property of addition, $5 + 2 = 2 + 5$.

Completing the square A method of solving a quadratic equation that involves turning one side of the equation into a perfect square trinomial.

Constant A number that does not have a variable part. In the expression $5x + 2$, 2 is a constant.

Correlation coefficient A number represented by r that measures how well a curve of best fit matches the points in a scatter plot. When the correlation coefficient is very close to 1 or to -1, the curve is a very good fit.

D

Degree of a polynomial The exponent on the highest power of a polynomial. In the polynomial $x^3 - 2x^2 + 5x - 2$, the degree is 3 since the highest power is a 3.

Difference of perfect squares Factoring a quadratic binomial in the form $x^2 - a^2$ into $(x - a)(x + a)$. For example, $x^2 - 9 = (x - 3)(x + 3)$.

Distributive property of multiplication over addition The rule that allows expressions of the form $a(b + c)$ to become $a \cdot b + a \cdot c$. For example, $2(3x + 5) = 6x + 10$.

Division property of equality A property of algebra that states that when both sides of a true equation are divided by the same non-zero number, the equation continues to be true. To solve the equation $2x = 8$, divide both sides of the equation by 2 using the division property of equality.

Domain The numbers that can be input into a function. When the function is defined as a set of ordered pairs or as a graph, the domain is the set of x-coordinates.

Dot plot A graphical way of representing a data set where each piece of data is represented with a dot.

E

Elimination method A way of solving a system of linear equations by combining the two equations in such a way as to eliminate one of the variables. In the set of equations,
$x + 2y = 12$
$3x - 2y = -4$
the y variable is eliminated by adding the two equations together.

Equation Two mathematical expressions with an equals sign between them. $3x + 2 = 8$ is an equation.

Exponential equation An equation in which the variable is an exponent. $2 \cdot 3^x = 18$ is an exponential equation.

Exponential function A function in which the variable is an exponent. $f(x) = 2 \cdot 3^x$ is an exponential function.

Expression Numbers and variables that are combined with the operations from math.

F

Factoring a polynomial Finding two polynomials that can be multiplied to become another polynomial. The polynomial $x^2 + 5x + 6$ can be factored into $(x + 2)(x + 3)$.

Factors The polynomials that evenly divide into a polynomial. The factors of $x^2 + 5x + 6$ are $(x + 2)$ and $(x + 3)$.

First quartile The number in a data set that is bigger than just 25% of the numbers in the set.

FOIL A way of multiplying two binomials of the form $(a + b)(c + d)$ where (F)irst the a and c are multiplied, then the (O)uters a and d are multiplied, then the (I)nners b and c are multiplied, and finally the (L)asts b and d are multiplied. Then the four results are added together. The product of $(x + 2)$ and $(x + 3)$ is $x^2 + 3x + 2x + 6 = x^2 + 5x + 6$ by this method.

Function Something that takes numbers as inputs and outputs numbers. Functions are often labeled with the letters f or g. The notation $f(2) = 7$ means that when the number 2 is input into function f, it outputs the number 7.

G

Geometric sequence A number sequence in which the ratio between two consecutive terms (what you get when you divide one term by the term before it) is a constant. The sequence 2, 6, 18, 54, 162, ... is a geometric sequence because the ratio between consecutive terms is always 3.

Graph A visual way to describe the solution set to an equation. Each solution to the equation corresponds to an ordered pair that is graphed as a point on the coordinate plane. Each of the ordered pairs that satisfy an equation produce a point on the coordinate plane and the collection of all the points is the graph of the equation.

Greatest common factor The largest expression that divides evenly into two or more monomials. The greatest common factor of $6x^2$ and $8x^3$ is $2x^2$.

Growth rate In an exponential expression $a \cdot b^x$, the b is the growth rate. For example, in the equation $y = 500 \cdot 1.05^x$, the growth rate is 1.05.

H

Histogram A way of representing data with repeated values. Each value is represented by a bar whose height corresponds to the number of times that value is repeated. There are no spaces between the bars.

I

Increasing A function is increasing on an interval if making the input value larger also makes the y output larger. On a graph, an increasing function "goes up" from left to right.

Inequality Like an equation, but there is a $<$, $>$, $\leq$, or $\geq$ sign between the two expressions. $x + 2 > 5$ is a one-variable inequality. $y \leq 2x + 6$ is a two-variable inequality.

Interquartile range The difference between the number that is the third quartile of a data set and the number that is the first quartile of a data set.

Isolating a variable A variable is isolated when it is by itself on one side of an equation. In the equation $x + 2 = 5$, the x is not yet isolated. Subtracting 2 from both sides of the equation transforms the original equation into $x = 3$ with the x now isolated.

L

Like terms Terms that have the same variable part. They can be combined by adding or subtracting. $2x^2$ and $3x^2$ are like terms. $2x^2$ and $3x^3$ are not like terms.

Line of best fit A line that comes closest to the set of points in a scatter plot.

Linear equation An equation in which the greatest exponent is a 1. The equation $2x + 3 = 7$ is a linear equation.

Linear function A function in which the greatest exponent is a 1. The function $f(x) = 2x + 3$ is a linear function.

M

Mean The average of the numbers in a data set. Calculate the mean by adding all the numbers and dividing the total by the number of numbers in the set.

Median The middle number in a data set after it has been arranged from least to greatest. If there are an even number of numbers in the data set, the median is found by adding the two middle numbers and dividing by 2.

Mode The most frequent number in a data set.

Monomial A mathematical expression that has a coefficient and/or a variable part. $3x^2$ is a monomial. $3 + x^2$ is not a monomial.

Multiplication property of equality A property of algebra that states that when equal values are multiplied by both sides of a true equation, the equation continues to be true. To solve the equation $(1/2)x = 5$, multiply both sides of the equation by 2, using the multiplication property of equality.

O

Ordered pair Two numbers written in the form (x, y). An ordered pair can be a solution to a two-variable equation. For example, $(2, 5)$ is one solution to the equation $y = 2x + 1$. Ordered pairs can be graphed on the coordinate axes by locating the point with the x-coordinate equal to the x value and the y-coordinate equal to the y value.

P

Parabola A "U"-shaped curve that is the graph of the solution set of a quadratic equation.

Perfect square trinomial A quadratic trinomial of the form $x^2 + bx + (b/2)^2$, which can be factored into $(x + b/2)^2$. For example, $x^2 + 6x + 9 = (x + 3)^2$.

Piecewise function A function that has multiple rules for determining output values from input values, depending on what the input values are. If the function $f(x)$ is defined as

$$f(x) = \begin{cases} 2x+1 & \text{if } x < 0 \\ x^2 & \text{if } x \geq 0 \end{cases}$$

then $f(-3) = 2(-3) + 1 = -5$ and $f(5) = 5^2 = 25$.

Polynomial The sum of one or more monomials. Each term of the polynomial has the form ax^n. $3x^4 + 2x^3 - 3x^2 + 6x - 1$ is a polynomial.

Q

Quadratic equation An equation in which the highest power on a variable is a 2. $x^2 + 5x + 6 = 0$ is a quadratic equation.

Quadratic formula The formula $x = \dfrac{(-b \pm \sqrt{(b^2 - 4ac)})}{2a}$. This formula can be used to find the two solutions to the quadratic equation $ax^2 + bx + c = 0$.

Quadratic function A function in which the highest power on a variable is 2. $f(x) = x^2 + 5x + 6$ is a quadratic function.

Quadratic polynomial A polynomial in which the highest power on a variable is 2. $x^2 + 5x + 6$ is a quadratic polynomial.

R

Range The set of values that can be output from a function is the range of that function.

Recursively defined sequence A way to define a sequence in which the first term or terms of the sequence are given and a formula is given for calculating the next term based on the previous term or terms.

$a_1 = 5$
$a_n = 3 + a_{n-1}$ for $n > 1$

is a recursive definition for the sequence 5, 8, 11, 14, ...

Glossary of Algebra I Terms

Regression Finding a curve that best fits a scatter plot. Three types of regression are linear, quadratic, and exponential.

Residual plot A set of points that represent how far points on a graph deviate from a curve of best fit.

Roots The roots of an equation are the values that solve that equation. The roots of $x^2 + 5x + 6 = 0$ are -3 and -2.

S

Sequence A list of numbers that usually has some kind of pattern.

Slope The slope of a line is a measure of the line's steepness. The equation for the slope of a line that passes through the two points (x_1, y_1) and (x_2, y_2) is $m = \dfrac{y_2 - y_1}{x_2 - x_1}$.

Slope-Intercept form An equation in the form $y = mx + b$ where m and b are numbers in slope intercept form. $y = 2x - 1$ is in slope intercept form. When a two-variable equation is in slope intercept form, the graph of the equation has a y-intercept of $(0, b)$ and a slope of m.

Solution set The set of numbers or ordered pairs that satisfies an equation. The solution set of $x + 2 = 5$ is $\{3\}$. The solution set of $x + y = 10$ has an infinite number of ordered pairs in its solution set, including $(2, 8)$, $(3, 7)$, and $(4, 6)$.

Substitution method A method for solving a system of equations in which one variable is isolated in one of the equations, and the expression equal to that variable is substituted for it in the other equation.

Subtraction property of equality A property of algebra that states that when equal values are subtracted from both sides of a true equation, the equation continues to be true. To solve the equation $x + 2 = 5$, subtract 2 from both sides of the equation by using the subtraction property of equality.

System of equations Two or more equations with two or more unknowns to solve for. An example of a system of equations with a solution of $(8, 2)$ is
$x + y = 10$
$x - y = 6$

T

Third quartile The number in a data set that is greater than just 75% of the numbers in the set.

Translation A translation of a graph is when the points on it are each shifted the same amount in the same direction. Examples of translations are vertical translations, horizontal translations, and combinations of vertical and horizontal translations.

Trinomial A polynomial with three terms. The polynomial $x^2 + 5x + 6$ is a trinomial.

Glossary of Algebra I Terms

V

Variable A letter, often an x, y, or z, that represents a value in a mathematical expression. In an algebraic equation, the variable is often the unknown that needs to be solved for.

Vertex The turning point of a parabola is its vertex. If the parabola opens upward, the vertex is the minimum point. If the parabola opens downward, the vertex is the maximum point.

Vertical line test A way of testing to see if a graph can represent a function. If at least one vertical line can pass through at least two points on the graph, the graph fails the vertical line test and cannot represent a function. If there are no vertical lines that can pass through at least two points, then the graph can be the graph of a function.

X

x-Intercept The location where a curve crosses the x-axis. The y-coordinate of the x-intercept is 0.

Y

y-Intercept The location where a curve crosses the y-axis. The x-coordinate of the y-intercept is 0. In slope-intercept form, $y = mx + b$, the y-intercept is located at $(0, b)$.

Z

Zeros The zeros of a function f are the numbers that can be input into the function so that 0 is output from the function. For example, the function $f(x) = 2x - 6$ has the number 3 as its only zero since $f(3) = 2(3) - 6 = 6 - 6 = 0$.

THE ALGEBRA I REGENTS EXAMINATION

The Regents Examination in Algebra I is a three-hour exam that is divided into four parts with a total of 37 questions. All 37 questions must be answered. Part I consists entirely of regular multiple-choice questions. Parts II, III, and IV each contain a set of questions that must be answered directly in the question booklet. You are required to show how you arrived at the answers for the questions in Parts II, III, and IV. The accompanying table shows how the exam breaks down.

Question Type	Number of Questions	Credit Value
Part I: Multiple choice	24	$24 \times 2 = 48$
Part II: 2-credit open ended	8	$8 \times 2 = 16$
Part III: 4-credit open ended	4	$4 \times 4 = 16$
Part IV: 6-credit open ended	1	$1 \times 6 = 6$
	Total = 37 questions	Total = 86 points

How Is the Exam Scored?

- Each of your answers to the 24 multiple-choice questions in Part I will be scored as either right or wrong.
- Solutions to questions in Parts II, III, and IV that are not completely correct may receive partial credit according to a special rating guide that is provided by the New York State Education Department. In order to receive full credit for a correct answer to a question in Parts II, III, or IV, you must show or explain how you arrived at your answer by indicating the key steps taken, including appropriate formula substitutions, diagrams, graphs, and charts. A correct numerical answer with no work shown will receive only 1 credit.
- The raw scores for the four parts of the test are added together. The maximum total raw score for the Algebra I Regents Examination is 86 points. Using a special conversion chart that is provided by the New York State Education Department, your total raw score will be equated to a final test score that falls within the usual 0 to 100 scale.

What Type of Calculator Is Required?

Graphing calculators are *required* for the Algebra I Regents Examination. During the administration of the Regents exam, schools are required to make a graphing calculator available for the exclusive use of each student. You will need to use your calculator to work with trigonometric functions of angles, find roots of numbers, and perform routine calculations.

Knowing how to use a graphing calculator gives you more options when deciding how to solve a problem. Rather than solving a problem algebraically with pen and paper, it may be easier to solve the same problem using a graph or table created by a graphing calculator. A graphical or numerical solution using a calculator can also be used to help confirm an answer obtained by solving the problem algebraically.

Are Any Formulas Provided?

The Algebra I Regents Examination test booklet will include a reference sheet containing the formulas in the accompanying table. Keep in mind that you may be required to know other formulas that are not included in this sheet. Please see pages 450 and 451 for these formulas.

What Else Should I Know?

- Do not omit any questions from Part I. Since there is no penalty for guessing, make certain that you record an answer for each of the 24 multiple-choice questions.
- If the method of solution is not stated in the problem, choose an appropriate method (numerical, graphical, or algebraic) with which you are most comfortable.
- If you solve a problem in Parts II, III, or IV using a trial-and-error approach, show the work for at least three guesses with appropriate checks. Should the correct answer be reached on the first trial, you must further illustrate your method by showing that guesses below and above the correct guess do not work.
- Avoid rounding errors when using a calculator. Unless otherwise directed, the (pi) key on a calculator should be used in computations involving the constant π rather than the common rational approximation of 3.14 or $\frac{22}{7}$. When performing a sequence of calculations in which the result of one calculation is used in a second calculation, do not round off. Instead, use the full power/display of the calculator by performing a "chain" calculation, saving intermediate results in the calculator's memory. Unless otherwise specified, rounding, if required, should be done only when the *final* answer is reached.
- Check that each answer is in the requested form. If a specific form is not required, answers may be left in any equivalent form, such as $\sqrt{75}$, $5\sqrt{3}$, or 8.660254038 (the full power/display of the calculator).
- If a problem requires using a formula that is not provided in the question, check the formula reference sheet in the test booklet to see if it is listed. Clearly write any formula you use before making any appropriate substitutions. Then evaluate the formula in step-by-step fashion.
- For any problem solved in Parts II, III, and IV using a graphing calculator, you must indicate how the calculator was used to obtain the answer such as by copying graphs or tables created by your calculator together with the equations used to produce them. When copying graphs, label each graph with its equation, state the dimensions of the viewing window, and identify the intercepts and any points of intersection with their coordinates. Whenever appropriate, indicate the rationale of your approach.

Examination June 2018
Algebra I

HIGH SCHOOL MATH REFERENCE SHEET

Conversions

1 inch = 2.54 centimeters
1 meter = 39.37 inches
1 mile = 5280 feet
1 mile = 1760 yards
1 mile = 1.609 kilometers

1 kilometer = 0.62 mile
1 pound = 16 ounces
1 pound = 0.454 kilogram
1 kilogram = 2.2 pounds
1 ton = 2000 pounds

1 cup = 8 fluid ounces
1 pint = 2 cups
1 quart = 2 pints
1 gallon = 4 quarts
1 gallon = 3.785 liters
1 liter = 0.264 gallon
1 liter = 1000 cubic centimeters

Formulas

Triangle	$A = \frac{1}{2}bh$
Parallelogram	$A = bh$
Circle	$A = \pi r^2$
Circle	$C = \pi d$ or $C = 2\pi r$
General Prisms	$V = Bh$
Cylinder	$V = \pi r^2 h$
Sphere	$V = \frac{4}{3}\pi r^3$

Formulas (continued)

Cone	$V = \frac{1}{3}\pi r^2 h$
Pyramid	$V = \frac{1}{3}Bh$
Pythagorean Theorem	$a^2 + b^2 = c^2$
Quadratic Formula	$x = \dfrac{-b \pm \sqrt{b^2 - 4ac}}{2a}$
Arithmetic Sequence	$a_n = a_1 + (n-1)d$
Geometric Sequence	$a_n = a_1 r^{n-1}$
Geometric Series	$S_n = \dfrac{a_1 - a_1 r^n}{1 - r}$ where $r \neq 1$
Radians	1 radian = $\dfrac{180}{\pi}$ degrees
Degrees	1 degree = $\dfrac{\pi}{180}$ radians
Exponential Growth/Decay	$A = A_0 e^{k(t - t_0)} + B_0$

Examination June 2018

PART I

Answer all 24 questions in this part. Each correct answer will receive 2 credits. No partial credit will be allowed. For each statement or question, write in the space provided the numeral preceding the word or expression that best completes the statement or answers the question. [48 credits]

1 The solution to $4p + 2 < 2(p + 5)$ is

(1) $p > -6$ (3) $p > 4$
(2) $p < -6$ (4) $p < 4$ 1 ____

2 If $k(x) = 2x^2 - 3\sqrt{x}$, then $k(9)$ is

(1) 315   (3) 159
(2) 307   (4) 153   2 ____

3 The expression $3(x^2 + 2x - 3) - 4(4x^2 - 7x + 5)$ is equivalent to

(1) $-13x - 22x + 11$ (3) $19x^2 - 22x + 11$
(2) $-13x^2 + 34x - 29$ (4) $19x^2 + 34x - 29$ 3 ____

4 The zeros of the function $p(x) = x^2 - 2x - 24$ are

(1) -8 and 3 (3) -4 and 6
(2) -6 and 4 (4) -3 and 8 4 ____

5 The box plot below summarizes the data for the average monthly high temperatures in degrees Fahrenheit for Orlando, Florida.

The third quartile is
(1) 92   (3) 83
(2) 90   (4) 71

6 Joy wants to buy strawberries and raspberries to bring to a party. Strawberries cost $1.60 per pound and raspberries cost $1.75 per pound. If she only has $10 to spend on berries, which inequality represents the situation where she buys x pounds of strawberries and y pounds of raspberries?

(1) $1.60x + 1.75y \leq 10$ (3) $1.75x + 1.60y \leq 10$
(2) $1.60x + 1.75y \geq 10$ (4) $1.75x + 1.60y \geq 10$

7 On the main floor of the Kodak Hall at the Eastman Theater, the number of seats per row increases at a constant rate. Steven counts 31 seats in row 3 and 37 seats in row 6. How many seats are there in row 20?

(1) 65   (3) 69
(2) 67   (4) 71

Examination June 2018

8 Which ordered pair below is *not* a solution to
   $f(x) = x^2 - 3x + 4$?

   (1) (0, 4) (3) (5, 14)
   (2) (1.5, 1.75) (4) (−1, 6)   8 __4__

9 Students were asked to name their favorite sport from a list of basketball, soccer, or tennis. The results are shown in the table below.

	Basketball	Soccer	Tennis
Girls	42	58	20
Boys	84	41	5

   What percentage of the students chose soccer as their favorite sport?

   (1) 39.6% (3) 50.4%
   (2) 41.4% (4) 58.6%   9 __2__

10 The trinomial $x^2 - 14x + 49$ can be expressed as

   (1) $(x - 7)^2$ (3) $(x - 7)(x + 7)$
   (2) $(x + 7)^2$ (4) $(x - 7)(x + 2)$ 10 __1__

11 A function is defined as {(0, 1), (2, 3), (5, 8), (7, 2)}. Isaac is asked to create one more ordered pair for the function. Which ordered pair can he add to the set to keep it a function?

   (1) (0, 2) (3) (7, 0)
   (2) (5, 3) (4) (1, 3)   11 __4__

Examination June 2018

12 The quadratic equation $x^2 - 6x = 12$ is rewritten in the form $(x + p)^2 = q$, where q is a constant. What is the value of p?

(1) −12 (3) −3
(2) −9   (4) 9

12 ____

13 Which of the quadratic functions below has the *smallest* minimum value?

$h(x) = x^2 + 2x - 6$ $k(x) = (x + 5)(x + 2)$
         (1)   (3)

(2) [graph of g(x)]   (4) [table of f(x) values:
                                  x | f(x)
                                  −1 | −2
                                  0 | −5
                                  1 | −6
                                  2 | −5
                                  3 | −2]

13 ____

14 Which situation is *not* a linear function?

(1) A gym charges a membership fee of $10.00 down and $10.00 per month.
(2) A cab company charges $2.50 initially and $3.00 per mile.
(3) A restaurant employee earns $12.50 per hour.
(4) A $12,000 car depreciates 15% per year.

14 ____

455

15 The Utica Boilermaker is a 15-kilometer road race. Sara is signed up to run this race and has done the following training runs:

   I.   10 miles
   II. 44,880 feet
   III. 15,560 yards

Which run(s) are at least 15 kilometers?

(1) I, only
(2) II, only
(3) I and III
(4) II and III

16 If $f(x) = x^2 + 2$, which interval describes the range of this function?

(1) $(-\infty, \infty)$
(2) $[0, \infty)$
(3) $[2, \infty)$
(4) $(-\infty, 2]$

17 The amount Mike gets paid weekly can be represented by the expression $2.50a + 290$, where a is the number of cell phone accessories he sells that week. What is the constant term in this expression and what does it represent?

(1) $2.50a$, the amount he is guaranteed to be paid each week
(2) $2.50a$, the amount he earns when he sells a accessories
(3) 290, the amount he is guaranteed to be paid each week
(4) 290, the amount he earns when he sells a accessories

Examination June 2018

18 A cubic function is graphed on the set of axes below.

Which function could represent this graph?

(1) $f(x) = (x - 3)(x - 1)(x + 1)$
(2) $g(x) = (x + 3)(x + 1)(x - 1)$
(3) $h(x) = (x - 3)(x - 1)(x + 3)$
(4) $k(x) = (x + 3)(x + 1)(x - 3)$

19 Mrs. Allard asked her students to identify which of the polynomials below are in standard form and explain why.

I. $15x^4 - 6x + 3x^2 - 1$
II. $12x^3 + 8x + 4$
III. $2x^5 + 8x^2 + 10x$

Which student's response is correct?

(1) Tyler said I and II because the coefficients are decreasing.
(2) Susan said only II because all the numbers are decreasing.
(3) Fred said II and III because the exponents are decreasing.
(4) Alyssa said II and III because they each have three terms.

20 Which graph does *not* represent a function that is always increasing over the entire interval –2 < x < 2?

(1) (3)
(2) (4)

21 At an ice cream shop, the profit, $P(c)$, is modeled by the function $P(c) = 0.87c$, where c represents the number of ice cream cones sold. An appropriate domain for this function is

(1) an integer ≤ 0 (3) a rational number ≤ 0
(2) an integer ≥ 0 (4) a rational number ≥ 0

22 How many real-number solutions does $4x^2 + 2x + 5 = 0$ have?

(1) one (3) zero
(2) two (4) infinitely many   22 _____

23 Students were asked to write a formula for the length of a rectangle by using the formula for its perimeter, $p = 2\ell + 2w$. Three of their responses are shown below.

I. $\ell = \dfrac{1}{2}p - w$

II. $\ell = \dfrac{1}{2}(p - 2w)$

III. $\ell = \dfrac{p - 2w}{2}$

Which responses are correct?

(1) I and II, only   (3) I and III, only
(2) II and III, only (4) I, II, and III   23 _____

24 If $a_n = n(a_{n-1})$ and $a_1 = 1$, what is the value of a_5?

(1) 5   (3) 120
(2) 20   (4) 720   24 _____

Examination June 2018

PART II

Answer all 8 questions in this part. Each correct answer will receive 2 credits. Clearly indicate the necessary steps, including appropriate formula substitutions, diagrams, graphs, charts, etc. For all questions in this part, a correct numerical answer with no work shown will receive only 1 credit.   [16 credits]

25 Graph $f(x) = \sqrt{x+2}$ over the domain $-2 \leq x \leq 7$.

Examination June 2018

26 Caleb claims that the ordered pairs shown in the table below are from a nonlinear function.

x	f(x)
0	2
1	4
2	8
3	16

Yes Caleb is correct

State if Caleb is correct. Explain your reasoning.

~~No Caleb is not correct because every input has only one output~~

27 Solve for x to the *nearest tenth*: $x^2 + x - 5 = 0$.

$x = 1.8$ and $x = -2.8$

~~$x^2 + x - 5 = 0$~~
~~$+5$~~
~~$x^2 + x = 5$~~

$x^2 = 5$

461

28 The graph of the function $p(x)$ is represented below. On the same set of axes, sketch the function $p(x + 2)$.

29 When an apple is dropped from a tower 256 feet high, the function $h(t) = -16t^2 + 256$ models the height of the apple, in feet, after t seconds. Determine, algebraically, the number of seconds it takes the apple to hit the ground.

30 Solve the equation below algebraically for the exact value of x.

$$6 - \frac{2}{3}(x+5) = 4x$$

$x = 0.225$ ~~(crossed out)~~

$\frac{4}{7}$

[boxed work, crossed out:] $6 - \frac{2x}{3} + -3 \cdot 3 = 4x$

$2 \cdot 7 - \frac{2x}{3} = 4x$
-3

$\frac{2 \cdot 7}{12} = \frac{12x}{12}$

31 Is the product of $\sqrt{16}$ and $\frac{4}{7}$ rational or irrational? Explain your reasoning.

Irrational because the decimal does not end or repeat so no one knows what number comes next.

Examination June 2018

32 On the set of axes below, graph the piecewise function:

$$f(x) = \begin{cases} -\dfrac{1}{2}x, & x < 2 \\ x, & x \geq 2 \end{cases}$$

Examination June 2018

PART III

Answer all 4 questions in this part. Each correct answer will receive 4 credits. Clearly indicate the necessary steps, including appropriate formula substitutions, diagrams, graphs, charts, etc. For all questions in this part, a correct numerical answer with no work shown will receive only 1 credit. [16 credits]

33 A population of rabbits in a lab, $p(x)$, can be modeled by the function $p(x) = 20(1.014)^x$, where x represents the number of days since the population was first counted.

Explain what 20 and 1.014 represent in the context of the problem.

Determine, to the *nearest tenth*, the average rate of change from day 50 to day 100.

34 There are two parking garages in Beacon Falls. Garage A charges $7.00 to park for the first 2 hours, and each additional hour costs $3.00. Garage B charges $3.25 per hour to park.

When a person parks for at least 2 hours, write equations to model the cost of parking for a total of x hours in Garage A and Garage B.

Determine algebraically the number of hours when the cost of parking at both garages will be the same.

35 On the set of axes below, graph the following system of inequalities:

$$2y + 3x \leq 14$$
$$4x - y < 2$$

Determine if the point (1, 2) is in the solution set.
Explain your answer.

36 The percentage of students scoring 85 or better on a mathematics final exam and an English final exam during a recent school year for seven schools is shown in the table below.

Percentage of Students Scoring 85 or Better	
Mathematics, x	English, y
27	46
12	28
13	45
10	34
30	56
45	67
20	42

Write the linear regression equation for these data, rounding all values to the *nearest hundredth*.

State the correlation coefficient of the linear regression equation, to the *nearest hundredth*. Explain the meaning of this value in the context of these data.

PART IV

Answer the question in this part. A correct answer will receive 6 credits. Clearly indicate the necessary steps, including appropriate formula substitutions, diagrams, graphs, charts, etc. A correct numerical answer with no work shown will receive only 1 credit. [6 credits]

37 Dylan has a bank that sorts coins as they are dropped into it. A panel on the front displays the total number of coins inside as well as the total value of these coins. The panel shows 90 coins with a value of $17.55 inside of the bank.

If Dylan only collects dimes and quarters, write a system of equations in two variables or an equation in one variable that could be used to model this situation.

Using your equation or system of equations, algebraically determine the number of quarters Dylan has in his bank.

Question 37 is continued on the next page.

Question 37 continued.

Dylan's mom told him that she would replace each one of his dimes with a quarter. If he uses all of his coins, determine if Dylan would then have enough money to buy a game priced at $20.98 if he must also pay an 8% sales tax. Justify your answer.

Answers June 2018
Algebra I

Answer Key

PART I
1. (3)
2. (4)
3. (1)
4. (2)
5. (1)
6. (1)
7. (2)
8. (1)
9. (4)
10. (2)
11. (2)
12. (1)
13. (3)
14. (3)
15. (4)
16. (3)
17. (2)
18. (4)
19. (3)
20. (3)
21. (3)
22. (2)
23. (4)
24. (2)

PART II
25.

26. Caleb is correct. It is a nonlinear function.
27. $x = 1.8$ and $x = -2.8$
28. See graph
29. 4 seconds
30. $\frac{4}{7}$
31. The product is rational.
32. See graph

PART III
33. 20 is the starting population. 1.014 is the growth factor. The average rate of change is 0.8.
34. For Garage A, $A = 7 + 3(x - 2)$. For Garage B, $B = 3.25x$. Both garages cost the same when you park for 4 hours.
35. (1, 2) is not in the solution set.
36. $y = 0.96x + 23.95$ The correlation coefficient is 0.92, which is a strong positive correlation.

PART IV
37. $0.10d + 0.25q = 17.55$
    $d + q = 90$

    Dylan has 57 quarters. No, he would not be able to buy the game.

In **Parts II–IV**, you are required to show how you arrived at your answers. For sample methods of solutions, see Barron's *Regents Exams and Answers* book for Algebra I.

Examination
June 2019
Algebra I

HIGH SCHOOL MATH REFERENCE SHEET

Conversions

1 inch = 2.54 centimeters
1 meter = 39.37 inches
1 mile = 5280 feet
1 mile = 1760 yards
1 mile = 1.609 kilometers

1 kilometer = 0.62 mile
1 pound = 16 ounces
1 pound = 0.454 kilogram
1 kilogram = 2.2 pounds
1 ton = 2000 pounds

1 cup = 8 fluid ounces
1 pint = 2 cups
1 quart = 2 pints
1 gallon = 4 quarts
1 gallon = 3.785 liters
1 liter = 0.264 gallon
1 liter = 1000 cubic centimeters

Formulas

Triangle	$A = \frac{1}{2}bh$
Parallelogram	$A = bh$
Circle	$A = \pi r^2$
Circle	$C = \pi d$ or $C = 2\pi r$
General Prisms	$V = Bh$
Cylinder	$V = \pi r^2 h$
Sphere	$V = \frac{4}{3}\pi r^3$

Formulas (continued)

Cone	$V = \frac{1}{3}\pi r^2 h$
Pyramid	$V = \frac{1}{3}Bh$
Pythagorean Theorem	$a^2 + b^2 = c^2$
Quadratic Formula	$x = \frac{-b \pm \sqrt{b^2 - 4ac}}{2a}$
Arithmetic Sequence	$a_n = a_1 + (n-1)d$
Geometric Sequence	$a_n = a_1 r^{n-1}$
Geometric Series	$S_n = \frac{a_1 - a_1 r^n}{1-r}$ where $r \neq 1$
Radians	1 radian = $\frac{180}{\pi}$ degrees
Degrees	1 degree = $\frac{\pi}{180}$ radians
Exponential Growth/Decay	$A = A_0 e^{k(t-t_0)} + B_0$

Examination June 2019

PART I

Answer all 24 questions in this part. Each correct answer will receive 2 credits. No partial credit will be allowed. For each statement or question, write in the space provided the numeral preceding the word or expression that best completes the statement or answers the question. [48 credits]

1 The expression $w^4 - 36$ is equivalent to
  (1) $(w^2 - 18)(w^2 - 18)$ (3) $(w^2 - 6)(w^2 - 6)$
  (2) $(w^2 + 18)(w^2 - 18)$ (4) $(w^2 + 6)(w^2 - 6)$ 1 ____

2 If $f(x) = 4x + 5$, what is the value of $f(-3)$?
  (1) -2 (3) 17
  (2) -7 (4) 4   2 ____

3 Which relation is *not* a function?

x	y
-10	-2
-6	2
-2	6
1	9
5	13

(1)

(3)

$3x + 2y = 4$

(2)

(4)   3 ____

4 Given: $f(x) = (x - 2)^2 + 4$
$g(x) = (x - 5)^2 + 4$

When compared to the graph of $f(x)$, the graph of $g(x)$ is

(1) shifted 3 units to the left
(2) shifted 3 units to the right
(3) shifted 5 units to the left
(4) shifted 5 units to the right

4 __2__

5 Students were asked to write $6x^5 + 8x - 3x^3 + 7x^7$ in standard form.

Shown below are four student responses.

Anne: $7x^7 + 6x^5 - 3x^3 + 8x$
Bob: $-3x^3 + 6x^5 + 7x^7 + 8x$
Carrie: $8x + 7x^7 + 6x^5 - 3x^3$
Dylan: $8x - 3x^3 + 6x^5 + 7x^7$

Which student is correct?

(1) Anne   (3) Carrie
(2) Bob   (4) Dylan

5 __1__

6 The function f is shown in the table below.

x	f(x)
0	1
1	3
2	9
3	27

Which type of function best models the given data?

(1) exponential growth function
(2) exponential decay function
(3) linear function with positive rate of change
(4) linear function with negative rate of change

6 __3__

7 Which expression results in a rational number?

(1) $\sqrt{2} \cdot \sqrt{18}$

(2) $5 \cdot \sqrt{5}$

(3) $\sqrt{2} + \sqrt{2}$

(4) $3\sqrt{2} + 2\sqrt{3}$

7 __1__

8 A polynomial function is graphed below.

Which function could represent this graph?

(1) $f(x) = (x + 1)(x^2 + 2)$
(2) $f(x) = (x - 1)(x^2 - 2)$
(3) $f(x) = (x - 1)(x^2 - 4)$
(4) $f(x) = (x + 1)(x^2 + 4)$

8 __3__

9 When solving $p^2 + 5 = 8p - 7$, Kate wrote $p^2 + 12 = 8p$. The property she used is

(1) the associative property
(2) the commutative property
(3) the distributive property
(4) the addition property of equality

9 __4__

10 David wanted to go on an amusement park ride. A sign posted at the entrance read "You must be greater than 42 inches tall and no more than 57 inches tall for this ride." Which inequality would model the height, x, required for this amusement park ride?

(1) $42 < x \leq 57$
(2) $42 > x \geq 57$
(3) $42 < x$ or $x \leq 57$
(4) $42 > x$ or $x \geq 57$

11 Which situation can be modeled by a linear function?

(1) The population of bacteria triples every day.
(2) The value of a cell phone depreciates at a rate of 3.5% each year.
(3) An amusement park allows 50 people to enter every 30 minutes.
(4) A baseball tournament eliminates half of the teams after each round.

12 Jenna took a survey of her senior class to see whether they preferred pizza or burgers. The results are summarized in the table below.

	Pizza	Burgers
Male	23	42
Female	31	26

Of the people who preferred burgers, approximately what percentage were female?

(1) 21.3
(2) 38.2
(3) 45.6
(4) 61.9

13 When $3a + 7b > 2a - 8b$ is solved for a, the result is

(1) $a > -b$
(2) $a < -b$
(3) $a < -15b$
(4) $a > -15b$

477

14 Three functions are shown below.

A: $g(x) = -\frac{3}{2}x + 4$

B: $f(x) = (x + 2)(x + 6)$

C:

[graph of downward-opening parabola labeled h(x)]

Which statement is true?

(1) B and C have the same zeros.
(2) A and B have the same y-intercept.
(3) B has a minimum and C has a maximum.
(4) C has a maximum and A has a minimum.

14 __3__

15 Nicci's sister is 7 years less than twice Nicci's age, a. The sum of Nicci's age and her sister's age is 41. Which equation represents this relationship?

(1) $a + (7 - 2a) = 41$
(2) $a + (2a - 7) = 41$
(3) $2a - 7 = 41$
(4) $a = 2a - 7$

15 __3__

16 The population of a small town over four years is recorded in the chart below, where 2013 is represented by $x = 0$. [Population is rounded to the nearest person]

Year	2013	2014	2015	2016
Population	3810	3943	4081	4224

The population, $P(x)$, for these years can be modeled by the function $P(x) = ab^x$, where b is rounded to the nearest thousandth. Which statements about this function are true?

    I. $a = 3810$
    II. $a = 4224$
    III. $b = 0.035$
    IV. $b = 1.035$

(1) I and III
(2) I and IV
(3) II and III
(4) II and IV

17 When written in factored form, $4w^2 - 11w - 3$ is equivalent to

(1) $(2w + 1)(2w - 3)$
(2) $(2w - 1)(2w + 3)$
(3) $(4w + 1)(w - 3)$
(4) $(4w - 1)(w + 3)$

18 Which ordered pair does *not* represent a point on the graph of $y = 3x^2 - x + 7$?

(1) $(-1.5, 15.25)$
(2) $(0.5, 7.25)$
(3) $(1.25, 10.25)$
(4) $(2.5, 23.25)$

19 Given the following three sequences:

    I.   2, 4, 6, 8, 10...
    II. 2, 4, 8, 16, 32...
    III. $a, a + 2, a + 4, a + 6, a + 8...$

Which ones are arithmetic sequences?

(1) I and II, only
(2) I and III, only
(3) II and III, only
(4) I, II, and III

Examination June 2019

20 A grocery store sells packages of beef. The function $C(w)$ represents the cost, in dollars, of a package of beef weighing w pounds. The most appropriate domain for this function would be

(1) integers
(2) rational numbers
(3) positive integers
(4) positive rational numbers

21 The roots of $x^2 - 5x - 4 = 0$ are

(1) 1 and 4
(2) $\dfrac{5 \pm \sqrt{41}}{2}$
(3) -1 and -4
(4) $\dfrac{-5 \pm \sqrt{41}}{2}$

22 The following table shows the heights, in inches, of the players on the opening-night roster of the 2015–2016 New York Knicks.

| 84 | 80 | 87 | 75 | 77 | 79 | 80 | 74 | 76 | 80 | 80 | 82 | 82 |

The population standard deviation of these data is approximately

(1) 3.5
(2) 13
(3) 79.7
(4) 80

23 A population of bacteria can be modeled by the function $f(t) = 1000(0.98)^t$, where t represents the time since the population started decaying, and $f(t)$ represents the population of the remaining bacteria at time t. What is the rate of decay for this population?

(1) 98%
(2) 2%
(3) 0.98%
(4) 0.02%

24 Bamboo plants can grow 91 centimeters per day. What is the approximate growth of the plant, in inches per hour?

(1) 1.49
(2) 3.79
(3) 9.63
(4) 35.83

PART II

Answer all 8 questions in this part. Each correct answer will receive 2 credits. Clearly indicate the necessary steps, including appropriate formula substitutions, diagrams, graphs, charts, etc. For all questions in this part, a correct numerical answer with no work shown will receive only 1 credit. [16 credits]

25 Solve algebraically for x: $-\frac{2}{3}(x + 12) + \frac{2}{3}x = -\frac{5}{4}x + 2$

Examination June 2019

26 If $C = G - 3F$, find the trinomial that represents C when $F = 2x^2 + 6x - 5$ and $G = 3x^2 + 4$.

27 Graph the following piecewise function on the set of axes below.

$$f(x) = \begin{cases} |x|, & -5 \leq x < 2 \\ -2x + 10, & 2 \leq x \leq 6 \end{cases}$$

28 Solve $5x^2 = 180$ algebraically.

29 A blizzard occurred on the East Coast during January, 2016. Snowfall totals from the storm were recorded for Washington, D.C. and are shown in the table below.

Washington, D.C.	
Time	Snow (inches)
1 A.M.	1
3 A.M.	5
6 A.M.	11
12 noon	33
3 P.M.	36

Which interval, 1 A.M. to 12 noon or 6 A.M. to 3 P.M., has the greatest rate of snowfall, in inches per hour? Justify your answer.

30 The formula for the volume of a cone is $V = \frac{1}{3}\pi r^2 h$.
Solve the equation for h in terms of V, r, and π.

$$h = \frac{3V}{\pi r^2}$$

31 Given the recursive formula:

$$a_1 = 3$$
$$a_n = 2(a_{n-1} + 1)$$

State the values of a_2, a_3, and a_4 for the given recursive formula.

$a_2 = 8$

$a_3 = 18$

$a_4 = 38$

32 Determine and state the vertex of $f(x) = x^2 - 2x - 8$ using the method of completing the square.

$(1, -9)$

PART III

Answer all 4 questions in this part. Each correct answer will receive 4 credits. Clearly indicate the necessary steps, including appropriate formula substitutions, diagrams, graphs, charts, etc. For all questions in this part, a correct numerical answer with no work shown will receive only 1 credit. [16 credits]

33 A school plans to have a fundraiser before basketball games selling shirts with their school logo. The school contacted two companies to find out how much it would cost to have the shirts made. Company A charges a $50 set-up fee and $5 per shirt. Company B charges a $25 set-up fee and $6 per shirt.

Write an equation for Company A that could be used to determine the total cost, A, when x shirts are ordered. Write a second equation for Company B that could be used to determine the total cost, B, when x shirts are ordered.

Determine algebraically and state the *minimum* number of shirts that must be ordered for it to be cheaper to use Company A.

Examination June 2019

34 Graph $y = f(x)$ and $y = g(x)$ on the set of axes below.

$$f(x) = 2x^2 - 8x + 3$$
$$g(x) = -2x + 3$$

Determine and state all values of x for which $f(x) = g(x)$.

35 The table below shows the number of hours ten students spent studying for a test and their scores.

Hours Spent Studying (x)	0	1	2	4	4	4	6	6	7	8
Test Scores (y)	35	40	46	65	67	70	82	88	82	95

Write the linear regression equation for this data set. Round all values to the *nearest hundredth*.

y = 7.79x + 34.27   r = .9

State the correlation coefficient of this line, to the *nearest hundredth*.

.98

Explain what the correlation coefficient suggests in the context of the problem.

The longer people study the higher the score

Examination June 2019

36 A system of inequalities is graphed on the set of axes below.

State the system of inequalities represented by the graph.

$y \le 2x - 2$ $y \le 3x + 3$

State what region A represents.

solution set of the inequalities

State what the entire gray region represents.

solution set for $y \le 2x - 2$

PART IV

Answer the question in this part. A correct answer will receive 6 credits. Clearly indicate the necessary steps, including appropriate formula substitutions, diagrams, graphs, charts, etc. A correct numerical answer with no work shown will receive only 1 credit. [6 credits]

37 When visiting friends in a state that has no sales tax, two families went to a fast-food restaurant for lunch. The Browns bought 4 cheeseburgers and 3 medium fries for $16.53. The Greens bought 5 cheeseburgers and 4 medium fries for $21.11.

Using c for the cost of a cheeseburger and f for the cost of medium fries, write a system of equations that models this situation.

$4c + 3f = 16.53$

$5c + 4f = 21.11$

The Greens said that since their bill was $21.11, each cheeseburger must cost $2.49 and each order of medium fries must cost $2.87 each. Are they correct? Justify your answer.

they are not correct because it would be $23.93 not 21.11

Using your equations, algebraically determine both the cost of one cheeseburger and the cost of one order of medium fries.

one cheese burger = 2.79

one order of frys = 1.79

Answers June 2019
Algebra I

Answer Key

PART I

1. (4)	5. (1)	9. (4)	13. (4)	17. (3)	21. (2)
2. (2)	6. (1)	10. (1)	14. (3)	18. (3)	22. (1)
3. (4)	7. (1)	11. (3)	15. (2)	19. (2)	23. (2)
4. (2)	8. (3)	12. (2)	16. (2)	20. (4)	24. (1)

PART II
25. $x = 8$
26. $-3x^2 - 18x + 19$
27.

28. $\{-6, 6\}$
29. 1 A.M. to 12 noon because $2\frac{10}{11} > 2\frac{7}{9}$
30. $h = \dfrac{3v}{\pi r^2}$
31. 8, 18, 38
32. $(1, -9)$

PART III
33. $A = 5x + 50$, $B = 6x + 25$; $x = 26$
34.

$x = 0, x = 3$

35. $y = 7.79x + 34.27$, $r = .98$
36. $y \leq 2x - 2$, $y < -3x + 3$, A represents the solution set of the system of inequalities. The gray region is the solution set to $y \leq 2x - 2$.

PART IV
37. $4c + 3f = 16.53$, $5c + 4f = 21.11$; they are not correct because it would be $23.93, not $21.11; one cheeseburger is $2.79, and one order of fries is $1.79

In **Parts II–IV**, you are required to show how you arrived at your answers. For sample methods of solutions, see *Barron's Regents Exams and Answers*: book for Algebra I.

Index

A

$(a - b)(a + b)$ pattern, multiplying of, 42–43
Absolute value, 123–124, 148–150
Addition, of polynomials, 32–33
Addition property of equality, 1–2
Algebra
  equations used in. *See* Equations
  properties of, 1–7
Algebra Reagents exam
  computational errors, 403
  conceptual errors, 403
  credits distribution by topic area, 449
  duration of, 401, 449
  formulas provided, 450–451
  partial credit on, 402–403
  points needed to pass, 410
  reasoning, 402–403
  sample examinations, 455–537
  scoring of, 450
  structure of, 402, 449
  time management for, 401
  tips for, 451–452
ANS feature, of graphing calculator, 406–408
Arithmetic sequences, 329, 331–333
Axis of symmetry, 174

B

Base case, 331
Best fit
  exponential curve of, 371–372
  line of. *See* Line of best fit
  parabola of, 368–370
Binomials
  definition of, 19
  factoring a trinomial into the product of, 46–47
  monomials multiplied by, 29
  multiplying of
    with distributive property, 40
    with FOIL shortcut, 40–41
  squaring, 42
Box plots, 395

C

Calculator. *See* Graphing calculator
Closed form defined sequences, 338–342
Coefficients
  definition of, 8
  negative, applying the quadratic formula with, 80–82
  two-step quadratic equations with, 58
Common factor, 45
Commutative property of addition or multiplication, 3–4
Completing the square, solving quadratic equations by, 66–71
Computational errors, 403
Conceptual errors, 403
Constant polynomials, 20
Constants
  description of, 7
  two-step quadratic equations with, 56–58
Correlation coefficient
  calculation of, 358–360
  definition of, 358
  negative, 362
  scatter plots used to compare, 361–362
Cubic polynomials, 20

D

Data representation
  box plots, 395
  dot plots, 393
  histograms, 394
Difference, 261
Distributive property
  of multiplication over addition, 3
  multiplying binomials with, 40
Division, of monomials, 24
Division property of equality, 2
Domain of function, 288–289, 297–298

E

Equations
  exponential. *See* Exponential equations
  for functions, 304–306
  identifying the type of
    by looking at graph of solution set, 254–256
    from table without graphing, 261–262
  isolating variables in, 14–15
  linear. *See* Linear equations
  multistep, 36–38
  one-step, 7–8
  quadratic. *See* Quadratic equations
  simplifying one side of, 36

Index

solving of, with variables on both sides of the equals sign, 37
three-step, 12
two-step, 11–12
Exponential curve of best fit, 371–372
Exponential decay, 249, 283
Exponential equations
  distinguishing characteristics of, 255
  interpreting of, 282–283
  real-world problems involving, 268–272, 279–285
  solutions to, 243
  solving for exponent in, 243–245
  table of values for identifying, 256–257
  two-variable, graphing solution sets to, 246–254
Exponential growth, 249, 282
Exponents
  in exponential equations, solving for, 243–245
  multiplying expressions involving, 22

F

Factoring
  greatest common factor, 45–46
  multiple-choice questions, 49–50
  perfect squares, 48–49
  perfect square trinomial into the square of a binomial, 48
  polynomials, 45–50
  trinomial into the product of two binomials, 46–47
Factors, 76–78
First quartile, 384–386
FOIL shortcut, multiplying binomials with, 40–41
Formulas, 450–451
Functions
  defining functions based on other functions, 305
  definition of, 287
  domain of, 288–289, 297–298
  equations for, 304–306
  graphing calculator with
    determining function values, 306–310
    piecewise functions, 312–313
  graphing of
    description of, 292–298
    horizontal translations, 317–319
    from its equation, 305–306
    piecewise functions, 312–313
    vertical translations, 316
  horizontal translations, 317–319
  as list of ordered pairs, 287–289
  negative, determining intervals for, 313–314
  piecewise, 310–313
  positive, determining intervals for, 313–314
  range of, 288–289, 297–298
  transformations, 316–319
  vertical translations, 316

G

Geometric sequences, 330
Graphing calculator
  ANS feature, 406–408
  description of, 450
  for determining function values, 306–310
  for graphing piecewise functions, 312–313
  for graphing solution sets
    to exponential equations, 247–249
    to linear equations, 146–157, 160–162
    to quadratic equations, 179–183
    to square root equations, 183
  for graphing systems of inequalities, 233
  for graphing table of values, 257–260
  for graphing two-variable inequalities, 223–224
  measures of central tendency, 386–388
  memory of, 404–405
  parentheses, 405–406
  for solving linear-quadratic systems of equations, 199–201
  for solving quadratic equations, 188–192
  types of, 403–404
Graphs/graphing
  of the equation, 120
  equations for
    description of, 158–162
    horizontal or vertical lines, 122–123
  of functions
    description of, 292–298
    horizontal translations, 317–319
    from its equation, 305–306
    piecewise functions, 312–313
    vertical translations, 316
  linear equations involving absolute value, 123–124
  quadratic equations, 205–210
  solution sets to exponential equations, 246–254
  solution sets to linear equations
    with graphing calculator, 146–157

Index

in slope-intercept form, 138–141
table of values for, 117–121
two-intercept method for, 121–122
solution sets to quadratic equations
graphing calculator for, 179–183
table of values for, 171–173
solution sets to square root equations
graphing calculator for, 183
table of values for, 173
solving systems of linear equations by, 124–125
systems of linear inequalities, 230–241
two-variable linear inequalities
involving > or < signs, 221–222
involving ≤ or ≥ signs, 218–220
Greatest common factor factoring, 45–46

H
Histograms, 394

I
Interquartile range, 385
Intersect feature, graphing systems of linear equations using, 150–156
Irrational numbers, 68–69

L
Like terms, 26–27, 37
Line
slope of. *See* Slope
word problems finding equation of a, 165–167
Linear equations
creating of, 273–278
distinguishing characteristics of, 254
interpreting of, 273–278
from real-world scenarios, 273–278
systems of. *See* Systems of linear equations
Linear inequalities
one-variable, 213–217
systems of, graphing of, 230–241
two-variable, graphing of
on graphing calculator, 223–224
involving > or < signs, 221–222
involving ≤ or ≥ signs, 218–220
Linear polynomials, 20
Linear-quadratic systems of equations
solving with algebra, 197–198
solving with graphing calculator, 199–201
solving with table of values, 197
Line of best fit
description of, 343–347

real-world scenario questions answered with, 347–348
residual plots, 348–351

M
Maximum vertex of parabola, 178
Mean, 383
Measures of central tendency
definition of, 383
graphing calculator for determining, 386–388
interquartile range, 385
mean, 383
median, 384, 386
mode, 384
quartiles, 384–386
Median, 384, 386
Minimum vertex of parabola, 178
Mode, 384
Monomials
definition of, 19
dividing of, 24
multiplying of
by binomials, 29
with coefficients and variable parts, 22–23
by polynomials with more than two terms, 29–30
Multiplication
of $(a - b)(a + b)$ pattern, 42–43
of monomials, 22–23
Multiplication property of equality, 2–3
Multiplicative inverse, 3, 8

N
Negative correlation coefficients, 362
Negative functions, 313–314
Negative reciprocal, 131

O
One-step algebra equations
one-step solutions, 7–8
zero-step solutions, 7
One-variable linear inequalities, 213–217
Ordered pairs
description of, 89–90
function represented as list of, 287–289
list of, function represented as, 287–289

P
Parabola
definition of, 171
vertex of, 174–178

Index

Parabola of best fit, 368–370
Parallel lines, 131
Parentheses, with graphing calculator, 405–406
Partial credit, 402–403
Percent decrease problems, 280–282
Percent increase problems, 268–269, 279–281
Perfect square pattern, 51–52
Perfect squares factoring, 48–49
Perfect square trinomial
   factoring of, into the square of a binomial, 48
   recognizing the pattern of, 53
Picture patterns, sequences for, 334, 340
Piecewise functions, 310–313
Polynomials
   addition of, 32–33
   constant, 20
   cubic, 20
   definition of, 19
   degree of, 19–20
   determining of, by number of terms, 19
   factoring, 45–50
   linear, 20
   monomials multiplied using, 29–30
   multiplying of, by polynomials, 40–43
   quadratic, 20
   subtraction of, 33–34
Positive functions, 313–314

Q

Quadratic equations
   distinguishing characteristics of, 255
   graphing of, for real-world applications, 205–210
   solving of
     by completing the square, 66–71
     by factoring, 71–75
     graphing calculator for, 188–192
     by guess and check, 63–65
     with quadratic formula, 79–82
     by taking the square root of both sides of the equation, 55–62
   word problems involving, 83–88
Quadratic formula, 79–82
Quadratic polynomials, 20
Quadratic trinomial, solving equations by factoring, 72–74
Quartiles, 384–386
Quotient, 261

R

Range of function, 288–289, 297–298
Rational numbers, 68–69
Real-world problems
   domain and range of function, 289
   exponential equations, 268–272, 279–285
   linear equations, 273–278
   line of best fit used to answer questions about, 347–348
   quadratic equations, 205–210
   sequences, 334, 340
Reasoning, 402–403
Reciprocal
   definition of, 3, 8
   negative, 131
Recursively defined sequences, 331–337
Reference sheet, 409, 453–454
Regression curves
   correlation coefficient. *See* Correlation coefficient
   line of best fit, 343–348
Relative two-way frequency table, 391
Residual plots, 348–351
Roots
   factors and, 76–78
   nature of, 68–69

S

Scatter plots, for comparing correlation coefficients, 361–362
Second quartile, 384
Sequence notation, 329
Sequences
   arithmetic, 329, 331–333
   closed form defined, 338–342
   definition of, 329
   geometric, 330, 332–333
   for picture patterns, 334, 340
   real-world scenarios, 334, 340
   recursively defined, 331–337
Slope
   calculating of, 130–131
   definition of, 129
   negative, 129
   positive, 129–131
   in time/distance graph, 131–132
Slope-intercept form
   definition of, 138
   finding the equation in, when y-intercept and another point are known, 141–142

Index

graphing solution set of a linear equation in, 138–141
Square root equations, 60, 183
Squaring binomials, 42
Statistics
  data representation methods, 393–398
  measures of central tendency. *See* Measures of central tendency
Subtraction, of polynomials, 33–34
Subtraction property of equality, 2
Symmetry, axis of, 174
Systems of linear equations
  combining two equations to form new equation, 95–96
  with intersect feature, 150–156
  setting up, 111–112
  solving of
    by changing both equations, 104–107
    by changing one of the equations, 102–104
    by graphing, 124–125
    with guess and check, 89–94
  solving word problems with, 111–112
  substitution method for solving, 98–99
  subtracting two equations to eliminate a variable, 98
  without solutions or infinite number of solutions, 107
Systems of linear inequalities, 230–241

T

Table of values
  calculator used to create, 148
  exponential equations identified by looking at, 256–257
  graphing calculator used to graph, 257–260
  for graphing the solution set
    of a linear equation, 117–121
    to quadratic equation, 171–173
    to square root equations, 173
  for solving linear-quadratic systems of equations, 197
Test-taking strategies
  graphing calculator, 403–408
  partial credit, 402–403
  time management, 401
Third quartile, 384–386
Three-step algebra equations, 12
TI-84, 403–404
Time/distance graphs
  for changing speeds, 132–133

interpreting slope in, 131–132
Time management, 401
TI-Nspire, 403
Trinomials
  definition of, 19
  factoring of
    into the product of two binomials, 46–47
    recognizing of, into two binomials pattern, 52–53
  perfect square
    factoring of, into the square of a binomial, 48
    recognizing the pattern of, 53
  x term of, 32
Two-intercept method, for graphing the solution set of a linear equation, 121–122
Two-step algebra equations, 11–12
Two-step quadratic equations
  with coefficients, 58
  with constants, 56–58
  with multiple variables, 59
Two-way frequency table, 388–390

V

Variables
  description of, 1
  isolating, in equations with multiple variables, 14–15
  subtracting two equations to eliminate, 98
  two-step quadratic equations with, 59
Vertex of parabola, 174–178
Vertical line test, 295

W

Word problems
  combining like terms, 37–38
  finding equation of a line, 165–167
  quadratic equations, 83–88
  solving, with systems of equations, 111–112

X

x-coordinate, 89–90
x-intercept, 121
x term, 32

Y

y-coordinate, 89–90
y-intercept, 121, 140, 142, 158

Z

Zeros feature, 188–191

BARRON'S

Regents Exams and Answers

Algebra I 2020

GARY M. RUBINSTEIN
B.A. Mathematics
Tufts University
M.S. Computer Science
University of Colorado

Copyright © 2020, 2018, 2017, 2016, 2015 by Kaplan, Inc., d/b/a Barron's Educational Series

All rights reserved.

No part of this book may be reproduced or distributed in any form or by any means without the written permission of the copyright owner.

Published by Kaplan, Inc., d/b/a Barron's Educational Series
750 Third Avenue
New York, NY 10017
www.barronseduc.com

ISBN: 978-1-5062-5383-1

Printed in Canada

10 9 8 7 6 5 4 3 2 1

Kaplan, Inc., d/b/a Barron's Educational Series print books are available at special quantity discounts to use for sales promotions, employee premiums, or educational purposes. For more information or to purchase books, please call the Simon & Schuster special sales department at 866-506-1949.

Contents

Preface   v

How to Use This Book   1

Test-Taking Tips and Strategies   3

Some Key Algebra I Facts and Skills   15
   1. Properties of Algebra and Solving Linear Equations with Algebra.... 15
   2. Polynomial Arithmetic... 23
   3. Quadratic Equations 31
   4. Systems of Linear Equations................................... 38
   5. Graphs of Linear Equations 48
   6. Graphs of Quadratic Equations 66
   7. Linear Inequalities 80
   8. Exponential Equations 94
   9. Creating and Interpreting Equations 103
  10. Functions.. 110
  11. Sequences.. 123
  12. Regression Curves 130
  13. Statistics 148

Glossary of Terms   158

Contents

Regents Examinations, Answers, and Self-Analysis Charts — 165

    June 2017 Exam 167
    August 2017 Exam 226
    June 2018 Exam 281
    August 2018 Exam 341
    June 2019 Exam 392
    August 2019 Exam 442

Preface

In June of 2014, New York State adopted a more difficult Algebra I course and changed the Regents exam to reflect this new course. On the Algebra I Regents exam, students are expected not only to solve algebra problems and answer algebra-related questions but also to *explain their reasoning*. This is a major shift in the testing of algebra concepts, so a thorough review is even more critical for success.

This book will introduce you to the concepts covered throughout the Algebra I Regents course as well as to the different types of problems you will encounter on the Regents exam at the end of the year.

WHO SHOULD USE THIS BOOK?

Students and teachers alike can use this book as a resource for preparing for the Algebra Regents exam.

Students will find this book to be a great study tool because it contains a review of Algebra I concepts, useful examples, and practice problems of varying difficulty that can be practiced throughout the school year to reinforce what they are learning in class. The most ideal way to prepare for the Algebra Regents exam is to work through the practice problems in the review sections and then the recently administered Regents exams at the end of the book.

Teachers can use this book as a tool to help structure an Algebra I course that will culminate with the Regents exam. The topics in the book are arranged by priority, so the sections in the beginning of the book are the ones from which more of the questions on the test are drawn. There are 13 sections, dedicated to all topics for Algebra I, each with practice exercises and solutions.

WHY IS THIS BOOK A HELPFUL RESOURCE?

Becoming familiar with the specific types of questions on the Algebra I Regents exam is crucial to performing well on this test. There are questions in which the math may be fairly easy but the way in which the question is asked makes the question seem much more difficult. For example, the question "Find all zeros of the function $f(x) = 2x + 6$" is a fancy way of asking the much simpler sounding "Solve for x if $2x + 6 = 0$." *Knowing exactly what the questions are asking is a big part of success on this test.*

Algebra has been around for thousands of years, and its basic concepts have never changed. So fundamentally, this algebra curriculum is not very different from the algebra taught in schools two years ago, ten years ago, or twenty years ago. But the exam, with more complicated ways of asking questions and presenting problems, requires a specifically presented study plan that's more important than ever.

Gary Rubinstein

How to Use This Book

This book is designed to help you get the most out of your review for the new Regents exam in Algebra I. Use this book to improve your understanding of the Algebra I topics and improve your grade.

TEST-TAKING TIPS

The first section in this book contains test-taking tips and strategies to help prepare you for the Algebra I Regents exam. This information is valuable, so be sure to read it carefully and refer to it during your study time. Remember: no single problem-solving strategy works for all problems—you should have a toolbox of strategies to pick from as you're facing unfamiliar or difficult problems on the test.

PRACTICE WITH KEY ALGEBRA I FACTS AND SKILLS

The second section in this book provides you with key Algebra I facts, useful skills, and practice problems with solutions. It provides you with a quick and easy way to refresh the skills you learned in class.

REGENTS EXAMS AND ANSWERS

The final section of the book contains actual Algebra I Regents exams that were administered in June 2017, August 2017, June 2018, August 2018, June 2019, and August 2019. These exams and thorough answer explanations are probably the most useful tool for your review, as they let you know what's most important. By answering the questions on these exams, you will be able to identify your strengths and weaknesses and then concentrate on the areas in which you may need more study.

Remember, the answer explanations in this book are more than just simple solutions to the problems—they contain facts and explanations that are crucial to success in the Algebra I course and on the Regents exam. Careful review of these answers will increase your chances of doing well.

SELF-ANALYSIS CHARTS

Each of the Algebra I Regents exams ends with a Self-Analysis Chart. This chart will further help you identify weaknesses and direct your study efforts where needed. In addition, the chart classifies the questions on each exam into an organized set of topic groups. This feature will also help you to locate other questions on the same topic in the other Algebra I exams.

IMPORTANT TERMS TO KNOW

The terms that are listed in the glossary are the ones that have appeared most frequently on past Integrated Algebra and the most recently Algebra I exams. All terms and their definitions are conveniently organized for a quick reference.

Test-Taking Tips and Strategies

Knowing the material is only part of the battle in acing the new Algebra I Regents exam. Things like improper management of time, careless errors, and struggling with the calculator can cost valuable points. This section contains some test-taking strategies to help you perform your best on test day.

Tip 1
Time Management

SUGGESTIONS

- *Don't rush.* The Algebra I Regents exam is three hours long. While you are officially allowed to leave after 90 minutes, you really should stay until the end of the exam. Just as it wouldn't be wise to come to the test an hour late, it is almost as bad to leave a test an hour early.
- *Do the test twice.* The best way to protect against careless errors is to do the entire test twice and compare the answers you got the first time to the answers you got the second time. For any answers that don't agree, do a "tie breaker" third attempt. Redoing the test and comparing answers is much more effective than simply looking over your work. Students tend to miss careless errors when looking over their work. By redoing the questions, you are less likely to make the same mistake.
- *Bring a watch.* What will happen if the clock is broken? Without knowing how much time is left, you might rush and make careless errors. Yes, the proctor will probably write the time elapsed on the board and update it every so often, but its better safe than sorry.

Test-Taking Tips and Strategies

The TI-84 graphing calculator has a built-in clock. Press the [MODE] to see it. If the time is not right, go to SET CLOCK and set it correctly. The TI-Nspire does not have a built-in clock.

TIP 2

Know How to Get Partial Credit

SUGGESTIONS

- *Know the structure of the exam.* The Algebra Regents exam has 37 questions. The first 24 of those questions are multiple-choice worth two points each. There is no partial credit if you make a mistake on one of those questions. Even the smallest careless error, like missing a negative sign, will result in no credit for that question. Parts Two, Three, and Four are free-response questions with no multiple-choice. Besides giving a numerical answer, you may be asked to explain your reasoning. Part Two has eight free-response questions worth two points each. The smallest careless error will cause you to lose one point, which is half the value of the question. Part Three has four free-response questions worth four points each. These questions generally have multiple parts. Part Four has one free-response question worth six points. This question will have multiple parts.

- *Explain your reasoning.* When a free-response question asks to "Justify your answer," "Explain your answer," or "Explain how you determined your answer," the grader is expecting a few clearly written sentences. For these, you don't want to write too little since the grader needs to see that you understand why you did the different steps you did to solve the equation. You also don't want to write too much because if anything you write is not accurate, points can be deducted.

Here is an example followed by two solutions. The first would not get full credit, but the second would.

Test-Taking Tips and Strategies

Example 1
Use algebra to solve for x in the equation $\frac{2}{3}x + 1 = 11$. Justify your steps.

Solution 1 (part credit):

$\frac{2}{3}x + 1 = 11$ $-1 = -1$ $\frac{2}{3}x = 10$ $x = 15$	I used algebra to get the x by itself. The answer was $x = 15$.

Solution 2 (full credit):

$\frac{2}{3}x + 1 = 11$ $-1 = -1$ $\frac{2}{3}x = 10$ $\frac{3}{2} \cdot \frac{2}{3}x = \frac{3}{2} \cdot 10$ $1x = 15$ $x = 15$	I used the subtraction property of equality to eliminate the $+1$ from the left-hand side. Then to make it so the x had a 1 in front of it, I used the multiplication property of equality and multiplied both sides of the equation by the reciprocal of $\frac{2}{3}$, which is $\frac{3}{2}$. Then since $1 \cdot x = x$, the left-hand side of the equation just became x and the right-hand side became 15.

- *Computational errors vs. conceptual errors*
  In the Part Three and Part Four questions, the graders are instructed to take off one point for a "computational error" but half credit for a "conceptual error." This is the difference between these two types of errors.
  If a four-point question was $x - 1 = 2$ and a student did it like this,

$$x - 1 = 2$$
$$+1 = +1$$
$$x = 4$$

the student would lose one point out of 4 because there was one computational error since $2 + 1 = 3$ and not 4.

Had the student done it like this,

$$x - 1 = 2$$
$$-1 = -1$$
$$x = 1$$

the student would lose half credit, or 2 points, since this error was conceptual. The student thought that to eliminate the -1, he should subtract 1 from both sides of the equation.

Either error might just be careless, but the conceptual error is the one that gets the harsher deduction.

TIP 3
Know Your Calculator

SUGGESTIONS

- *Which calculator should you use?* The two calculators used for this book are the TI-84 and the TI-Nspire. Both are very powerful. The TI-84 is somewhat easier to use for the functions needed for this test. The TI-Nspire has more features for courses in the future. The choice is up to you. This author prefers the TI-84 for the Algebra Regents. Graphing calculators come with manuals that are as thick as the book you are holding. There are also plenty of video tutorials online for learning how to use advanced features of the calculator. To become an expert user, watch the online tutorials or read the manual.

- *Clearing the memory.* You may be asked at the beginning of the test to clear the memory of your calculator. When practicing for the test, you should clear the memory too so you are practicing under test-taking conditions.

    This is how you clear the memory.

For the TI-84:

Press [2ND] and then [+] to get to the MEMORY menu. Then press [7] for Reset.

Test-Taking Tips and Strategies

Use the arrows to go to [ALL] for All Memory. Then press [1].

```
RAM ARCHIVE ALL
1:All Memory...
```

Press [2] for Reset.

```
RESET MEMORY
1:No
2:Reset
Resetting ALL
will delete all
data, programs &
Apps from RAM &
Archive.
```

The calculator will be reset as if in brand new condition. The one setting that you may need to change is to turn the diagnostics on if you need to calculate the correlation coefficient.

For the TI-Nspire:

The TI-Nspire must be set to Press-To-Test mode when taking the Algebra Regents. Turn the calculator off by pressing [ctrl] and [home]. Press and hold [esc] and then press [home].

```
Press-to-Test
Prevent access to 3D graphing functionality
and pre-existing Scratchpad data, documents
and folders.
  Angle Settings: Degree
  Select additional restrictions:
    ☑ Limit geometry functions
    ☑ Disable function grab and move
              Enter Press-to-Test  Cancel
```

While in Press-to-Test mode, certain features will be deactivated. A small green light will blink on the calculator so a proctor can verify the calculator is in Press-to-Test mode.

To exit Press-to-Test mode, use a USB cable to connect the calculator to another TI-Nspire. Then from the home screen on the calculator in Press-to-Test mode, press [doc], [9] and select Exit Press-to-Test.

Test-Taking Tips and Strategies

- *Use parentheses*

  The calculator always uses the order of operations where multiplication and division happen before addition and subtraction. Sometimes, though, you may want the calculator to do the operations in a different order.

  Suppose at the end of a quadratic equation, you have to round $x = \dfrac{-1 + \sqrt{5}}{2}$ to the nearest hundredth. If you enter $(-)$ (1) $(+)$ (2ND) (x^2) (5) $(/)$ (2), it displays

  ```
  -1+√(5/2
          .5811388301
  ```

  ```
  -1+√5     0.618034
  ───
   2
  ```

  which is not the correct answer.

  One reason is that for the TI-84 there needs to be a closing parenthesis (or on the TI-Nspire, press [right arrow] to move out from under the radical sign) after the 5 in the square root symbol. Without it, it calculated $-1 + \sqrt{\dfrac{5}{2}}$. More needs to be done, though, since

  ```
  -1+√(5/2
          .5811388301
  -1+√(5)/2
          .1180339887
  ```

  ```
  -1+√(5/2)    0.581139
  
  -1+√5/2      0.118034
  ```

  still is not correct. This is the solution to $-1 + \sqrt{\dfrac{5}{2}}$.

  To get this correct, there also needs to be parentheses around the entire numerator, $-1 + \sqrt{5}$.

Test-Taking Tips and Strategies 9

```
-1+√(5/2
        .5811388301
-1+√(5)/2
        .1180339887
(-1+√(5))/2
        .6180339887
```

	Scratchpad ▽	
$-1+\sqrt{\frac{5}{2}}$		
$\frac{-1+\sqrt{5}}{2}$		0.118034
$\frac{-1+\sqrt{5}}{2}$		0.618034
		3/99

This is the correct answer.

On the TI-Nspire, fractions like this can also be done with [templates].

	Scratchpad ▽			Scratchpad ▽	
			$\frac{-1+\sqrt{5}}{2}$		0.618034
		0/99			1/99

- *Using the ANS feature*

The last number calculated with the calculator is stored in something called the ANS variable. This ANS variable will appear if you start an expression with a $+$, $-$, $\times$, or $\div$. When an answer has a lot of digits in it, this saves time and is also more accurate.

If for some step in a problem you need to calculate the decimal equivalent of $\frac{1}{7}$, it will look like this on the TI-84:

```
1/7
        .1428571429
```

For the TI-Nspire, if you try the same thing, it leaves the answer as $\frac{1}{7}$. To get the decimal approximation, press [ctrl] and [enter] instead of just [enter].

Now if you want to multiply this by 3, just press [×], and the calculator will display "Ans*"; press [3] and [enter].

The ANS variable can also help you do calculations in stages. To calculate $x = \dfrac{-1 + \sqrt{5}}{2}$ without using so many parentheses as before, it can be done by first calculating $-1 + \sqrt{5}$ and then pressing [÷] and [2] and Ans will appear automatically.

Test-Taking Tips and Strategies

The ANS variable can also be accessed by pressing [2ND] and [−] at the bottom right of the calculator. If after calculating the decimal equivalent of $\frac{1}{7}$ you wanted to subtract $\frac{1}{7}$ from 5, for the TI-84 press [5], [−], [2ND], [ANS], and [ENTER]. For the TI-Nspire press [5], [−], [ctrl], [ans], and [enter].

TIP 4

Use the Reference Sheet

SUGGESTIONS

- In the back of the Algebra Regents booklet is a reference sheet that contains 17 conversion facts, such as inches to centimeters and quarts to pints, and also 17 formulas. Many of these conversion facts and formulas will not be needed for an individual test, but the quadratic formula and the arithmetic sequence formula are the two that will come in the handiest.

High School Math Reference Sheet

1 inch = 2.54 centimeters	1 kilometer = 0.62 mile	1 cup = 8 fluid ounces
1 meter = 39.37 inches	1 pound = 16 ounces	1 pint = 2 cups
1 mile = 5280 feet	1 pound = 0.454 kilogram	1 quart = 2 pints
1 mile = 1760 yards	1 kilogram = 2.2 pounds	1 gallon = 4 quarts
1 mile = 1.609 kilometers	1 ton = 2000 pounds	1 gallon = 3.785 liters
		1 liter = 0.264 gallon
		1 liter = 1000 cubic centimeters

Triangle	$A = \frac{1}{2}bh$	Pythagorean Theorem	$a^2 + b^2 = c^2$
Parallelogram	$A = bh$	Quadratic Formula	$x = \frac{-b \pm \sqrt{b^2 - 4ac}}{2a}$
Circle	$A = \pi r^2$	Arithmetic Sequence	$a_n = a_1 + (n-1)d$
Circle	$C = \pi d$ or $C = 2\pi r$	Geometric Sequence	$a_n = a_1 r^{n-1}$
General Prisms	$V = Bh$	Geometric Series	$S_n = \frac{a_1 - a_1 r^n}{1 - r}$ where $r \neq 1$
Cylinder	$V = \pi r^2 h$	Radians	1 radian = $\frac{180}{\pi}$ degrees
Sphere	$V = \frac{4}{3}\pi r^3$	Degrees	1 degree = $\frac{\pi}{180}$ radians
Cone	$V = \frac{1}{3}\pi r^2 h$	Exponential Growth/Decay	$A = A_0 e^{k(t - t_0)} + B_0$
Pyramid	$V = \frac{1}{3}Bh$		

- *How Many Points Do You Need to Pass?*

The Algebra Regents exam is scored out of a possible 86 points. Unlike most tests given in the year by your teacher, the score is not then turned into a percent out of 86. Instead each test has a conversion sheet that varies from year to year. For example, for the June 2014 test, the conversion sheet looked like this.

Raw Score	Scale Score	Raw Score	Scale Score	Raw Score	Scale Score
86	100	57	75	28	64
85	99	56	74	27	63
84	97	55	74	26	62
83	96	54	74	25	61
82	95	53	73	24	60
81	94	52	73	23	59
80	92	51	73	22	58
79	91	50	72	21	56
78	90	49	72	20	55
77	89	48	72	19	54
76	88	47	72	18	52
75	87	46	71	17	50
74	86	45	71	16	49
73	85	44	71	15	47
72	84	43	70	14	45
71	83	42	70	13	42
70	82	41	70	12	40
69	82	40	70	11	38
68	81	39	69	10	35
67	80	38	69	9	32
66	79	37	69	8	30
65	79	36	68	7	26
64	78	35	68	6	23
63	78	34	67	5	20
62	77	33	67	4	16
61	77	32	66	3	12
60	76	31	66	2	9
59	76	30	65	1	4
58	75	29	64	0	0

On this test, 30 points became a 65, 57 points became a 75, and 73 points became an 85. This means that for this examination a student who got 30 out of 86, which is just 35% of the possible points, would get a 65 on this exam. 57 out of 86 is 66%, but this scaled to a 75. 73 out of 86, however, is actually 85% and became an 85. So in the past there has been a curve on the exam for lower scores, though the scaling is not released until after the exam.

Some Key Algebra I Facts and Skills

1. PROPERTIES OF ALGEBRA AND SOLVING LINEAR EQUATIONS WITH ALGEBRA

1.1 ONE-STEP ALGEBRA PROBLEMS

An algebra problem, like $x - 5 = 2$, is one that can be solved by changing both sides of the equation until the variable x is isolated. There are four main properties that can be used in solving algebra problems.

- The **addition property** of equality

    The equation $x - 5 = 2$ is solved by adding 5 to both sides of the equation. When you add to both sides of an equation, you are using the addition property of equality.

    $$\begin{aligned} x - 5 &= 2 \quad &\text{the given equation} \\ +5 &= +5 \quad &\text{addition property of equality} \\ x &= 7 \quad &\text{x is isolated. The solution is 7.} \end{aligned}$$

- The **subtraction property** of equality

    The equation $x + 2 = 7$ is solved by subtracting 2 from both sides of the equation. When you subtract from both sides of an equation, you are using the subtraction property of equality.

    $$\begin{aligned} x + 2 &= 7 \quad &\text{the given equation} \\ -2 &= -2 \quad &\text{subtraction property of equality} \\ x &= 5 \quad &\text{x is isolated. The solution is 5.} \end{aligned}$$

- The **division** property of equality

    The equation $2x = 10$ is solved by dividing both sides of the equation by 2. When you divide both sides of the equation, you are using the division property of equality.

    $2x = 10$ the given equation

    $\dfrac{2x}{2} = \dfrac{10}{2}$ division property of equality

    $x = 5$ x is isolated. The solution is 5.

- The **multiplication** property of equality

    The equations $\dfrac{x}{5} = 3$ and $\dfrac{2}{3}x = 8$ can be solved by multiplying both sides of the equation by the same number.

    $\dfrac{x}{5} = 3$ the given equation

    $5 \cdot \dfrac{x}{5} = 5 \cdot 3$ multiplication property of equality

    $x = 15$ x is isolated. The solution is 5.

    $\dfrac{2}{3}x = 8$ the given equation

    $\dfrac{3}{2} \cdot \dfrac{2}{3}x = \dfrac{3}{2} \cdot 8$ multiplication property of equality

    $x = 12$ x is isolated. The solution is 12.

1.2 TWO-STEP ALGEBRA PROBLEMS

When an equation has the form $mx + b = y$, it takes two steps to solve for x. The first step is to eliminate the b, which is called the **constant**. The second step is to eliminate the m, which is called the **coefficient**. The b is eliminated with either the addition or the subtraction property of equality. The m is eliminated with either the division or the multiplication property of equality.

Properties of Algebra and Solving Linear Equations with Algebra   17

- For the equation $3x - 7 = 11$

$$3x - 7 = 11 \quad \text{the given equation}$$
$$+7 = +7 \quad \text{addition property of equality}$$
$$\frac{3x}{3} = \frac{18}{3} \quad \text{division property of equality}$$
$$x = 6 \quad x \text{ is isolated. The solution is 6.}$$
$$\frac{2}{3}x + 5 = 11 \quad \text{the given equation}$$
$$-5 = -5 \quad \text{subtraction property of equality}$$
$$\frac{2}{3}x = 6$$
$$\frac{3}{2} \cdot \frac{2}{3}x = \frac{3}{2} \cdot 6 \quad \text{multiplication property of equality}$$
$$x = 9 \quad x \text{ is isolated. The solution is 9.}$$

1.3 COMBINING LIKE TERMS BEFORE SOLVING

If the equation has multiple x terms or multiple constants, the equation should first be simplified by combining like terms. After all like terms have been combined, the question will usually be a two-step algebra problem and can be solved with the methods from Section 1.2.

$$2x + 5 + 3x - 2 = 23 \quad \text{the given equation}$$
$$2x + 3x + 5 - 2 = 23 \quad \text{Terms are rearranged so the } x \text{ terms are together and the constants are together. This step is optional.}$$
$$5x + 3 = 23 \quad \text{Like terms have been combined.}$$
$$-3 = -3 \quad \text{subtraction property of equality}$$
$$\frac{5x}{5} = \frac{20}{5} \quad \text{division property of equality}$$
$$x = 4 \quad x \text{ is isolated. The solution is 4.}$$

1.4 VARIABLES ON BOTH SIDES OF THE EQUATION

When there are x terms on both sides of the equation, the addition property of equality or the subtraction property of equality can be used to change the equation into one where the x terms are all on the same side of the equation.

$$5x - 3 = 12 + 2x \quad \text{the given equation}$$
$$-2x = -2x \quad \text{subtraction property of equality eliminates the } x \text{ term from the right-hand side of the equation}$$
$$3x - 3 = 12$$
$$+3 = +3 \quad \text{addition property of equality}$$
$$\frac{3x}{3} = \frac{15}{3} \quad \text{division property of equality}$$
$$x = 5 \quad x \text{ is isolated. The solution is 5.}$$

1.5 EQUATIONS WITH MORE THAN ONE VARIABLE

An equation with more than one variable can be solved the same way as an equation with just one variable. The **solution** will not be a number in these problems, but an **expression** with variables and numbers in it.

$$ax + 2 = c \quad \text{the given equation}$$
$$-2 = -2 \quad \text{subtraction property of equality}$$
$$ax = c - 2 \quad \text{the } c \text{ and the 2 cannot be combined since they are unlike terms}$$
$$\frac{ax}{a} = \frac{c - 2}{a} \quad \text{division property of equality}$$
$$x = \frac{c - 2}{a} \quad x \text{ is isolated. The solution for } x \text{ is not a number but an expression in terms of } c \text{ and } a. \text{ The answer is } \frac{c - 2}{a}.$$

Practice Exercises

1. Antonio started the question $2x + 1 = 11$ by writing $2x = 10$. Which property justifies this step?
   (1) Commutative property of addition
   (2) Distributive property of multiplication over addition
   (3) Addition property of equality
   (4) Subtraction property of equality

2. Mila used the multiplication property to justify the first step in solving an equation. The original equation was $\frac{x}{2} + 4 = 10$. What could the equation have been transformed into after this step?
   (1) $\frac{x}{2} = 6$
   (2) $\frac{x}{2} + 6 + 12$
   (3) $x + 8 = 20$
   (4) $x + 2 = 5$

3. What is the solution set for the equation $x - 6 = 7$?
   (1) {1}
   (2) {6}
   (3) {11}
   (4) {13}

4. What value of x makes the equation $3x + 7 = 22$ true?
   (1) 1
   (2) 3
   (3) 5
   (4) 7

5. Find the solution set for the equation $5(x + 4) = 35$.
   (1) {1}
   (2) {2}
   (3) {3}
   (4) {4}

Some Key Algebra I Facts and Skills

6. Solve for d in terms of c, e, and f.
   $cd - e = f$

   (1) $\dfrac{f-e}{c}$
   (2) $\dfrac{f+e}{c}$
   (3) $\dfrac{f}{c} + e$
   (4) $\dfrac{f}{c} - e$

7. Solve for m in terms of a, b, and c.
   $b - ma = c$

   (1) $\dfrac{b-c}{-a}$
   (2) $(c-b) - a$
   (3) $\dfrac{c-b}{-a}$
   (4) $(c+b) - a$

8. Solve for r in terms of c and π.
   $c = 2\pi r$

   (1) $\dfrac{2c}{\pi}$
   (2) $\dfrac{2\pi}{c}$
   (3) $\dfrac{c}{2\pi}$
   (4) $\dfrac{2}{c\pi}$

Properties of Algebra and Solving Linear Equations with Algebra

Solutions

1. The first step of the process is to subtract 1 from each side of the equation. This is called the subtraction property of equality.

   The correct choice is **(4)**.

2. Though it is more common to begin this question by subtracting 4 from both sides of the equation, in this case she does it by multiplying both sides of the equation by 2. The left-hand side becomes $x + 8$ and the right-hand side becomes 20.

   The correct choice is **(3)**.

3. Isolate the x by adding 6 to both sides of the equation. The equation then becomes $x = 13$.

   The correct choice is **(4)**.

4. Subtract 7 from both sides of the equation to get $3x = 15$. Divide both sides of the equation by 3 to get $x = 5$.

   The correct choice is **(3)**.

5. One way to solve this equation is to first distribute the 5 through the left-hand side to get the equation $5x + 20 = 35$, then subtract 20 from both sides of the equation to get $5x = 15$, and finally to divide both sides of the equation by 5 to get $x = 3$. Another way is to first divide both sides of the equation by 5 to get $x + 4 = 7$, and then subtract 4 from both sides of the equation to get $x = 3$.

   The correct choice is **(3)**.

6. First add e to both sides of the equation to get $cd = f + e$. Then divide both sides of the equation by c to get $d = (f+e)/c$.

   The correct choice is **(2)**.

7. First subtract b from both sides of the equation to get $-ma = c - b$. Then divide both sides by $-a$ to get $m = \dfrac{c-b}{-a}$.

   The correct choice is **(3)**.

8. Divide both sides of the equation by 2π to get $\dfrac{c}{2\pi} = r$.

   The correct choice is **(3)**.

2. POLYNOMIAL ARITHMETIC

2.1 CLASSIFYING POLYNOMIALS

A **polynomial** is an expression like $2x + 5$ or $3x^2 - 5x + 3$. The **terms** of a polynomial are separated by $+$ or $-$ signs. The polynomial $2x + 5$ has two terms. The polynomial $3x^2 - 5x + 3$ has three terms. The terms of a polynomial have a **coefficient** and a **variable part**. The term $3x^2$ has a coefficient of 3 and a variable part of x^2. A term with no variable part is called a **constant**.

- A polynomial with three terms is called a **trinomial**.
- A polynomial with two terms is called a **binomial**.
- A polynomial with one term is called a **monomial**.

2.2 MULTIPLYING AND DIVIDING MONOMIALS

- To *multiply* one monomial by another, multiply the coefficients and multiply the variable parts by adding the exponents on the same variables.

$$8x^3 \cdot 2x$$

Multiply the coefficients $8 \cdot 2 = 16$.

Multiply the variable parts by adding the exponents $x^3 \cdot x^1 = x^4$. The solution is $16x^4$.

- To *divide* one monomial by another, divide the coefficients and divide the variable parts by subtracting the exponents on the same variables.

$$8x^3 \div 2x$$

Divide the coefficients. $8 \div 2 = 4$.

Divide the variable parts by subtracting the exponents

$$x^3 \div x^1 = x^2$$

The solution is $4x^2$.

This question can also be expressed as $\dfrac{8x^3}{2x} = 4x^2$.

2.3 COMBINING LIKE TERMS

Like terms are terms that have the same variable part. For example, $3x^2$ and $2x^2$ are like terms because the variable part for both is x^2. Like terms can be added or subtracted by adding or subtracting the coefficients and by not changing the variable part.

$$3x^2 + 2x^2 = 5x^2$$
$$3x^2 - x^2 = 3x^2 - 1x^2 = 2x^2$$

- To simplify $2x + 3 + 4x - 5$, combine the like terms with the variable part of x. $2x + 4x = 6x$. Also combine the constants $3 - 5 = -2$. This expression simplifies to $6x - 2$.

If terms are not like terms, they cannot be combined by adding or subtracting. $3x^2 + 5x^3$ cannot be combined because the exponents are different, so they are not like terms.

2.4 MULTIPLYING MONOMIALS AND POLYNOMIALS

- To *multiply* a polynomial by a monomial, use the **distributive property**.

$$2(3x + 5) = 2 \cdot 3x + 2 \cdot 5 = 6x + 10$$

This works for more complex monomials and polynomials also.

$$2x^2(5x^2 - 7x + 3) = 2x^2 \cdot 5x^2 + 2x^2 \cdot -7x + 2x^2 \cdot 3$$
$$= 10x^4 - 14x^3 + 6x^2$$

2.5 ADDING AND SUBTRACTING POLYNOMIALS

- To *add* two polynomials, remove the parentheses from both and combine like terms.

$$(5x + 2) + (3x - 4) = 5x + 2 + 3x - 4 = 8x - 2$$

- To **subtract** two polynomials, remove the parentheses of the polynomial on the left, then negate all the terms of the polynomial on the right, and remove the parentheses before combining like terms.

$$(5x + 2) - (3x - 4) = 5x + 2 - 3x + 4 = 2x + 6$$

2.6 MULTIPLYING BINOMIALS

- To *multiply* binomials, use the <u>FOIL</u> process.

$$(2x + 3)(5x - 2)$$

The F stands for firsts. Multiply $2x \cdot 5x$, the first term in each of the parentheses. $10x^2$.
The O stands for outers. Multiply $2x \cdot -2$, the terms on the far left and on the far right. $-4x$.
The I stands for inners. Multiply $3 \cdot 5x$, the two terms in the middle. $15x$.
The L stands for lasts. Multiply $3 \cdot -2$, the second term in each of the parentheses. -6
These four answers become $10x^2 - 4x + 15x - 6$. Combine like terms to get $10x^2 + 11x - 6$ in simplified form.

2.7 FACTORING POLYNOMIALS

Factoring a number, like 15, is when two numbers are found that can be multiplied to become that number, $15 = 3 \cdot 5$. Factoring polynomials is more involved and there are certain patterns to be aware of.

- **Greatest Common Factor Factoring**

The terms of some polynomials have a greatest common factor. This can be factored out like a reverse use of the distributive property.
In $6x^2 + 8x$, the terms have a common factor of $2x$. Write $2x$ outside the parentheses and divide each term by $2x$ to determine what goes inside the parentheses.

$$2x(3x + 4)$$

- **Difference of Perfect Squares Factoring**

The expression $a^2 - b^2$ can be factored into $(a - b)(a + b)$. This works anytime both terms of a binomial are perfect squares and there is a minus sign between the two terms.

$$x^2 - 9 = x^2 - 3^2 = (x - 3)(x + 3)$$

- **Reverse FOIL**

A **trinomial** like $x^2 + 8x + 15$ can be factored if there are two numbers that have a sum equal to the coefficient of the x term, 8, and a product equal to the constant 15. Since $3 + 5 = 8$ and $3 \cdot 5 = 15$,

$$x^2 + 8x + 15 = (x + 3)(x + 5).$$

For $x^2 + 3x - 10$, the numbers that have a sum of 3 and a product of -10 are -2 and 5.

$$x^2 + 3x - 10 = (x - 2)(x + 5)$$

2.8 MORE COMPLICATED FACTORING

Sometimes none of the factoring patterns seems to match the polynomial that needs to be factored. When this happens, see if it is possible to rewrite it in a way that resembles the pattern better.

- The polynomial $x^4 - 9$ can be rewritten as $(x^2)^2 - 3^2$, which now has the difference of perfect squares pattern.

$$x^4 - 9 = (x^2)^2 - 3^2 = (x^2 - 3)(x^2 + 3)$$

- The polynomial $x^4 + 8x^2 + 15$ can be rewritten as

$$(x^2)^2 + 8(x^2) + 15$$

$$x^4 + 8x^2 + 15 = (x^2)^2 + 8(x^2) + 15 = (x^2 + 3)(x^2 + 5)$$

Practice Exercises

1. Classify this polynomial $5x^2 + 3$.
   - (1) Monomial
   - (2) Binomial
   - (3) Trinomial
   - (4) None of the above

2. Classify this polynomial $7x^2 - 3x + 2$.
   - (1) Monomial
   - (2) Binomial
   - (3) Trinomial
   - (4) None of the above

3. Multiply $3x^3 \cdot 4x^5$.
   - (1) $7x^8$
   - (2) $7x^{15}$
   - (3) $12x^8$
   - (4) $12x^{15}$

4. Which expression is equivalent to $2x^2 + 5x^2$?
   - (1) $10x^2$
   - (2) $7x^2$
   - (3) $7x^4$
   - (4) The expression cannot be simplified any further.

5. Simplify $6x(2x + 3)$.
   - (1) $8x^2 + 18x$
   - (2) $12x^2 + 18x$
   - (3) $12x^2 + 3$
   - (4) $20x$

28 Some Key Algebra I Facts and Skills

6. Simplify $3x(5x^2 - 2x + 3)$.
    (1) $15x^3 - 2x + 3$
    (2) $15x^3 - 6x^2 + 9x$
    (3) $5x^2 + x + 3$
    (4) $15x^3 + 6x^2 - 9x$

7. Simplify $(3x - 4) - (5x - 3)$.
    (1) $-2x - 7$
    (2) $-2x - 1$
    (3) $2x - 1$
    (4) $2x - 7$

8. $(2x + 3)(3x - 1) =$
    (1) $6x^2 - 3$
    (2) $6x^2 + 11x - 3$
    (3) $6x^2 + 7x - 3$
    (4) $6x^2 - 7x - 3$

9. Factor $x^2 - 2x - 15$.
    (1) $(x - 3)(x - 5)$
    (2) $(x + 3)(x - 5)$
    (3) $(x - 15)(x + 1)$
    (4) $(x + 15)(x - 1)$

10. Factor $x^4 + 7x^2 + 12$.
    (1) $(x^2 + 6)(x^2 + 2)$
    (2) $(x^2 - 3)(x^2 - 4)$
    (3) $(x^2 + 3)(x^2 + 4)$
    (4) This cannot be factored.

Polynomial Arithmetic

Solutions

1. There are two terms, $5x^2$ and 3 separated by a + sign. A polynomial with two terms is called a binomial.

   The correct choice is **(2)**.

2. There are three terms, $7x^2$, $3x$, and 2 separated by + and − signs. A polynomial with three terms is called a trinomial.

   The correct choice is **(3)**.

3. To multiply two monomials, first multiply the coefficients, $3 \cdot 4 = 12$. Then multiply the variable parts. Remember that when multiplying variables, you add the exponents. $x^3 \cdot x^5 = x^{(3+5)} = x^8$. The solution is $12x^8$.

   The correct choice is **(3)**.

4. Since these are like terms with variable part x^2, they can be combined. The solution will also have a variable part of x^2 with a coefficient equal to the sum of the two coefficients. Since $2 + 5 = 7$, $2x^2 + 5x^2 = 7x^2$.

   The correct choice is **(2)**.

5. Using the distributive property it becomes $6x \cdot 2x + 6x \cdot 3 = 12x^2 + 18x$.

   The correct choice is **(2)**.

6. Using the distributive property it becomes $3x \cdot 5x^2 + 3x(-2x) + 3x(3) = 15x^3 - 6x^2 + 9x$.

   The correct choice is **(2)**.

7. Distribute the − sign through the parentheses on the right. The expression becomes $3x - 4 - 5x + 3$. Combine like terms to get $-2x - 1$.

    The correct choice is **(2)**.

8. Use the FOIL process. The firsts are $2x \cdot 3x = 6x^2$. The outers are $2x \cdot (-1) = -2x$. The inners are $3 \cdot 3x = 9x$. The lasts are $3 \cdot (-1) = -3$. Combine these four terms to get $6x^2 - 2x + 9x - 3$. Combine like terms to get $6x^2 + 7x - 3$.

    The correct choice is **(3)**.

9. To factor this quadratic trinomial, find two numbers that have a product of -15 and a sum of -2. The numbers are -5 and $+3$. The factors, then, are $(x - 5)(x + 3)$.

    The correct choice is **(2)**.

10. This trinomial can be written as $(x^2)^2 + 7(x^2) + 12$. It has the same structure, then, as a quadratic trinomial and can be factored by finding two numbers that have a product of 12 and a sum of 7. The two numbers are $+3$ and $+4$. The factors are $(x^2 + 3)$ and $(x^2 + 4)$.

    The correct choice is **(3)**.

3. QUADRATIC EQUATIONS

3.1 METHODS OF SOLVING QUADRATIC EQUATIONS

A **quadratic equation** is an equation that can be written in the form $ax^2 + bx + c = 0$. For example, $x^2 + 4x - 5 = 0$ is a quadratic equation. There are three ways to solve a quadratic equation.

- 1. Solve by factoring. If possible, factor the left-hand side of the equation.

$$x^2 + 4x - 5 = 0$$
$$(x + 5)(x - 1) = 0$$

Since the only way that two things can have a product of zero is if at least one of them is zero, this means that either $(x + 5)$ or $(x - 1)$ must equal zero.

$$
\begin{array}{lll}
x + 5 = 0 & \text{or} & x - 1 = 0 \\
-5 = -5 & & +1 = +1 \\
x = -5 & \text{or} & x = 1
\end{array}
$$

$x = -5$ or $x = 1$ are the solutions to this quadratic equation.

- 2. Solve by completing the square. First eliminate the constant from the left-hand side by adding or subtracting.

$$x^2 + 4x - 5 = 0$$
$$+5 = +5$$
$$x^2 + 4x = 5$$

Next, *divide* the coefficient of the x by 2, square that answer, and add that number to both sides of the equation.

$$\left(\frac{4}{2}\right)^2 = 2^2 = 4$$

$$x^2 + 4x + 4 = 5 + 4$$
$$x^2 + 4x + 4 = 9$$

The left-hand side of the equation will factor.

$$(x + 2)(x + 2) = 9$$
$$(x + 2)^2 = 9$$

- Take the square root of both sides of the equation, putting a ± in front of the square root of the right-hand side.

$$\sqrt{(x+2)^2} = \pm\sqrt{9}$$
$$x + 2 = \pm 3$$
$$-2 = -2$$
$$x = -2 \pm 3$$
$$x = -2 + 3 \text{ or } x = -2 - 3$$
$$x = 1 \text{ or } x = -5$$

- 3. Solve with the quadratic formula. Any quadratic equation of the form $ax^2 + bx + c = 0$ can be solved with the equation

$$x = \frac{-b \pm \sqrt{b^2 - 4ac}}{2a}$$

For $x^2 + 4x - 5 = 0$, $a = 1, b = 4, c = -5$.

$$x = \frac{-4 \pm \sqrt{4^2 - 4 \cdot 1 \cdot (-5)}}{2 \cdot 1} = \frac{-4 \pm \sqrt{16 + 20}}{2} = \frac{-4 \pm \sqrt{36}}{2} = \frac{-4 \pm 6}{2}$$

$$x = \frac{-4 + 6}{2} = \frac{2}{2} = 1 \text{ or } x = \frac{-4 - 6}{2} = -\frac{10}{2} = -5$$

3.2 THE RELATIONSHIP BETWEEN FACTORS AND ZEROS

If a quadratic equation has factors $(x - p)$ and $(x - q)$ then the roots of the equation are p and q. If the equation has roots (or zeros) p and q, the factors are $(x - p)$ and $(x - q)$.

For example, the factors of the equation $x^2 - 7x + 10$ are $(x - 5)$ and $(x - 2)$. Therefore, the zeros of the equation are 5 and 2.

If the roots of a quadratic equation are -2 and 8, then the factors are $(x - (-2))$ and $(x - 8)$. The $(x - (-2))$ can be expressed as $(x + 2)$.

Quadratic Equations

3.3 WORD PROBLEMS INVOLVING QUADRATIC EQUATIONS

Some real-world scenarios can be modeled with quadratic equations.

- **Area Problems**

The width of a rectangle is 3 units more than the length. If the area of the rectangle is 70 square units, what are the length and width of the rectangle?

$$A = 70 \quad l + 3$$
$$l$$

Since area is length times width, for this scenario

$$70 = l \cdot w$$
$$70 = l \cdot (l + 3)$$
$$70 = l^2 + 3l$$
$$0 = l^2 + 3l - 70$$

Any of the methods can be used to solve for the answers $l = -10$ and $l = 7$. Since the length must be positive, the answer is 7 units.

- **Projectile Problems**

The height of a projectile after t seconds can be modeled with a quadratic equation. If the equation for the height of a baseball is $h = -16t^2 + 48t + 64$, when will the baseball land on the ground?

When the ball lands on the ground, the height will be 0.
$0 = -16t^2 + 48t + 64$ is the equation.

- This can be solved by any of the methods for the answers $t = -1$ and $t = 4$. Since the amount of time must be positive, the solution is 4 seconds.

Practice Exercises

1. Find all solutions to $(x+2)^2 = 64$.
   - (1) $\sqrt{62}, -\sqrt{62}$
   - (2) 6, −10
   - (3) 6
   - (4) −10

2. Use completing the square to find both solutions for x in the equation $x^2 + 8x + 16 = 9$.
   - (1) −1, −7
   - (2) −1, −2
   - (3) −2, −3
   - (4) −3, −4

3. Find all solutions to $x^2 + 6x = 0$.
   - (1) 0
   - (2) −6
   - (3) 0, −6
   - (4) 0, 6

4. Solve $x^2 + 10x + 24 = 0$ by factoring.
   - (1) 4, −6
   - (2) −4, −6
   - (3) −4, 6
   - (4) 4, 6

5. If the roots of an equation are 3 and −6, what could the equation be?
   - (1) $(x-3)(x-6) = 0$
   - (2) $(x-3)(x+6) = 0$
   - (3) $(x+3)(x-6) = 0$
   - (4) $(x+3)(x+6) = 0$

Quadratic Equations

6. If the roots of a polynomial are 1 and −8, what could be the factors?
   (1) $(x - 1)$ and $(x + 8)$
   (2) $(x - 1)$ and $(x - 8)$
   (3) $(x + 1)$ and $(x + 8)$
   (4) $(x + 1)$ and $(x - 8)$

7. Solve $x^2 + 4x - 7 = 0$ with the quadratic formula.
   (1) 1.3
   (2) $2 \pm \sqrt{11}$
   (3) $2 \pm \sqrt{12}$
   (4) $-2 \pm \sqrt{11}$

8. The width of a rectangle is 10 inches longer than its length. If the area of the rectangle is 56 square inches, which equation could be used to determine its length (l)?
   (1) $l(l + 10) = 56$
   (2) $l(l - 10) = 56$
   (3) $2l + 2(l + 10) = 56$
   (4) $2l - 2(l + 10) = 56$

9. The height of a projectile in feet at time t is determined by the equation $h = -16t^2 + 128t + 320$. At what time will the projectile be 560 feet high?
   (1) 4 seconds
   (2) 5 seconds
   (3) 6 seconds
   (4) 7 seconds

10. The height of a projectile in feet at time t is determined by the equation $h = -16t^2 + 112t + 128$. At what time will the projectile be 0 feet high?
    (1) 5 seconds
    (2) 6 seconds
    (3) 7 seconds
    (4) 8 seconds

Solutions

1. Take the square root of both sides to get $x + 2 = \pm 8$. The equation $x + 2 = 8$ has solution $x = 6$. The equation $x + 2 = -8$ has solution $x = -10$.

   The correct choice is **(2)**.

2. Since the constant, 16, is already equal to the square of half the coefficient $\left(\frac{8}{2}\right)^2$, the left-hand side of the equation is already a perfect square trinomial.

   It can be factored as $(x + 4)^2 = 9$. Then take the square root of both sides to get $x + 4 = \pm 3$. The equation $x + 4 = 3$ has solution $x = -1$. The equation $x + 4 = -3$ has solution $x = -7$.

   The correct choice is **(1)**.

3. Factor out an x to get the equation $x(x + 6) = 0$. This equation is true when either $x = 0$ or when $x + 6 = 0$, which becomes $x = -6$.

   The correct choice is **(3)**.

4. The two numbers that have a product of 24 and a sum of 10 are $+4$ and $+6$. The factors are $(x + 4)(x + 6)$. The solutions to the equation $(x + 4)(x + 6) = 0$ are when $x + 4 = 0$, which becomes $x = -4$ and also when $x + 6 = 0$, which becomes $x = -6$.

   The correct choice is **(2)**.

5. When a is a root of an equation, $(x - a)$ is a factor. If the roots are 3 and -6, the factors can be $(x - 3)$ and $(x - (-6)) = (x + 6)$.

   The correct choice is **(2)**.

6. If 1 is a root, $(x - 1)$ is a factor. If -8 is a root $(x - (-8)) = (x + 8)$ is a factor.

   The correct choice is **(1)**.

7. $a = 1, b = 4, c = -7$.

$$x = \frac{-4 \pm \sqrt{4^2 - 4(1)(-7)}}{2} = \frac{-4 \pm \sqrt{16 + 28}}{2} = \frac{-4 \pm \sqrt{44}}{2}$$

$$= \frac{-4 \pm 2\sqrt{11}}{2} = -2 \pm \sqrt{11}$$

The correct choice is **(4)**.

8. The area of a rectangle is $l \cdot w$. If l is the length and $l + 10$ is the width, the area is $l(l + 10)$. If the area is known to be 56, the equation that can be used to solve for l is $l(l + 10) = 56$.

The correct choice is **(1)**.

9. When 560 is substituted for h, the equation becomes $560 = -16t^2 + 128t + 320$. Subtract 560 from both sides of the equation to get $0 = -16t^2 + 128t - 240$. Divide both sides by -16 to get $0 = t^2 - 8t + 15$. The right-hand side factors and the equation becomes $0 = (t - 3)(t - 5)$ with solutions $t = 3$ and $t = 5$. Of the choices listed, only 5 seconds is correct.

The correct choice is **(2)**.

10. Substituting $h = 0$ into the equation, it becomes $0 = -16t^2 + 112t + 128$. Divide both sides by -16 to get $0 = t^2 - 7t - 8$. The right side factors so the equation becomes $0 = (t - 8)(t + 1)$. The solutions are $t = 8$ and $t = -1$. Since the amount of time must be positive, the -1 is rejected.

The correct choice is **(4)**.

4. SYSTEMS OF LINEAR EQUATIONS

4.1 WHAT IS A SYSTEM OF LINEAR EQUATIONS?

A **system of equations** is a set of two equations that each have two variables. The system is *linear* if there are no exponents greater than one on any of the variables.

$$2x + 3y = 21$$
$$5x - 2y = 5$$

is a system of linear equations.

The solution to a system of equations is the set of **ordered pairs** that satisfy both equations at the same time. For the system above, the solution is the ordered pair (3, 5) since

$$2(3) + 3(5) = 6 + 15 = 21$$
$$5(3) - 2(5) = 15 - 10 = 5$$

There are two main techniques for solving systems of equations.

4.2 SOLVING A SYSTEM OF EQUATIONS WITH THE SUBSTITUTION METHOD

If one of the two equations is in the form $y = mx + b$, use **the substitution method**.

$$y = 2x + 3$$
$$2x + 4y = 42$$

- Substitute the expression $2x + 3$ for the y in the second equation.

$$2x + 4(2x + 3) = 42$$

Solve for x.

$$2x + 8x + 12 = 42$$
$$10x + 12 = 42$$
$$-12 = -12$$
$$\frac{10x}{10} = \frac{30}{10}$$
$$x = 3$$

Systems of Linear Equations

- Substitute 3 for x into either of the equations and solve for y.

$$y = 2(3) + 3$$
$$y = 6 + 3$$
$$y = 9$$

The solution is the ordered pair (3, 9).

4.3 SOLVING A SYSTEM OF EQUATIONS WITH THE ELIMINATION METHOD

When both equations are in the form $ax + by = c$, use the **elimination method**. The elimination method is when the two equations are combined in a way that one of the variables is eliminated.

In the system of equations below, the y term will be eliminated if the two equations are added. When the coefficient of one of the variables in one equation is the same number with the opposite sign of the same variable in the other equation, add the equations to eliminate that variable. In this case, the $+2y$ has the opposite coefficient as the $-2y$.

$$3x + 2y = 19$$
$$+4x - 2y = 16$$
$$\frac{7x}{7} = \frac{35}{7}$$
$$x = 5$$

- Substitute 5 for x into either of the equations

$$3(5) + 2y = 19$$
$$15 + 2y = 19$$
$$-15 = -15$$
$$\frac{2y}{2} = \frac{4}{2}$$
$$y = 2$$

The solution is the ordered pair (5, 2).

For some systems of equations, one or both of the equations need to be changed so the elimination will happen.

In the system

$$2x - 4y = 4$$
$$3x + 2y = 14$$

Since 4 is a multiple of 2, multiply both sides of the bottom equation by 2 and the coefficients of the y variables will be the same number with opposite signs.

$$2x - 4y = 4$$
$$2(3x + 2y) = 2(14)$$
$$2x - 4y = 4$$
$$+6x + 4y = 28$$
$$\frac{8x}{8} = \frac{32}{8}$$
$$x = 4$$

$$2(4) - 4y = 4$$
$$8 - 4y = 4$$
$$-8 = -8$$
$$\frac{-4y}{-4} = \frac{-4}{-4}$$
$$y = 1$$

The solution is the ordered pair (4, 1).

In the system

$$3x - 5y = 17$$
$$2x + 4y = 4$$

Systems of Linear Equations

- To eliminate the y, determine the least common multiple of 4 and 5, which is 20. Multiply both sides of both equations so that one of the coefficients on the y is -20 and the other is $+20$. To do this, multiply both sides of the top equation by 4 and multiply both sides of the bottom equation by 5.

$$4(3x - 5y) = 4(17)$$
$$5(2x + 4y) = 5(4)$$

$$12x - 20y = 68$$
$$+10x + 20y = 20$$
$$\overline{}$$
$$\frac{22x}{22} = \frac{88}{22}$$
$$x = 4$$

- Substitute 4 for x in either of the original equations and solve for y.

$$3(4) - 5y = 17$$
$$12 - 5y = 17$$
$$-12 = -12$$
$$\frac{-5y}{-5} = \frac{5}{-5}$$
$$y = -1$$

The solution to the system of equations is the ordered pair $(4, -1)$.

4.4 WORD PROBLEMS INVOLVING SYSTEMS OF EQUATIONS

Some real-world scenarios can be modeled with a system of linear equations. Here is a typical example.

If five slices of pizza and three drinks cost $21 and two slices of pizza and five drinks cost $16, how much is it for just one slice of pizza?

- Let x be the cost of a slice of pizza and y be the cost of a drink.

The system of equations is

$$5x + 3y = 21$$
$$2x + 5y = 16$$

- To eliminate the y, make the $3y$ and the $5y$ into $15y$ and $-15y$, respectively, by multiplying both sides of the top equation by 5 and both sides of the bottom equation by -3. Then add the two equations, and solve for x.

$$5(5x + 3y) = 5(21)$$
$$-3(2x + 5y) = -3(16)$$

$$25x + 15y = 105$$
$$+\ -6x - 15y = -48$$
$$\overline{}$$
$$\frac{19x}{19} = \frac{57}{19}$$
$$x = 3$$

Since the question just asked for the price of a slice of pizza, it is not necessary to also find the value of y. A slice of pizza costs $3.

Practice Exercises

1. (2, 5) is a solution to which equation?
   - (1) $x + 2y = 9$
   - (2) $2x + y = 9$
   - (3) $8x - y = 9$
   - (4) $3x + 4y = 25$

2. Which equation has the same solution set as the equation $2x + 3y = 5$?
   - (1) $8x + 12y = 20$
   - (2) $8x + 12y = 15$
   - (3) $6x + 9y = 12$
   - (4) $4x + 6y = 8$

3. Solve the system of equations
$$3x + 2y = 17$$
$$4x - 2y = 4$$
   - (1) (4, 3)
   - (2) (5, 1)
   - (3) (3, 4)
   - (4) (1, 6)

4. Solve the system of equations
$$y = 5x + 3$$
$$2x + 6y = 50$$
   - (1) (1, 8)
   - (2) (8, 1)
   - (3) (−1, −8)
   - (4) (−8, −1)

5. Solve the system of equations

$$2x + 3y = -1$$
$$-2x + 5y = -23$$

(1) $(-4, 3)$ (3) $(4, -3)$
(2) $(4, 3)$ (4) $(-4, -3)$

6. Solve the system of equations

$$y = 3x - 2$$
$$4x - 2y = -4$$

(1) $(10, 4)$ (3) $(-10, -4)$
(2) $(4, 10)$ (4) $(-4, -10)$

7. In order to eliminate the x from this system of equations,

$$12x - 3y = 21$$
$$-2x + 6y = 2$$

you could

(1) Multiply both sides of the first equation by 2.
(2) Multiply both sides of the second equation by 6.
(3) Multiply both sides of the first equation by −2.
(4) Multiply both sides of the second equation by 1/2.

8. Solve the system of equations

$$8x - 2y = 28$$
$$4x + 3y = 6$$

(1) (3, 2)
(2) (−3, 2)
(3) (−3, −2)
(4) (3, −2)

9. Which system of equations can be used to model the following scenario?

There are 50 animals. Some of the animals have two legs and the rest of them have four legs. In total there are 172 legs.

(1) $x + y = 172$
    $2x + 4y = 50$
(2) $x + 50 = y$
    $2x + 172 = 4y$
(3) $y + 50 = x$
    $4y + 172 = 2x$
(4) $x + y = 50$
    $2x + 4y = 172$

10. A pet store has 30 animals. Some are cats and the rest are dogs. The cats cost $50 each. The dogs cost $100 each. If the total cost for all 30 animals is $1,900, how many cats are there?

(1) 8
(2) 20
(3) 22
(4) 24

Solutions

1. Substitute 2 for x and 5 for y into each of the choices. Choice (1) becomes $2 + 2(5) = 9$, which is not true. Choice (2) becomes $2(2) + 5 = 9$, which is true.

   The correct choice is **(2)**.

2. If both sides of an equation are multiplied by the same number, the new equation has the same solution set as the original equation. If both sides of the equation $2x + 3y = 5$ are multiplied by 4, it becomes $8x + 12y = 20$, which is choice (1). The other choices could be obtained by multiplying the right-hand side of the original equation by one number and the left-hand side of the original equation by another number, which will not produce an equation with the same solution set as the original.

   The correct choice is **(1)**.

3. Since the top equation has a $+2y$ and the bottom equation has a $-2y$, the equations can be added together, and the y terms will drop out leading to the equation $7x = 21$ or $x = 3$. To solve for y, substitute 3 for x into either of the original equations, like $3(3) + 2y = 17$, $9 + 2y = 17$, $2y = 8$, which leads to $y = 4$.

   The correct choice is **(3)**.

4. Since the y is isolated in the top equation, substitute $5x + 3$ for y in the bottom equation. It becomes $2x + 6(5x + 3) = 50$, $2x + 30x + 18 = 50$, $32x + 18 = 50$, $32x = 32$, $x = 1$. Only one choice has an x-coordinate of 1.

   The correct choice is **(1)**.

5. Since the top equation has a $+2x$ and the bottom equation has a $-2x$, the equations can be added together to get $8y = -24$, $y = -3$. Substitute -3 for y into either equation to solve for x. $2x + 3(-3) = -1$, $2x - 9 = -1$, $2x = 8$, $x = 4$.

   The correct choice is **(3)**.

Systems of Linear Equations

6. Since the y is isolated in the top equation, substitute $3x - 2$ for y in the bottom equation. $4x - 2(3x - 2) = -4$, $4x - 6x + 4 = -4$, $-2x + 4 = -4$, $-2x = -8$, $x = 4$. To solve for y, substitute 4 for x into either equation. $y = 3(4) - 2 = 12 - 2 = 10$.

    The correct choice is **(2)**.

7. The x will be eliminated after combining two equations when the coefficient of the x in one of the equations is the opposite of the coefficient of the x in the other equation. For choice (1), if both sides of the top equation are multiplied by 2, it would become $24x - 6y = 42$. $+24$ is not the opposite of -2. For choice (2) if both sides of the second equation are multiplied by 6, it becomes $-12x + 36y = 12$. Since -12 is the opposite of $+12$, this is the best answer.

    The correct choice is **(2)**.

8. Multiply both sides of the bottom equation by -2 to get $-8x - 6y = -12$. Add this to the top equation to eliminate the x and get $-8y = 16$. Divide both sides by -8 to get $y = -2$. Substitute -2 for y into one of the original equations. $8x - 2(-2) = 28$, $8x + 4 = 28$, $8x = 24$, $x = 3$.

    The correct choice is **(4)**.

9. If x is the number of two-legged animals and y is the number of four-legged animals, the number of animals is $x + y$ and the number of legs is $2x + 4y$. The system, then, is $x + y = 50$ and $2x + 4y = 172$.

    The correct choice is **(4)**.

10. If x is the number of dogs and y is the number of cats, the system of equations is
$$x + y = 30$$
$$100x + 50y = 1,900$$

    To eliminate the x, multiply both sides of the top equation by -100 to get $-100x - 100y = -3,000$. Add this to the bottom equation to get $-50y = -1,100$ or $y = 22$.

    The correct choice is **(3)**.

5. GRAPHS OF LINEAR EQUATIONS

5.1 GRAPHING THE SOLUTION SET OF A LINEAR EQUATION BY MAKING A TABLE OF VALUES

The equation $x + y = 10$ has an infinite number of ordered pairs that satisfy it. One way to organize the information before creating a graph is to make a table of values.

This is a chart with three ordered pairs satisfying the equation $x + y = 10$. For a linear equation, only two ordered pairs are needed, but it is wise to do an extra ordered pair in case one of your first two is incorrect.

x	y
2	8
3	7
9	1

Plot the ordered pair (2, 8) on the coordinate plane by locating the point that is two units to the right of the y-axis and 8 units above the x-axis. One way to do this is to start at the origin point where the two axes intersect and move to the right two units from there and then up 8 units.

Do the same for the other two ordered pairs on the chart (3, 7) and (9, 1).

Graphs of Linear Equations

Draw a line through the three points. If the three points do not all lie on the same line, one of your ordered pairs is incorrect.

MATH FACTS

The line on a graph contains an infinite number of points. Each point corresponds to an ordered pair that is part of the solution set for the equation, and each ordered pair that is part of the solution set for the equation corresponds to a point on the line.

5.2 GRAPHING A LINEAR EQUATION USING THE INTERCEPT METHOD

When a linear equation is written in the form $ax + by = c$, the graph of the equation can be created quickly by finding the x-intercept and the y-intercept.

For the equation $2x - 3y = 12$:

- To find the y-intercept, substitute 0 for x and solve for y.

$$2(0) - 3y = 12$$
$$\frac{-3y}{-3} = \frac{12}{-3}$$
$$y = -4$$

The y-intercept is $(0, -4)$.

- To find the *x*-intercept, substitute 0 for y and solve for x.

$$2x - 3(0) = 12$$
$$2x = 12$$
$$x = 6$$

The *x*-intercept is (6, 0).

- Plot (0, −4) and (6, 0) on a set of coordinate axes and draw the line that passes through both points.

5.3 CALCULATING AND INTERPRETING SLOPE

The *slope* of a line is a number that measures how steep it is. A horizontal line has a slope of 0. A line with a positive slope goes up as it goes to the right. A line with a negative slope goes down as it goes to the right. The variable used for slope is the letter m.

5.4 GRAPHING A LINEAR EQUATION IN SLOPE-INTERCEPT FORM

When the linear equation is in the form $y = mx + b$, it can be graphed very quickly. The y-intercept is $(0, b)$, and from that point to another point move one unit to the right and m units up.

For the equation $y = 2x - 3$, the y-intercept is $(0, -3)$. From that point, move 1 unit to the right and 2 units up. Draw the line through the two points $(0, -3)$ and $(1, -1)$.

Some Key Algebra I Facts and Skills

- If m is a fraction, n/d, it is more accurate to move d units to the right and n units up from the y-intercept.

- For the equation $y = \frac{2}{3}x + 2$, the y-intercept is (0, 2). From that point move 3 to the right and 2 up to get to the point (3, 4). Draw a line through these two points.

Graphs of Linear Equations

If the *m* value is negative, move down instead of up to get from the *y*-intercept to the next point.

5.5 GRAPHING LINEAR EQUATIONS ON THE GRAPHING CALCULATOR

An equation in slope-intercept form can be quickly graphed on the graphing calculator.

- For the TI-84, press [Y=], enter the equation, and press [ZOOM] and [6].

- For the TI-Nspire, go to the home screen and select [B] for the Graph Scratchpad. Enter the equation on the entry line after $f1(x)=$ and press [enter].

- If a linear equation with two variables is not originally in the slope-intercept form, $y = mx + b$, algebra can be used to rewrite it in slope-intercept form before graphing.

5.6 EQUATIONS OF VERTICAL AND HORIZONTAL LINES

The graph of the equation $y = k$ is a horizontal line through the point $(0, k)$. The graph of the equation $x = h$ is a vertical line through the point $(h, 0)$.

5.7 SOLVING SYSTEMS OF LINEAR EQUATIONS GRAPHICALLY

The solution to a system of equations is the set of coordinates of the point where the graphs of the two lines intersect.

- The solution to the system of equations

$$y = 2x - 1$$
$$y = -\frac{2}{3}x + 7$$

can be found by carefully producing the two graphs on the same set of axes. The coordinates of the intersection point $(3, 5)$ means the solution to the system of equations is $x = 3, y = 5$.

Graphs of Linear Equations

The graphing calculator can also determine the intersection point of two lines. On the TI-84, graph both lines and press [2ND], [TRACE], and [5] to find the intersection. On the TI-Nspire, graph both lines and press [menu], [6], and [4] to find the intersection.

5.8 DETERMINING THE SLOPE OF A LINE

If two points on a line are (x_1, y_1) and (x_2, y_2), then the slope of the line can be determined by the formula

$$m = \frac{y_2 - y_1}{x_2 - x_1}$$

The slope of the line passing through (3, 5) and (7, 8) is

$$m = \frac{8 - 5}{7 - 3} = \frac{3}{4}$$

5.9 INTERPRETING THE SLOPE OF A LINE

When a graph represents a real-world scenario, the slope is the rate that the x value is changing with relation to the y value. In a distance-time graph, the slope of a line segment corresponds to the speed that the moving object is traveling.

In the distance-time graph above, a car's time traveled is represented by the x-coordinate, and the distance it has covered at that time is represented by the y-coordinate. The slope of the segment connecting (0, 0) and (5, 300) is 60 so the car was traveling at a speed of 60 mph for the first 5 hours. The slope of the segment connecting (5, 300) and (8, 540) is 80 so the car was traveling at a speed of 80 mph for the last 3 hours.

5.10 FINDING THE EQUATION OF A LINE THROUGH TWO GIVEN POINTS

If two points are known, use the slope-intercept formula to find the slope of the line. Substitute the value you calculated for m and the x- and y-coordinates of either of the given two points into the equation $y = mx + b$ and solve for b. Substitute the values calculated for m and b into the equation $y = mx + b$.

- For the line through the two points $(3, 5)$ and $(9, 1)$, m is the slope of the line $= \dfrac{1-5}{9-3} = -\dfrac{4}{6} = -\dfrac{2}{3}$.
- Using the point $(3, 5)$, substitute $x = 3$, $y = 5$, $m = -\dfrac{2}{3}$ into the equation $y = mx + b$ and solve for b.

$$5 = -\frac{2}{3}(3) + b$$
$$5 = -2 + b$$
$$+2 = +2$$
$$7 = b$$

$m = -\dfrac{2}{3}$ and $b = 7$ so the equation is $y = -\dfrac{2}{3}x + 7$.

Practice Exercises

1. What is the x-intercept of the graph of the solution set of the equation $2x + 5y = 20$?

(1) (10, 0)
(2) (4, 0)
(3) (0, 10)
(4) (0, 4)

2. This is a graph of the solution set of which equation?

(1) $x = 8$
(2) $x = -8$
(3) $y = 8$
(4) $y = -8$

Graphs of Linear Equations

3. Below is the graph of the solution set of an equation. Based on this graph, which ordered pair does not seem to be part of the solution set of the equation $y = \frac{1}{3}x + 2$?

(1) (3, 3)
(2) (6, 4)
(3) (6, 8)
(4) (9, 5)

4. What is the slope of the line that passes through $(-2, 1)$ and $(8, 5)$?

(1) $\frac{2}{5}$
(2) $-\frac{2}{5}$
(3) $\frac{5}{2}$
(4) $-\frac{5}{2}$

Some Key Algebra I Facts and Skills

5. Below is a distance-time graph for a bicycle trip. During which time interval is the cyclist going the fastest?

(1) 0 to 2 hours
(2) 2 to 3 hours
(3) 3 to 5 hours
(4) 5 to 8 hours

6. What is the slope of the line defined by the equation $y = -3x + 4$?

(1) 3
(2) -3
(3) 4
(4) -4

7. This is the graph of the solution set of which equation?

(1) $y = 5x + \dfrac{1}{3}$
(2) $y = 5x + 3$
(3) $y = 3x + 5$
(4) $y = \dfrac{1}{3}x + 5$

Graphs of Linear Equations

8. Find the equation of the line passing through the two points (0, −7) and (5, 8).
   (1) $y = 3x + 7$
   (2) $y = 3x - 7$
   (3) $y = \frac{1}{3}x + 7$
   (4) $y = \frac{1}{3}x - 7$

9. Find the equation of the line passing through the two points (4, −2) and (12, 4).
   (1) $y = \frac{4}{3}x - 5$
   (2) $y = \frac{4}{3}x + 5$
   (3) $y = \frac{3}{4}x - 5$
   (4) $y = \frac{3}{4}x + 5$

10. Find the equation of the line through the points (3, 5) and (3, 8).
    (1) $x = -3$
    (2) $x = 3$
    (3) $y = -3$
    (4) $y = 3$

Solutions

1. The *x*-intercept has a *y*-coordinate of zero. Substitute zero for *y* to get $2x + 5(0) = 20$, $2x = 20$, $x = 10$. The *x*-intercept is $(10, 0)$.

   The correct choice is **(1)**.

2. The points on this vertical line all have *x*-coordinates of 8. The equation is $x = 8$.

   The correct choice is **(1)**.

3. When the four points are plotted on the same graph as the line, only $(6, 8)$ is not on the line.

   The correct choice is **(3)**.

4. Using the slope formula $m = \dfrac{y_2 - y_1}{x_2 - x_1}$, with the points $(-2, 1)$ and $(8, 5)$ becomes $m = \dfrac{5 - 1}{8 - (-2)} = \dfrac{4}{10} = \dfrac{2}{5}$.

   The correct choice is **(1)**.

5. The interval between 0 and 2 has a slope of $30/2 = 15$. The interval between 2 and 3 has a slope of $10/1 = 10$. The interval between 3 and 5 has a slope of $40/2 = 20$. The interval between 5 and 8 has a slope of $54/3 = 18$. Since the slope represents the speed the bicycle is going, the interval between 3 and 5 hours is the fastest. It can also be seen from the graph that the interval between 3 and 5 hours looks the steepest.

   The correct choice is **(3)**.

6. When the equation for a line is in $y = mx + b$ form, the m is the slope. For this equation, the coefficient of the x is -3 so the slope of the line is -3.

   The correct choice is **(2)**.

7. According to the graph, the y-intercept is $(0, 5)$. The slope of the line through $(0, 5)$ and $(3, 6)$ is $m = \dfrac{6-5}{3-0} = \dfrac{1}{3}$. In $y = mx + b$ form, the m is the slope and the b is the y-coordinate of the y-intercept so the equation is $y = \dfrac{1}{3}x + 5$.

   The correct choice is **(4)**.

8. The slope of the line is $m = \dfrac{8-(-7)}{5-0} = \dfrac{15}{5} = 3$. Choose one of the points for the x and the y value and 3 for the m value. Using the point $(5, 8)$ the equation $y = mx + b$ becomes $8 = 3(5) + b$, $8 = 15 + b$, $b = -7$. The equation is $y = 3x - 7$.

   The correct choice is **(2)**.

9. The slope of the line is $m = \dfrac{4-(-2)}{12-4} = \dfrac{6}{8} = \dfrac{3}{4}$. Pick one of the points for x and y and substitute $\dfrac{3}{4}$ for m into $y = mx + b \cdot 4 = \dfrac{3}{4}$. $12 + b$, $4 = 9 + b$, $b = -5$. The equation is $y = \dfrac{3}{4}x - 5$.

   The correct choice is **(3)**.

10. Since the x-coordinates are the same, the line is a vertical line. Vertical lines have equations $x = $ constant. Since the x-coordinates are both 3, the equation is $x = 3$.

    The correct choice is **(2)**.

6. GRAPHS OF QUADRATIC EQUATIONS

6.1 GRAPHING A QUADRATIC EQUATION WITH A CHART

A **quadratic equation** can be written in the form $y = ax^2 + bx + c$. The graph of a quadratic equation is always a **parabola**, which resembles the letter U. When the a value is positive, the parabola looks like a right-side-up U. When the a value is negative, the parabola looks like an upside-down U.

- The simplest way to create the graph of a quadratic equation is to choose at least five consecutive x values and create a chart.

For the equation $y = x^2 - 2x - 3$, the chart could look like this:

x	y
-2	$(-2)^2 - 2(-2) - 3 = 4 + 4 - 3 = 5$
-1	$(-1)^2 - 2(-1) - 3 = 1 + 2 - 3 = 0$
0	$(0)^2 - 2(0) - 3 = -3$
1	$(1)^2 - 2(1) - 3 = 1 - 2 - 3 = -4$
2	$(2)^2 - 2(2) - 3 = 4 - 4 - 3 = -3$

- The graph of these five ordered pairs looks like this:

- *Connect* the points with a U-shaped parabola.

The low point of the parabola is called the **vertex**. In this example the vertex is $(1, -4)$. The two x-intercepts of the parabola are $(-1, 0)$ and $(3, 0)$. The y-intercept of this parabola is $(0, -3)$.

6.2 GRAPHING A PARABOLA BY FINDING THE VERTEX AND INTERCEPTS

The coordinates of the vertex, the y-intercept, and the x-intercepts can be calculated. These four points help produce an accurate graph of the parabola.

- The y-intercept of the parabola for the equation $y = ax^2 + bx + c$ is $(0, c)$.
- For the equation $y = x^2 - 2x - 3$, the y-intercept is $(0, -3)$.

The x-coordinate of the vertex of the parabola for the equation $y = ax^2 + bx + c$ is $x = -\frac{b}{2a}$. To get the y-coordinate, substitute the $-\frac{b}{2a}$ value in for x in the equation $y = ax^2 + bx + c$ to solve for y.

- For the equation $y = x^2 - 2x - 3$, the x-coordinate of the vertex is $x = -\frac{-2}{2(1)} = \frac{2}{2} = 1$. The y-coordinate is $y = 1^2 - 2(1) - 3 = 1 - 2 - 3 = -4$. So the vertex is $(1, -4)$.

The x-intercepts of the parabola for the equation $y = ax^2 + bx + c$ can be found by solving the quadratic equation $0 = ax^2 + bx + c$.

For the equation $y = x^2 - 2x - 3$, this becomes

$$0 = x^2 - 2x - 3$$

This quadratic equation can be solved by factoring.

$$0 = (x + 1)(x - 3)$$
$$x + 1 = 0 \text{ or } x - 3 = 0$$
$$-1 = -1 \quad +3 = +3$$
$$x = -1 \quad \text{or} \quad x = 3$$

The x-intercepts are $(-1, 0)$ and $(3, 0)$.

These four points help make an accurate sketch of the parabola.

Graphs of Quadratic Equations

6.3 GRAPHING QUADRATIC EQUATIONS ON THE GRAPHING CALCULATOR

The graphing calculator can easily graph quadratic equations. On the TI-84 press [Y=] and enter the equation after "Y1=", and press [ZOOM] and [6]. On the TI-Nspire, from the home screen press [B] for the Graph Scratchpad. Then enter the equation on the entry line and press [enter].

The graphing calculator can also determine the vertex and the intercepts using the min/max feature and the zeros feature.

- For the TI-84, press [2ND], [TRACE], and [3].

- For the TI-Nspire, press [menu], [6], and [2] to find the minimum point.

6.4 USING THE GRAPHING CALCULATOR TO SOLVE QUADRATIC EQUATIONS

The solutions to the equation $ax^2 + bx + c = 0$ are also the x-coordinates of the x-intercepts of the parabola defined by $y = ax^2 + bx + c$.

- To solve the equation $x^2 - 2x - 3 = 0$ with the graphing calculator, graph $y = x^2 - 2x - 3$ and then use the zeros feature to find the x-intercepts.
- For the TI-84, enter the equation $Y1 = x^2 - 2x - 3$ and [ZOOM] and [6] to graph. Then press [2ND], [TRACE], and [2] to find each zero.

- For the TI-Nspire, enter the equation $f1(x) = x^2 - 6x + 8$ to graph it. Then press [menu], [6], and [1] to find each zero.

Graphs of Quadratic Equations

6.5 SOLVING A LINEAR-QUADRATIC SYSTEM OF EQUATIONS BY GRAPHING

When a system of equations has one linear equation and one quadratic equation, one way to find the solution is to graph the line and the parabola and find the coordinates of the intersection point, or intersection points. There can be up to two solutions.

- **For the System of Equations**

$$y = -2x + 1$$
$$y = x^2 - 2x - 3$$

- Create the graph for both equations either by hand or with the graphing calculator. The coordinates of the two intersection points are the two solutions to the system of equations. For this system the two solutions are $x = -2, y = 5$ and $x - 2, y = -3$.

This can also be done on the graphing calculator.

Some Key Algebra I Facts and Skills

- For the TI-84, graph the two equations and press [2ND], [TRACE], and [5] for each intersection point.

- For the TI-Nspire, graph the two equations and press [menu], [6], and [4] to find the intersection points.

6.6 WORD PROBLEMS INVOLVING THE GRAPH OF A QUADRATIC EQUATION

The height of a projectile thrown in the air can be modeled with a quadratic equation. The graph of this equation provides information about when the projectile reaches its highest point and when the projectile lands on the ground.

- If the equation is $h = -16t^2 + 48t + 160$, the y-coordinate of the vertex of the graph is the highest point the projectile reaches. The x-coordinate of the vertex is the amount of time it takes for the projectile to reach its highest point. The x-coordinate of the x-intercept is the amount of time it takes for the projectile to land. For this example, the highest point the projectile reaches is 196 feet high after 1.5 seconds. The projectile lands after 5 seconds.

Practice Exercises

1. Which is a point on the graph of the solution set of $y = x^2 + 5x - 2$?
   - (1) (3, 19)
   - (2) (3, 20)
   - (3) (3, 21)
   - (4) (3, 22)

2. What are the coordinates of the vertex of the parabola defined by the equation $y = x^2 - 4x - 1$?
   - (1) (−2, 5)
   - (2) (2, 5)
   - (3) (−2, −5)
   - (4) (2, −5)

3. $x = -4$ is the x-coordinate of the vertex for the parabola defined by which equation?
   - (1) $y = x^2 + 8x + 3$
   - (2) $y = x^2 - 8x + 3$
   - (3) $y = x^2 + 4x + 3$
   - (4) $y = x^2 - 4x + 3$

Graphs of Quadratic Equations   75

4. What could be the equation that determines this parabola?

(1) $y = x^2 - 6x - 8$ (3) $y = x^2 + 6x - 8$
(2) $y = x^2 - 6x + 8$ (4) $y = x^2 + 6x + 8$

5. Which is the graph of $y = -x^2 + 2x + 3$?

(1)

(2)

(3)

(4)

6. Which ordered pair is a solution to the system?

$$y = x^2$$
$$y = x + 2$$

(1) (1, 4)
(2) (2, 4)
(3) (3, 4)
(4) (4, 4)

7. Solve this system of equations using algebra.

$$y = x^2$$
$$y = 2x + 3$$

(1) (−1, 1) and (9, 3)
(2) (−1, 1) and (3, 9)
(3) (−1, 1) and (−3, 9)
(4) (1, −1) and (9, 3)

8. Which system of equations could be used to solve this graph?

(2, 4)
(0, 1)

(1) $y = x^2$
$y = \frac{2}{3}x + 1$

(2) $y = x^2$
$y = \frac{3}{2}x - 1$

(3) $y = x^2$
$y = \frac{3}{2}x + 1$

(4) $y = x^2$
$y = \frac{2}{3}x - 1$

Graphs of Quadratic Equations

9. The x-intercepts of the parabola defined by which equation are the solutions to the equation $x^2 + 5x = 15$?
   (1) $y = x^2 + 5x + 15$
   (2) $y = x^2 + 5x$
   (3) $y = x^2 + 5x - 15$
   (4) $y = x^2 - 5x - 15$

10. Solve for all values of x, rounded to the nearest hundredth, $x^2 + 10x + 23 = 0$.
    (1) $-3.41, -6.58$
    (2) $-3.59, -6.41$
    (3) $-3.31, -6.64$
    (4) $-3.62, -6.18$

Solutions

1. For each choice, substitute the x-coordinate in for x and the y-coordinate in for y to see which makes the equation true. 22 does equal $3^2 + 5(3) - 2$, so the solution is $(3, 22)$.

   The correct choice is **(4)**.

2. The x-coordinate of the vertex of a parabola $y = ax^2 + bx + c$ is $x = \frac{-b}{2a}$.

   Since a is 1 and b is -4, the x-coordinate of the vertex is $x = -\frac{(-4)}{2(1)} = 2$. To get the y-coordinate of the vertex, substitute 2 for x into the equation and solve for y. $y = 2^2 - 4(2) - 1 = 4 - 8 - 1 = -5$. The vertex is $(2, -5)$.

   The correct choice is **(4)**.

3. Graph each choice on the graphing calculator and use the minimum feature to find the vertex of each. The vertex of the parabola $y = x^2 + 8x + 3$ is $(-4, -15)$, which has an x-coordinate of -4.

   The correct choice is **(1)**.

4. Since the x-intercepts are $(2, 0)$ and $(4, 0)$, the equation for the parabola is $y = a(x - 2)(x - 4) = a(x^2 - 6x + 8)$. Since the y-intercept is $+8$, $a = 1$ and the equation is $y = x^2 - 6x + 8$.

   The correct choice is **(2)**.

5. The x-coordinate of the vertex must be $x = -\frac{2}{2(1)} = 1$.

   The correct choice is **(2)**.

6. Substitute x^2 for y in the bottom equation to get $x^2 = x + 2$, $x^2 - x - 2 = 0$, $(x - 2)(x + 1) = 0$, $x = 2$ or $x = -1$. The x-coordinates of the intersection points are 2 and -1. This can also be done with the graphing calculator. The answer is $(2, 4)$.

   The correct choice is **(2)**.

7. Substitute x^2 for y in the bottom equation to get $x^2 = 2x + 3$, $x^2 - 2x - 3 = 0$, $(x - 3)(x + 1) = 0$. $x = 3$ and $x = -1$ are the x-coordinates of the two intersection points.

   The correct choice is (2).

8. The parabola with vertex (0, 0) passing through (2, 4) is $y = x^2$. The line through (0, 1) and (2, 4) is $y = (3/2)x + 1$.

   The correct choice is (3).

9. This equation can be rewritten as $x^2 + 5x - 15 = 0$. Equations in this form can be solved by finding the x-intercepts of the parabola $y = x^2 + 5x - 15$.

   The correct choice is (3).

10. Using the graphing calculator, graph $y = x^2 + 10x + 23$ and use the "zero" feature to get $x = 3.59$ or $x = -6.41$.

    The correct choice is (2).

7. LINEAR INEQUALITIES

7.1 ONE-VARIABLE LINEAR INEQUALITIES

A linear inequality is like a linear equation, but instead of an = sign there is either a >, <, ≥, or ≥ sign. Solving a one-variable linear inequality is almost the same as solving a linear equality. The only difference is that when multiplying or dividing both sides by a negative number to eliminate the coefficient, the direction of the inequality sign must be reversed.

$$-2x + 3 < 11$$

- Eliminate the constant 3 by subtracting it from both sides of the inequality.

$$-2x + 3 < 11$$
$$-3 = -3$$
$$-2x < 8$$

- *Divide* both sides by -2 to isolate the x. Because you are dividing by a negative, the direction of the inequality sign must be reversed to keep the equation true. If the -2 were a $+2$, the direction of the inequality sign would not need to be reversed.

$$\frac{-2x}{-2} < \frac{8}{-2}$$
$$x > -4$$

The solution is $x > -4$.

7.2 GRAPHING TWO-VARIABLE INEQUALITIES

$y \leq 2x - 5$ is a two-variable inequality. To graph the two-variable inequality, first graph the line $y = 2x - 5$.

One side of the line needs to be shaded in. To determine which side of the line to shade, substitute the ordered pair (0, 0) into the inequality to test if it yields a true inequality.

$$0 < 2(0) - 5$$
$$0 < 0 - 5$$
$$0 < -5 \text{ is not true.}$$

Since (0, 0) does not satisfy the inequality, shade the side of the line that does not contain (0, 0).

- If the inequality sign is a < or a >, the line must be a dotted line. If it is a ≤ or a ≥, it must be a solid line. If (0, 0) is on the line, an ordered pair that is not on the line must be used to test which side to shade.
- For the graph of $y > 3x$, graph the line $y = 3x$ and make it a dotted line.

Since (0, 0) is on the line, test a point that is not on the line, like (2, 0).

$$0 > 3(2)$$

Since $0 > 6$ is not true, shade the side that does not contain (2, 0).

Linear Inequalities

7.3 GRAPHING SYSTEMS OF LINEAR INEQUALITIES

To graph the solution set to a system of linear inequalities, graph both inequalities on the same graph and locate the portion of the graph that is shaded twice.

$$y > 2x - 5$$
$$y < -\frac{2}{3}x + 2$$

The portion of the graph that has an S is the solution set. Any point in that region will satisfy the system of inequalities. (−3, 1) is an example of a point that is in this double-shaded region and will satisfy both inequalities.

7.4 GRAPHING INEQUALITIES ON THE GRAPHING CALCULATOR

The TI-84 and the TI-Nspire c an graph linear inequalities.

- On the TI-84, the shading can be set by moving the cursor to the icon to the left of the Y= and pressing [ENTER] until the icon shows the proper shading.

- On the TI-Nspire, delete the "=" from the equation in the entry line and choose the symbol to replace it with.

The graphing calculator can also graph systems of linear inequalities by graphing both inequalities on the same set of coordinate axes.

Practice Exercises

1. What is the solution set for $3x < -18$?
   - (1) $x < -6$
   - (2) $x > -6$
   - (3) $x \geq -6$
   - (4) $x \leq -6$

2. What is the solution set for $-4x \geq 20$?
   - (1) $x \geq -5$
   - (2) $x > -5$
   - (3) $x \leq -5$
   - (4) $x < -5$

3. What is the smallest integer that satisfies the equation $-6x < -18$?
   - (1) 3
   - (2) 4
   - (3) 5
   - (4) 6

Some Key Algebra I Facts and Skills

4. Which is the graph of $y < x + 4$?

(1)

(3)

(2)

(4)

5. Below is the graph of $x + y \leq 8$. Which is a point in the solution set?

(1) (1, 8)
(2) (2, 7)
(3) (4, 5)
(4) (3, 5)

Linear Inequalities

6. All of these ordered pairs are part of the shaded region for the graph of $2x + y \leq 12$ except

(1) (2, 8)
(2) (3, 6)
(3) (5, 3)
(4) (4, 4)

7. Which graph shows the solution to the system of inequalities?

$$y < 2x + 1$$
$$y > \frac{1}{3}x + 4$$

(1)

(2)

(3)

(4)

8. Which system of inequalities does the following graph show the solution for?

(1) $y \geq x - 2$
$y < -\frac{2}{3}x + 1$

(2) $y \leq x - 2$
$y > -\frac{2}{3}x + 1$

(3) $y < x - 2$
$y \geq -\frac{2}{3}x + 1$

(4) $y > x - 2$
$y \leq -\frac{2}{3}x + 1$

Linear Inequalities

9. Which graph has the solution set shaded in for the following system of inequalities?

$$y \leq -x + 6$$
$$y \geq \frac{1}{2}x - 1$$

(1)

(2)

(3)

(4)

10. Which is the graph of the following system of inequalities?

$$y \geq 0$$
$$x \leq 0$$

(1)

(2)

(3)

(4)

Linear Inequalities

Solutions

1. Divide both sides of the inequality by 3 to get $x < -6$.

   The correct choice is **(1)**.

2. Divide both sides of the inequality by -4. Switch the direction of the inequality sign because you divided by a negative to get $x \leq -5$.

   The correct choice is **(3)**.

3. Divide both sides of the inequality by -6 and switch the direction of the inequality sign to get $x > 3$. The smallest integer that satisfies the equation is 4. The answer is not 3 since it is not true that $3 > 3$.

   The correct choice is **(2)**.

4. Because it is a $<$ and not a $\leq$ sign, the line must be dotted. Test to see if $(0, 0)$ is in the solution set by checking if $0 < 0 + 4$ is true. Since it is true, the side of the line containing $(0, 0)$ must be shaded.

   The correct choice is **(4)**.

5. When all four choices are plotted on the graph, three of them are not in the shaded region. The point $(3, 5)$ is on the line, but since there is a $\leq$ sign, the line is part of the solution set.

   The correct choice is **(4)**.

6. Three of the points are on the line, which is part of the solution set because it is a $\leq$ sign. The point (5, 3) is not in the shaded area.

   The correct choice is **(3)**.

7. Both of the lines have to be dotted because of the inequality signs $<$ and $>$. This eliminates choices 1 and 3. Both choices 2 and 4 have the region below the line $y = 2x + 1$ shaded. The difference between choices 2 and 4 is that choice 2 has the region above $y = (1/3)x + 4$ shaded, and choice 4 has the region below $y = (1/3)x + 4$ shaded. To check which side is correct, substitute (0,0) into the equation $y > (1/3)x + 4$ to get $0 > (1/3)(0) + 4$. Since 0 is not greater than 4, (0, 0) is not part of the solution set to $y > (1/3)x + 4$. The side of the line $y = (1/3)x + 4$ that contains (0, 0) should not be shaded so the region above the line $y = (1/3)x + 4$ should be shaded.

   The correct choice is **(2)**.

Linear Inequalities

8. Since the line $y = x - 2$ is solid, choices 3 and 4 can be eliminated. Substituting $(0, 0)$ into both inequalities in choice 1 makes them both true while substituting $(0, 0)$ into both inequalities in choice 2 makes them both false.

   The correct choice is **(1)**.

9. To check choice 1, pick a point in the shaded region for that choice and check to see if it satisfies both inequalities. Since $(0, 0)$ is a point in the shaded region, check to see if $0 \leq -0 + 6$ and $0 \geq (1/2) \cdot 0 - 1$. $(0, 0)$ does make both inequalities true.

   The correct choice is **(1)**.

10. The points on or above the x-axis make $y \geq 0$ true. The points on or to the left of the y-axis make $x \leq 0$ true. The points that satisfy both inequalities must be above the x-axis and to the left of the y-axis.

    The correct choice is **(2)**.

8. EXPONENTIAL EQUATIONS

8.1 EVALUATING EXPONENTIAL EXPRESSIONS

An **exponential equation** is one where one of the variables is an exponent. The equation $y = 3 \cdot 2^x$ is a two-variable exponential equation. Exponential equations can be written in the form $y = a \cdot b^x$.

- When evaluating an *exponential expression*, raise the base to the exponent before multiplying. For example, to evaluate $y = 3 \cdot 2^x$ when $x = 4$ it becomes

$$y = 3 \cdot 2^4$$
$$y = 3 \cdot 16 \text{ (NOT } y = 6^4)$$
$$y = 48$$

8.2 EXPONENTIAL GROWTH VS. EXPONENTIAL DECAY

In the exponential equation $y = a \cdot b^x$, the b is called the **base**. When b is greater than 1, the equation is an example of *exponential growth* since $a \cdot b^x$ grows as x grows. When b is between 0 and 1 the equation is an example of *exponential decay* since $a \cdot b^x$ gets smaller as x grows.

$y = 3 \cdot 2^x$ is an example of exponential growth since $b = 2$, which is greater than 1.

$y = 3 \cdot (0.7)^x$ is an example of exponential decay since $b = 0.7$, which is between 0 and 1.

Exponential Equations

8.3 GRAPHS OF EXPONENTIAL EQUATIONS

A table of values for a two-variable exponential equation, like $y = 3 \cdot 2^x$, looks like this:

x	y
-3	$3 \cdot 2^{-3} = 3 \cdot \dfrac{1}{2^3} = 3 \cdot \dfrac{1}{8} = \dfrac{3}{8}$
-2	$3 \cdot 2^{-2} = 3 \cdot \dfrac{1}{2^2} = 3 \cdot \dfrac{1}{4} = \dfrac{3}{4}$
-1	$3 \cdot 2^{-1} = 3 \cdot \dfrac{1}{2^1} = 3 \cdot \dfrac{1}{2} = \dfrac{3}{2}$
0	$3 \cdot 2^0 = 3 \cdot 1 = 3$
1	$3 \cdot 2^1 = 3 \cdot 2 = 6$
2	$3 \cdot 2^2 = 3 \cdot 4 = 12$
3	$3 \cdot 2^3 = 3 \cdot 8 = 24$

Remember that raising a number to a negative power, like a^{-x}, is equivalent to $\dfrac{1}{a^x}$. For example, $2^{-3} = \dfrac{1}{2^3} = \dfrac{1}{8}$.

The graph for this equation has a shape like this:

- For exponential decay, like $y = 3 \cdot \left(\frac{1}{2}\right)^x$, the graph has a shape like this:

Exponential equations can also be graphed on the graphing calculator.

8.4 REAL-WORLD SCENARIOS INVOLVING EXPONENTIAL EQUATIONS

Many real-world scenarios can be modeled with exponential equations. The population of a country over time is generally an example of exponential growth. The temperature of food over time after being put into a freezer is an example of exponential decay.

Practice Exercises

1. If $x = 2$ and $y = 3^x$, solve for y.
   - (1) 8
   - (2) 9
   - (3) 10
   - (4) 11

2. If $x = 3$ and $y = 2 \cdot 3^x$, solve for y.
   - (1) 216
   - (2) 27
   - (3) 54
   - (4) 18

3. Which ordered pair is in the solution set of $y = 5 \cdot 2^x$?
   - (1) (0, 0)
   - (2) (2, 25)
   - (3) (3, 40)
   - (4) (3, 100)

Some Key Algebra I Facts and Skills

4. Which is the graph of $y = 2^x$?

(1)

(3)

(2)

(4)

5. Below is the graph of which equation?

(1) $y = \left(\frac{1}{3}\right)^x$

(2) $y = 10^x$

(3) $y = 4^x$

(4) $y = 3^x$

Exponential Equations

6. In what interval is the graph of $y = 1.5^x$ increasing?
   (1) Always
   (2) Never
   (3) When $x \geq 0$
   (4) When $x \leq 0$

7. Below is the graph of $y = b^x$. What is true about the value of b?

   (1) b must be greater than 1
   (2) b must be less than 1
   (3) b must be less than 0
   (4) b must be less than -1

8. What type of equation has a graph like the one below?

   (1) Linear
   (2) Exponential
   (3) Quadratic
   (4) None of the above

9. The population of a country can be modeled with the equation $P = 250 \cdot 1.07^t$, where P is the population in millions and t is the number of years since 2010. According to this model, rounded to the nearest ten million, what will the population of this country be in 2019?
   (1) 450,000,000
   (2) 460,000,000
   (3) 470,000,000
   (4) 480,000,000

10. A cup of tea that is 200 degrees is put into a room that is 80 degrees. The temperature of the tea can be calculated with the formula $t = 200 \cdot 0.9^m + 80$, where m is the number of minutes since the tea was put into the room. What will the temperature of the tea be after 15 minutes rounded to the nearest degree?
    (1) 116 degrees
    (2) 118 degrees
    (3) 120 degrees
    (4) 121 degrees

Solutions

1. Substitute 2 for x and the equation becomes $y = 3^2 = 9$.

   The correct choice is **(2)**.

2. Substitute 3 for x and the equation becomes $y = 2 \cdot 3^3 = 2 \cdot 27 = 54$.

   The correct choice is **(3)**.

3. Substitute each ordered pair into the equation to see which makes it true. For the ordered pair $(3, 40)$, $5 \cdot 2^3 = 5 \cdot 8 = 40$.

   The correct choice is **(3)**.

4. The graph for $y = 2^x$ must pass through the points $(0, 1)$, $(1, 2)$, and $(2, 4)$. It also passes through $(-1, 1/2)$ and $(-2, 1/4)$.

   The correct choice is **(3)**.

5. Since this graph passes through $(0, 1)$, $(1, 3)$, and $(2, 9)$, it is the equation $y = 3^x$.

   The correct choice is **(4)**.

6. An exponential graph with a base greater than 1 is always increasing.

   The correct choice is **(1)**.

7. When an exponential graph is decreasing, the base is between 0 and 1.

   The correct choice is **(2)**.

8. This graph has the standard shape of an exponential curve based on an equation $y = b^x$ with $b > 1$. An exponential curve like this starts off relatively flat for x values close to zero and then increases rapidly.

    The correct choice is **(2)**.

9. Since 2019 is 9 years past 2010, substitute 9 for t to get $P = 250 \cdot 1.07^9 = 459.6$. In millions, this is approximately 460,000,000.

    The correct choice is **(2)**.

10. Substitute 15 for m to get $t = 200 \cdot 0.9^{15} + 80 = 121.1782$ or 121.2, which rounds to 121.

    The correct choice is **(4)**.

9. CREATING AND INTERPRETING EQUATIONS

9.1 CREATING AND INTERPRETING LINEAR EQUATIONS

Many real-world situations can be modeled with a linear equation in the form $y = mx + b$. The m and the b have to be replaced with the appropriate values from the situation. In general, the b value is the starting value and the m value is the amount the total changes each time the x variable increases.

- If a carnival costs $10 admission and $3 for each ride, the 10 and the 3 can be used in a linear equation. Since the 10 is the starting amount, it would take the place of the b in the equation. Since 3 is the amount the total increases by for each new ride, it would take the place of m in the equation.

  The equation could be written $y = 3x + 10$, where y is the total and x is the number of rides. Instead of x and y, the total could be represented by the variable T, whereas the number of rides could be represented by the variable R to form the equation $T = 3R + 10$.

  When given an equation modeling a real-world situation, it is also possible to interpret what the values for m and b represent. The b represents the starting value, and the m represents the amount that the total changes for each increase in x.

- If the equation for the cost of a pizza with N toppings is $C = 2N + 12$, you could be asked to interpret what the 2 and the 12 represent. Since the 2 is in the place of the m in $y = mx + b$, it represents the cost of each topping. Since the 12 is in the place of the b in $y = mx + b$, it represents the cost of the pizza before any toppings are added.

9.2 CREATING AND INTERPRETING EXPONENTIAL EQUATIONS

An **exponential equation** is a good model for many real-world situations including population growth, compound interest, and liquid cooling. Exponential equations have the form $y = a \cdot (1 + r)^x$, where the a represents the starting value and the r represents the growth rate.

- If the population of a town is 10,000 people and the annual growth rate is 7%, then the equation that relates total population, P, to the number of years that have passed, T, is $P = 10{,}000 \cdot 1.07^T$. Since 10,000 is the starting value, it takes the place of the a and since the growth rate is .07, it takes the place of the r in the equation $y = a \cdot (1 + r)^x$.

- If an exponential equation that models a real-world situation is given, it is possible to interpret what the numbers in the equation represent.

- If, in another town, the equation relating their population to the number of years that have passed is $P = 20{,}000 \cdot 1.09^T$, you could be asked to interpret what the numbers 20,000 and .09 represent. In this case, the 20,000 represents the starting population and the .09 represents the growth rate.

- When a ball is dropped, each bounce is 80% as high as the previous bounce. If the ball is dropped from a window 50 feet above the ground, the equation that relates the height of the bounce (H) to the number of bounces (B) will be $H = 50 \cdot 0.80^B$. Since the starting height is 50, it takes the place of the a, and since 0.80 can be expressed as $(1 - 0.20)$, the r value is -0.20, which is the growth rate. A -0.20 growth rate can also be called a *decay* rate of 0.20.

Practice Exercises

1. It costs $10 to go to the movies and $3 for each bag of popcorn. Which equation relates the total cost (C) to the number of bags of popcorn purchased (P)?

   (1) $C = 3P + 10$
   (2) $C = 10P + 3$
   (3) $P = 3C + 10$
   (4) $P = 10C + 3$

2. A tablet computer costs $400 and $2 for each app. Which equation relates the total cost (C) to the number of apps purchased (A)?

   (1) $A = 2 + 400C$
   (2) $A = 400 + 2C$
   (3) $C = 2 + 400A$
   (4) $C = 400 + 2A$

3. A cable TV plan costs $80 a month plus $10 extra for each premium channel. Which equation relates the monthly bill (B) to the number of premium channels ordered (C)?

   (1) $B = 80C + 10$
   (2) $B = 10C + 80$
   (3) $C = 80B + 10$
   (4) $C = 10B + 80$

4. Lydia wants to buy a DVD player and some DVDs. The equation that relates the total cost for the DVD player and N DVDs is $P = 20N + 200$. What does the number 200 in the equation represent?

   (1) The cost of the DVD player
   (2) The cost of each DVD
   (3) The cost of all N DVDs
   (4) The total cost of the DVD player and all N DVDs

105

5. Amelia buys an empty sticker album and some sticker sheets. The equation that relates the total cost for the empty sticker album and N sticker sheets is $P = 0.75N + 3.00$. What does the number 0.75 in the equation represent?

   (1) The cost of the empty sticker album
   (2) The cost of each sticker sheet
   (3) The cost of all N sticker sheets
   (4) The total cost of the empty sticker album and all N sticker sheets

6. There were 900 birds in a forest. Each year the bird population increases by 12%. Which equation relates the bird population (P) to the number of years that have passed?

   (1) $P = 900(1.12)^t$
   (2) $P = 900(0.12)^t$
   (3) $P = 900(0.88)^t$
   (4) $t = 900(1.12)^P$

7. A bouncing ball is dropped from 20 feet high. After each bounce, the height of the next bounce is 65% as high as the last bounce. Which equation relates the height of the bounce (H) to the number of bounces that have happened (N)?

   (1) $H = 20(0.35)^N$
   (2) $H = 20(1.65)^N$
   (3) $H = 20(0.65)^N$
   (4) $N = 20(0.65)^H$

8. The population (P) of a town after t years can be modeled with the equation $P = 20{,}000(1.07)^t$. What does the 20,000 represent?

   (1) The growth rate
   (2) The percent increase each year
   (3) The population after t years
   (4) The starting population of the town

9. After Allie takes some medicine, each hour the number of milligrams of medicine (M) remaining in her body after t minutes can be modeled with the equation $M = 200(1 - 0.27)^t$. Which number represents the decay rate?
   (1) 0.27
   (2) 200
   (3) 0.73
   (4) 146

10. Mason puts money into a bank that offers interest compounded annually. The formula relating the amount of money in the bank (A) to the number of years it has been in the bank (t) is $A = 800(1.2)^t$. What is the interest rate the bank offers?
    (1) 1.2%
    (2) 2%
    (3) 20%
    (4) 120%

Some Key Algebra I Facts and Skills

Solutions

1. P bags of popcorn cost $3P$. The entrance price is $10 so the total price is $3P + 10$.

   The correct choice is **(1)**.

2. A apps cost $2A$. The computer costs $400 so the total price is $2A + 400$ or $400 + 2A$.

   The correct choice is **(4)**.

3. C premium channels cost $10C$. The monthly cost is $80 so the total price is $10C + 80$.

   The correct choice is **(2)**.

4. The 20 is the cost for each DVD and the 200 is the price of the DVD player. In general, the constant represents the fixed cost.

   The correct choice is **(1)**.

5. The 3 is the part that does not depend on the value of N. The $0.75N$ increases by 0.75 each time N increases by 1 so the 0.75 represents the cost of each sticker sheet.

   The correct choice is **(2)**.

6. In an exponential equation of the form $P = a \cdot (1 + r)^t$, the r value is the percent increase (or decrease if r is negative), and the a is the initial value. When $r = 0.12$ and $a = 900$, the equation is $P = 900(1.12)^t$.

   The correct choice is **(1)**.

Creating and Interpreting Equations

7. The height of each bounce is 0.65 multiplied by the height of the previous bounce. The initial height is 20 feet. After one bounce, the second is $20 \cdot 0.65$. After two bounces it has a height of $20 \cdot 0.65 \cdot 0.65$. In general, after N bounces, the height will be $H = 20 \cdot 0.65^N$.

   The correct choice is **(3)**.

8. In an equation of the form $P = a \cdot (1 + r)^t$, the r is the percent increase each year and the a is the initial value at $t = 0$. The 20,000 then represents the starting population of the town.

   The correct choice is **(4)**.

9. In an equation of the form $M = a \cdot (1 - r)^t$, the r is the decay rate. Since the $(1 - 0.27)$ is being raised to the t power in this equation, the decay rate is 0.27.

   The correct choice is **(1)**.

10. The interest rate is the percent increase each year. The percent increase in an equation of the form $A = P \cdot (1 + r)^t$ is the r variable. Since $1.2 = 1 + 0.2$, the r value is 0.2, which is 20%.

    The correct choice is **(3)**.

10. FUNCTIONS

10.1 DIFFERENT REPRESENTATIONS OF FUNCTIONS

A **function** is like a machine that takes a number as an input and outputs a number. Functions are often named with lowercase letters like f and g.

- If a function f takes the number 2 as an input and outputs the number 7, we say $f(2) = 7$. The number in the parentheses is the number that is input into the function. The number after the equals sign is the number that is output from the function.

A function can be represented in several different ways. Here are the most common ways:

1. As a list of ordered pairs
- If the function f is defined as $f = \{(1, 4), (2, 7), (3, 10), (4, 13), (5, 16)\}$, the numbers in the parentheses represent an input value and an output value for the function. In this example, the point $(1,4)$ in the definition means that if 1 is put into the function, 4 is output from the function, or $f(1) = 4$. Likewise, $f(2) = 7, f(3) = 10, f(4) = 13$, and $f(5) = 16$.

2. As an equation
- A function can be defined by an equation that allows you to calculate the output value for a given input value. An example is the definition $f(x) = 3x + 1$. With this definition, it is possible to calculate the output value for any input value. For example, to calculate $f(10)$, substitute the number 10 into the equation $f(x) = 3x + 1$ to become $f(10) = 3(10) + 1 = 30 + 1 = 31$ so $f(10) = 31$.

Functions

3. As a graph
  - If the function is defined as a graph, determine the value of $f(2)$ by finding a point on the graph that has an x-coordinate of 2. The y-coordinate of that point is the value of $f(2)$. Since the point with x-coordinate of 2 is (2, 7) the value of $f(2) = 7$.

10.2 DOMAIN AND RANGE OF FUNCTIONS

The possible input values of a function are called the *domain* of the function. The possible output values are called the *range* of the function.

For the function $f = \{(1, 4), (2, 7), (3, 10), (4, 13), (5, 16)\}$, the domain is the set of input values {1, 2, 3, 4, 5}. The range is the set of output values {4, 7, 10, 13, 16}.

- When a function is described as a graph, the domain is the set of x-coordinates of all the points on the graph, and the range is the set of y-coordinates of all the points on the graph.

For this graph, the domain is {1, 2, 3, 4}, and the range is {3, 5, 8}.

For this graph, the domain is $2 \leq x \leq 5$, and the range is $3 \leq y \leq 7$.

In a real-world situation, the domain is often a special subset of the real numbers. If the function has as its input the number of cars a salesman sells in a month, the domain would be the set of nonnegative integers $\{0, 1, 2, \ldots\}$ since you cannot sell fractions of a car or negative cars.

10.3 GRAPHING FUNCTIONS

The graph of the function $f(x) = x^2$ is the same as the graph $y = x^2$. All the methods for graphing by hand or with the graphing calculator described earlier can be used for functions.

10.4 HOW TO TELL WHEN A GRAPH OR A LIST CANNOT DEFINE A FUNCTION

In a function, each time a number from the domain is put into the function, the same value is output from the function. So if $f(2) = 7$, the number 7 will always be output from the function whenever 2 is put into the function.

- When a function is defined as a list of ordered pairs, then there will never be two ordered pairs with the same x-coordinate, but different y-coordinates.

$f = \{(1, 4), (2, 7), (2, 10), (3, 13)\}$ is not the definition of a function since $f(2)$ can be 7 or 10.

$g = \{(1, 4), (2, 7), (3, 7), (4, 13)\}$ is the definition of a function. The fact that $g(2) = 7$ and $g(3) = 7$ does not contradict the definition of a function. As long as there are no repeats of x-coordinates, there can be repeats of y-coordinates.

A graph will not be the graph of a function if there are two points that have the same x-coordinate, but different y-coordinates. Graphs that cannot be functions fail the **vertical line test**, which means that at least one vertical line will pass through two or more points. All the points it passes through will have the same x-coordinates but different y-coordinates.

These graphs cannot be graphs of functions since they fail the vertical line test at at least one location.

Some Key Algebra I Facts and Skills

These graphs can be graphs of functions since they pass the vertical line test at all locations.

Functions

10.5 GRAPHING TRANSFORMED FUNCTIONS

- If the graph of $y = f(x)$ is already known, the graph of $y = f(x + a)$, $y = f(x - a)$, $y = f(x) + a$, and $y = f(x) - a$ can be easily graphed by knowing the four basic transformations.

- If the graph of $y = f(x)$ looks like this:

then the graph of

1. $f(x) + a$ will be the graph of $f(x)$ with every point shifted *up* by a units. The graph of $f(x) + 2$ looks like this:

2. $f(x) - a$ will be the graph of $f(x)$ with every point shifted *down* by a units. The graph of $f(x) - 2$ looks like this:

Some Key Algebra I Facts and Skills

3. $f(x + a)$ will be the graph of $f(x)$ with every point shifted *left* by a units. The graph of $f(x + 2)$ looks like this:

4. $f(x - a)$ will be the graph of $f(x)$ with every point shifted *right* by a units. The graph of $f(x - 2)$ looks like this:

Practice Exercises

1. If a function f is defined as $f = \{(1, 2), (2, 3), (3, 1), (4, 4)\}$, what is $f(2)$?
   (1) 1
   (2) 2
   (3) 3
   (4) 4

2. Which of the following *cannot* be the definition of a function?
   (1) $f = \{(1, 5), (2, 7), (2, 8), (4, 9)\}$
   (2) $f = \{(1, 2), (2, 2), (3, 2), (4, 2)\}$
   (3) $f = \{(0, 0), (1, 1), (-1, 1), (2, 4), (-2, 4)\}$
   (4) $f = \{(6, 1)\}$

3. What is the domain of the function defined as $f = \{(1, 4), (3, 7), (4, 8), (5, 8)\}$?
   (1) $\{4, 7, 8\}$
   (2) $\{1, 3, 4, 5, 7, 8\}$
   (3) $\{1, 3, 4, 5\}$
   (4) $\{4\}$

4. Below is the graph of $y = f(x)$. What is the value of $f(3)$?

   (1) 1
   (2) 2
   (3) 3
   (4) 4

5. Which is the graph of a function?

(1)

(2)

(3)

(4)

6. If $g(x) = -x^2 + 7x + 1$ what is $g(2)$?
 (1) 11
 (2) 19
 (3) 27
 (4) 35

Functions

7. If below is the graph of $y = f(x)$, which is the graph of $y = f(x) - 5$?

(1)

(2)

(3)

(4)

8. If $f(x) = 5x - 2$, what is $f(x+1)$?
   (1) $5x - 2$
   (2) $5x - 3$
   (3) $5x + 1$
   (4) $5x + 3$

Some Key Algebra I Facts and Skills

9. Below is the graph of $f(x)$ on the left and $g(x)$ on the right. Which is equivalent to $g(x)$?

   (1) $f(x) + 4$
   (2) $f(x) - 4$
   (3) $f(x + 4)$
   (4) $f(x - 4)$

10. If the graph of $y = f(x)$ is a parabola with the vertex at (5, 1), what is the vertex of the graph of the parabola $y = f(x - 2)$?

    (1) (5, 3)
    (2) (5, −1)
    (3) (7, 1)
    (4) (3, 1)

Functions

Solutions

1. Since the ordered pair (2, 3) is in the set, $f(2) = 3$.

   The correct choice is **(3)**.

2. A set of ordered pairs cannot be the definition of a function if there are two ordered pairs with the same x-coordinate, but different y-coordinates. In choice 1, there are two ordered pairs with x-coordinates of 2, (2, 7) and (2, 8), so that cannot be the definition of a function.

   The correct choice is **(1)**.

3. The domain is the set of x-coordinates. In this example, the x-coordinates are 1, 3, 4, and 5.

   The correct choice is **(3)**.

4. There is a point at (3, 4) so $f(3) = 4$.

   The correct choice is **(4)**.

5. Choices 1, 2, and 4 all fail the vertical line test since at least one vertical line can pass through more than one point on them. Choice 3 can be the graph of a function since there is no possible vertical line that would pass through more than one point on it.

   The correct choice is **(3)**.

6. Substitute 2 for x to get $g(2) = -2^2 + 7 \cdot 2 + 1 = -4 + 14 + 1 = 11$.

   The correct choice is **(1)**. Note that $-2^2 = -4$ and not 4. If there were parentheses around -2, as in $(2)^2$, then it would equal 4.

7. The graph of $y = f(x) = -5$ is the same as the graph of $y = f(x)$ with each point shifted 5 units down.

   The correct choice is **(4)**.

8. Substitute $x + 1$ for x to get $f(x + 1) = 5(x + 1) - 2 = 5x + 5 - 2 = 5x + 3$.

   The correct choice is **(4)**.

9. The graph of $g(x)$ is the same as the graph of $y = f(x)$ with each point shifted 4 units to the right. $g(x)$ must be equivalent to $f(x - 4)$.

   The correct choice is **(4)**.

10. The graph of $f(x - 2)$ is the same as the graph of $f(x)$ with each point shifted two units to the right. If the vertex of the parabola for $f(x)$ is at $(5, 1)$, the vertex of the parabola for $f(x - 2)$ will be two units to the right of $(5, 1)$, which is at $(7, 1)$.

    The correct choice is **(3)**.

11. SEQUENCES

11.1 TYPES OF SEQUENCES

A **sequence** is a series of numbers that can be predicted by some kind of pattern. Two of the most common types of sequences are **arithmetic sequences** and **geometric sequences**.

- The sequence 2, 5, 8, 11, 14,... is an example of an arithmetic sequence since each term after the first can be obtained by adding the same number, 3, to the previous term.

- The sequence 2, 6, 18, 54, 162,... is an example of a geometric sequence since each term after the first can be obtained by multiplying the same number, 2, by the previous term.

- There are two notations that are used to describe the terms of a sequence. If the sequence is 2, 5, 8, 11, 14,... the first term can be described as either $a_1 = 2$ or $a(1) = 2$. The second notation is similar to function notation.

11.2 DESCRIBING A SEQUENCE WITH A DIRECT FORMULA

The terms of the sequence 2, 5, 8, 11,... can be described by the formula $a_n = 2 + 3(n - 1)$ or $a(n) = 2 + 3(n - 1)$. If you substitute $n = 1$ into the formula, it becomes $a_1 = 2 + 3(1 - 1) = 2 + 3(0) = 2 + 0 = 2$.

- For any arithmetic sequence, the direct formula for the nth term is $a_n = a_1 + d(n - 1)$ where a_1 is the first term of the sequence and d is the common difference between two consecutive terms. For the sequence 2, 5, 8, 11,..., $a_1 = 2$ and $d = 3$.

- For any geometric sequence, the direct formula for the nth term is $a_n = a_1 \cdot r^{n-1}$, where a_1 is the first term of the sequence and r is the common ratio between two consecutive terms. For the sequence 2, 6, 18, 54,..., $a_1 = 2$ and $r = 3$ since if you divide any term by the previous term you get 3.

11.3 DESCRIBING A SEQUENCE WITH A RECURSIVE FORMULA

Another way to describe the nth term of a sequence is to relate it to the previous term. The term before the a_n term is called the a_{n-1} term. A *recursive formula* for a sequence first defines the exact value of one or more of the terms and then describes how to obtain new terms from the previous ones.

- For the sequence 2, 5, 8, 11, 14, ... the recursive formula is

$$a_1 = 2$$
$$a_n = 3 + a_{n-1} \text{ for } n > 1$$

From the set of two equations, the entire sequence can be calculated. For example, when $n = 2$, $a_2 = a_1 + 3 = 2 + 3 = 5$. From this, the value of a_3 can be calculated, and this pattern can continue to obtain other n values. The recursive formulas are not convenient for large values of n, however.

- For the sequence 2, 6, 18, 54, 162, ..., the recursive formula is

$$a_1 = 2$$
$$a_n = 3 \cdot a_{n-1} \text{ for } n > 1$$

Practice Exercises

1. What type of sequence is 3, 7, 11, 15, ...?
   (1) Increasing arithmetic
   (2) Decreasing arithmetic
   (3) Increasing geometric
   (4) Decreasing geometric

2. What type of sequence is 4, 8, 16, 32, ...?
   (1) Increasing arithmetic
   (2) Decreasing arithmetic
   (3) Increasing geometric
   (4) Decreasing geometric

3. What is the next number in the sequence 20, 10, 5, $\frac{5}{2}$?
   (1) $\frac{5}{4}$
   (2) $\frac{5}{8}$
   (3) $\frac{5}{16}$
   (4) $\frac{5}{32}$

4. Find the value of a_2 in the sequence defined by
$$a_1 = 4$$
$$a_n = 3 + a_{n-1} \text{ for } n > 1$$
   (1) 3
   (2) 5
   (3) 7
   (4) 12

Some Key Algebra I Facts and Skills

5. The sequence 5, 11, 17, 23, ... can be generated by which definition?

   (1) $a_1 = 5$
   $a_n = 6a_{n-1}$ for $n > 1$

   (2) $a_1 = 5$
   $a_n = \frac{11}{5}a_{n-1}$ for $n > 1$

   (3) $a_1 = 5$
   $a_n = -6 + a_{n-1}$ for $n > 1$

   (4) $a_1 = 5$
   $a_n = 6 + a_{n-1}$ for $n > 1$

6. The sequence 3, 15, 75, 375, ... can be generated by which definition?

   (1) $a_1 = 3$
   $a_n = 5a_{n-1}$ for $n > 1$

   (2) $a_1 = 3$
   $a_n = 12 + a_{n-1}$ for $n > 1$

   (3) $a_1 = 3$
   $a_n = \frac{1}{5}a_{n-1}$ for $n > 1$

   (4) $a_1 = 3$
   $a_n = -12 + a_{n-1}$ for $n > 1$

7. A ball is dropped from a window 50 feet above the ground. Each bounce is $\frac{4}{5}$ the height of the previous bounce. Which definition would generate the height of the bounces?

   (1) $a_1 = 50$
   $a_n = -10 + a_{n-1}$ for $n > 1$

   (2) $a_1 = 50$
   $a_n = 10 + a_{n-1}$ for $n > 1$

   (3) $a_1 = 50$
   $a_n = \frac{4}{5}a_{n-1}$ for $n > 1$

   (4) $a_1 = 50$
   $a_n = \frac{5}{4}a_{n-1}$ for $n > 1$

8. What is the fifth term of the sequence generated by the definition $a_n = 10 - 4(n-1)$?
   - (1) 18
   - (2) 12
   - (3) 2
   - (4) −6

9. What definition would produce the sequence 4, 13, 22, 31, ...?
   - (1) $a_n = 4 + 9n$
   - (2) $a_n = 4 + 9(n-1)$
   - (3) $a_n = 9 + 4n$
   - (4) $a_n = 9 + 4(n-1)$

10. Which expression could be used to find the 20th term of the sequence 5, 9, 13, 17, 21, ...?
    - (1) $5 + 4(20)$
    - (2) $5 + 4(19)$
    - (3) $4 + 5(20)$
    - (4) $4 + 5(19)$

Solutions

1. Since $7 = 3 + 4$ and $11 = 7 + 4$ and $15 = 11 + 4$, this is an increasing arithmetic sequence.

   The correct choice is **(1)**.

2. Since $8 = 2 \cdot 4$ and $16 = 2 \cdot 8$ and $32 = 2 \cdot 16$, this is an increasing geometric sequence.

   The correct choice is **(3)**.

3. Each term is equal to $\frac{1}{2}$ multiplied by the previous term. So the next term is $\frac{1}{2} \cdot \frac{5}{2} = \frac{5}{4}$.

   The correct choice is **(1)**.

4. $a_2 = 3 + a_{2-1} = 3 + a_1 = 3 + 4 = 7$.

   The correct choice is **(3)**.

5. The first term is a_1, so $a_1 = 5$. Each term is equal to 6 plus the previous term, so $a_n = 6 + a_{n-1}$.

   The correct choice is **(4)**.

6. The first term is a_1, so $a_1 = 3$. Each term is equal to 5 multiplied by the previous term, so $a_n = 5 \cdot a_{n-1}$.

   The correct choice is **(1)**.

Sequences

7. The initial height is 50, so $a_1 = 50$. Each bounce is 4/5 the height of the previous bounce so $a_n = 4/5 \cdot a_{n-1}$.

   The correct choice is **(3)**.

8. Substitute 5 for n to get $a_5 = 10 - 4(5-1) = 10 - 4(4) = 10 - 16 = -6$.

   The correct choice is **(4)**.

9. Substitute 1 for n in each equation. Only the second equation gets $a_1 = 4$. Also, an arithmetic sequence has the form $a_n = a_1 + d(n-1)$, where d is the common difference and a_1 is the first term. Since $a_1 = 4$ and $d = 9$, the equation is $a_n = 4 + 9(n-1)$.

   The correct choice is **(2)**.

10. An arithmetic sequence with first term a_1 and common difference of d has the equation $a_n = a_1 + (n-1)d$. a_1 is 5 in this example, and $d = 4$. Therefore, the equation is $a_n = 5 + 4(n-1)$. Substitute 20 for n to get a $20 = 5 + 4(19)$.

    The correct choice is **(2)**.

12. REGRESSION CURVES

12.1 THE LINE OF BEST FIT

A **line of best fit** is a line that comes as close as possible to a set of points on a graph. In the scatterplot below, there is no line that could pass through all ten points. Of all the possible lines, though, there is one that is a better fit than the others and this is the line of best fit.

The equation for the line of best fit can be determined quickly on a graphing calculator.

Below is a scatter plot and the chart on which it was based.

Regression Curves 131

x	y
2	5
5	8
11	4
12	10
16	11
20	15

Instructions for the TI-84:

Press [STAT] and [1], and enter the x values into L1 and the y values into L2.

Press [STAT], [right], [4], and [ENTER] for the equation of the line of best fit.

```
LinReg
y=ax+b
a=.4776785714
b=3.578869048
```

Instructions for the TI-Nspire:

From the home screen select the Add Lists and Spreadsheet icon. In the header row, label column A with an x and column B with a y. Enter the x values into cells A1 to A6. Enter the y values into cells B1 to B6.

Press [menu], [4], [1], and [3] and select x in the X List field and y in the Y List field.

Press the OK button.

The equation for the line of best fit is $y = 0.477679x + 3.57887$.

12.2 THE CORRELATION COEFFICIENT

The **correlation coefficient**, denoted by the variable r, is a number between -1 and 1. When a line of best fit has a positive slope and passes exactly through each of the points, the correlation coefficient is 1. When a line of best fit has a negative slope and passes exactly through each of the points, the correlation coefficient is -1. All correlation coefficients are between -1 and 1.

- The line of best fit for the scatter plot below is close to 1 because it has a positive slope and comes close to the points on the scatter plot. In this case $r = 0.9$

Regression Curves

- The line of best fit for the following scatter plot is close to -1 because it has a negative slope and comes close to the points on the scatter plot, but not as close as the line in the previous example did to the points on that graph. In this case the $r = -0.8$

The graphing calculator can display the correlation coefficient. On the TI-84 it will be displayed along with the line of best fit only if the "diagnostics" are turned on. Press [2ND] and [0] to access the catalog and scroll to the DiagnosticOn command to do this. On the TI-Nspire it will display the correlation coefficient along with the equation of the line of best fit.

The correlation coefficient for the example from Section 12.1 was $r = 0.78552$.

12.3 RESIDUAL PLOTS

When a line is a good fit for a scatterplot, the points in the plot are close to the points on the line. One way to measure this is to calculate the *residuals* and to examine them on a *residual plot*.

Here is a scatterplot with five points (1, 2), (2, 4), (3, 4), (4, 6), (5, 7).

The line of best fit for this scatterplot is $y = 1.2x + 1$.

Only one of the points is on the line. For the others, two are above the line and two are below. Residuals measure how far above or below the line the points are. The information can be collected on a chart.

Regression Curves

x	y	1.2x + 1	Residuals (y column) − (1.2x + 1 column)
1	2	2.2	−0.2
2	4	3.4	0.6
3	4	4.6	−0.6
4	6	5.8	0.2
5	7	7	0

The two points above the line have positive residuals. The two points below the line have negative residuals.

A residual plot is a scatter plot with the x-coordinates from the x column in the chart and the y-coordinates from the residuals column in the chart.

MATH FACTS

When the original scatterplot resembles a line, the residual plot will be a random scattering of points with no apparent pattern. When the original scatterplot resembles some other kind of curve, the residual plot will not be a bunch of random points but will look like a line or some kind of curve like a line or a U shape.

Some Key Algebra I Facts and Skills

Example 1

Construct a residual plot for the scatter plot based on the following chart. What does this residual plot suggest about the original scatterplot?

Solution: The line of best fit is $y = 1.453x - 1.987$.

x	y	$1.453x - 1.987$	Residuals (y column) $- (1.453x - 1.987)$
1	0.42	-0.534	0.954
2	0.62	0.919	-0.299
3	1.22	2.372	-1.152
4	3.21	3.825	-0.615
5	6.39	5.278	1.112

Regression Curves

The residual plot looks like this:

Because this has a U shape and is not just a random scattering of points, it suggests that the original scatter plot did not resemble a line.

12.4 PARABOLAS AND EXPONENTIALS OF BEST FIT

Even though every scatter plot has a line of best fit, sometimes even the best possible line isn't a very good fit.

Some Key Algebra I Facts and Skills

If the points in a scatter plot look more like a parabola or an exponential curve, the most appropriate curve to model it may not be a line. For parabolas, there is a parabola of best fit, and for exponential curves, there is an exponential curve of best fit. Both can be calculated on the graphing calculator. Select quadratic regression for the parabola of best fit and exponential regression for the exponential curve of best fit.

For the TI-84:

Enter the data into L1 and L2. Then press [STAT], [right], and [5].

For the TI-Nspire:

To find the parabola of best fit, after entering the data into column A and column B, press [menu], [4], [1], and [6].

To find the exponential curve of best fit on the TI-84, enter the data into L1 and L2 and press [STAT], [right], and [0].

Regression Curves

To find the exponential curve of best fit on the TI-Nspire, enter the data into column A and column B, press [menu], [4], [1], and [A].

The parabola of best fit for the first scatter plot is $y = 0.1634989x^2 - 3.67983x + 24.2616$.

The exponential curve of best fit for the second scatterplot is $y = 1.30687 \cdot 1.13394^x$.

Practice Exercises

1. Calculate the equation for the line of best fit for the following set of data in $y = mx + b$ form. Round m and b to the nearest tenth.

x	y
1	3
2	5
3	4
4	6
5	8

(1) $y = 1.1x + 1.9$
(2) $y = 1.4x + 1.7$
(3) $y = 1.7x + 1.4$
(4) $y = 1.9x + 1.1$

Regression Curves

2. Calculate the equation for the line of best fit for the following set of data in $y = mx + b$ form. Round m and b to the nearest tenth.

x	y
10	33
20	20
30	10
40	14
50	6

(1) $y = 34.6x - 0.6$
(2) $y = -0.6x + 34.6$
(3) $y = 31.8x - 0.7$
(4) $y = -0.7x + 31.8$

3. What is the equation for the line of best fit for the points on this scatter plot?

(1) $y = -0.5x + 7$
(2) $y = 0.7x - 0.5$
(3) $y = -0.4x + 0.9$
(4) $y = 0.9x - 0.4$

4. Of these four choices, which line appears to be the best fit for this scatter plot?

Regression Curves

5. What is the correlation coefficient (r), rounded to the nearest hundredth, for the line of best fit for the data on the table below?

x	y
3	10
6	13
9	27
12	38
15	40

(1) 0.97
(2) 0.96
(3) 0.95
(4) 0.94

6. For which scatter plot is the correlation coefficient closest to 1?

(1)

(2)

(3)

(4)

7. Of the four choices, which is closest to the correlation coefficient for this scatter plot?

(1) 0.8
(2) −0.8
(3) 1
(4) −1

8. Find the equation of the parabola of best fit for the data on the table below.

x	y
1	2
2	2
3	3
4	4
5	6
6	8
7	10
8	14

(1) $y = 0.31x^2 - 67x + 3.53$
(2) $y = 0.24x^2 - 0.52x + 2.23$
(3) $y = 0.58x^2 - 0.41x + 4.12$
(4) $y = 0.73x^2 - 0.72x + 6.87$

Regression Curves 145

9. Find the equation of the exponential curve of best fit for the data on the table below.

x	y
1	2
2	2
3	3
4	4
5	6
6	8
7	10
8	14

(1) $y = 1.28 \cdot 1.35^x$
(2) $y = 1.35 \cdot 1.28^x$
(3) $y = 1.47 \cdot 1.39^x$
(4) $y = 1.39 \cdot 1.47^x$

10. For which scatter plot would an exponential curve of best fit be most appropriate?

Solutions

1. Enter the data into the graphing calculator and do linear regression to get the equation $y = 1.1x + 1.9$.

   The correct choice is **(1)**.

2. Enter the data into the graphing calculator and do linear regression to get the equation $y = -0.6x + 34.6$.

   The correct choice is **(2)**.

3. Enter 2, 4, 6, 8, 10 for the x values and 1, 5, 3, 7, 9 for the y values into the graphing calculator and do linear regression. The equation is $y = 0.9x - 0.4$.

   The correct choice is **(4)**.

4. Of the four choices, choice 3 seems to have the line closer to the points than the other three choices.

   The correct choice is **(3)**.

5. Enter the data into the graphing calculator and do linear regression to get an r value of 0.97.

   The correct choice is **(1)**.

6. To have an r value close to $+1$, the points must lie close to a line with a positive slope. Choices 1 and 2 both resemble lines with positive slopes. In choice 2 the points seem to fall closer to a straight line than choice 1.

   The correct choice is **(2)**.

Regression Curves

7. Since this scatter plot resembles a line with a positive slope, choices 2 and 4 can be eliminated because r must be positive. To have an r value of $+1$ the points would have to lie perfectly on a line, so choice 3 can be eliminated too.

    The correct choice is **(1)**.

8. Enter the data into the graphing calculator and do quadratic regression to get $y = 0.24x^2 - 0.52x + 2.23$.

    The correct choice is **(2)**.

9. Enter the data into the graphing calculator and do exponential regression to get $y = 1.28 \cdot 1.35^x$.

    The correct choice is **(1)**.

10. Of the four choices, choice 3 looks the most like an exponential curve. Choice 1 looks like a parabola. Choice 2 looks like a line. Choice 4 looks like a bunch of random points.

    The correct choice is **(3)**.

13. STATISTICS

13.1 MEAN, MEDIAN, AND MODE

In a set of numbers, the **mean**, also known as the **average**, is the sum of the numbers divided by how many numbers there are. For the set 70, 75, 78, 80, 82, the mean can be calculated as $\frac{70 + 75 + 78 + 80 + 82}{5} = \frac{385}{5} = 77$.

The **median** is the middle number (or average of the two middle numbers if there are an even amount of numbers) when the numbers are arranged from least to greatest. In the set 70, 75, 78, 80, 82, the median is the number 78. If the set were 70, 75, 78, 80, 82, 85, the median would be the average of 78 and 80, which is $\frac{78 + 80}{2} = 79$.

The **mode** is the number that appears most frequently. In the set 70, 75, 78, 80, 80, the mode is 80 since there are two 80s and only one of each of the other numbers.

13.2 FIRST QUARTILE AND THIRD QUARTILE

The median is greater than 50% of the numbers in the list. The number that is greater than just 25% of the numbers on the list is called the *first quartile*. The number that is greater than 75% of the numbers on the list is called the *third quartile*. To find the first quartile, find the median of all the numbers less than the median of the list. To find the third quartile, find the median of all the numbers greater than the median of the list. The interquartile range is the difference between the third quartile and the first quartile.

- For the list 40, 43, 45, 47, 48, 52, 57, 60, 61, 64, 68:
- The numbers less than the median 52 are 40, 43, 45, 47, 48. The median of these five numbers is 45, which is the first quartile.
- The numbers greater than the median 52 are 57, 60, 61, 64, 68. The median of these five numbers is 61, which is the third quartile. The interquartile range is $61 - 45 = 16$.

13.3 BOX PLOTS

A box plot is a picture that shows five different metrics, minimum, first quartile, median, third quartile, and maximum. To make a box plot, draw a segment connecting the maximum and the minimum. Then draw a rectangle with sides at the first quartile and third quartile. Finally, draw a line through the rectangle at the median.

- For the following numbers, the box plot looks like this:

    10, 10, 10, 12, 12, 12, 12, 17, 25, 25, 25, 25, 30, 30, 30

13.4 USING THE GRAPHING CALCULATOR TO DETERMINE MAXIMUM, MINIMUM, MEDIAN, FIRST QUARTILE, AND THIRD QUARTILE

For the TI-84:

The graphing calculator can calculate the five measures of central tendency. First enter all the numbers into L1 by pressing [STAT] and [1] for Edit.

To find the minimum, first quartile, median, third quartile, and maximum of the seven numbers 10, 4, 8, 12, 6, 16, 14, enter them into L1. Then press [STAT] and [1] for 1-Var Stats and press [ENTER].

Some Key Algebra I Facts and Skills

The screen will display

```
1-Var Stats
 x̄=10
 Σx=70
 Σx²=812
 Sx=4.320493799
 σx=4
↓n=7
```

The $\bar{x} = 10$ is for the mean. The $n = 7$ means that there were seven elements in the list. For the minimum, first quartile, median, third quartile, and maximum, press the down arrow five times.

```
1-Var Stats
↑n=7
 minX=4
 Q₁=6
 Med=10
 Q₃=14
 maxX=16
```

minX is for the minimum, Q1 is for the first quartile, Med is for the median, Q3 is for the third quartile, and maxX is for the maximum element.

For the TI-Nspire:

From the home screen, select the Add Lists & Spreadsheet icon. Name column A x and fill in cells A1 through A7 with the numbers 10, 4, 8, 12, 6, 16, 14.

Press [4], [1], and [1] for One-Variable Statistics.

Press [OK] since there is just one list. Set the X1 List to x since that was what the column with the data was named in the spreadsheet.

Some Key Algebra I Facts and Skills

In cells C2 through C13, the one-variable statistics will be displayed. The median is the $\bar{x}$-bar. n is for the number of numbers. MinX is the smallest number. Q1X is the first quartile. MedianX is the median. Q3X is the third quartile. MaxX is the largest number.

	A	B	C	D
	x		=OneVar($\triangleright$)	
1	10	Title	One-Var...	
2	4	$\bar{x}$	10.	
3	8	Σx	70.	
4	12	Σx^2	812.	
5	6	sx := Sn-1...	4.32049	

C1 ="One-Variable Statistics"

	A	B	C	D
	x		=OneVar($\triangleright$)	
6	16	$\sigma x := \sigma nX...$	4.	
7	14	n	7.	
8		MinX	4.	
9		Q₁X	6.	
10		MedianX...	10.	

C10 =10.

	A	B	C	D
	x		=OneVar($\triangleright$)	
11		Q₃X	14.	
12		MaxX	16.	
13		SSX := Σ...	112.	
14				
15				

C15

Practice Exercises

1. Find the mean of this set of numbers {4, 5, 8, 8, 8, 10, 10, 13, 15, 17, 23}.
   (1) 8
   (2) 9
   (3) 10
   (4) 11

2. Find the median of this set of numbers {4, 5, 8, 8, 8, 10, 10, 13, 15, 17, 23}.
   (1) 8
   (2) 10
   (3) 11
   (4) 15

3. Find the interquartile range of this set of numbers {4, 5, 8, 8, 8, 10, 10, 13, 15, 17, 23}.
   (1) 7
   (2) 8
   (3) 9
   (4) 10

4. For the first four days of a five-day vacation, the mean temperature was 80 degrees. What must the temperature be on the fifth day in order for the mean temperature to be 82 degrees?
   (1) 88
   (2) 89
   (3) 90
   (4) 91

5. For which data set is the median greater than the mean?
   (1) {4, 7, 10, 13, 16}
   (2) {8, 9, 10, 18, 19}
   (3) {8, 9, 10, 11, 12}
   (4) {1, 2, 10, 11, 12}

154 Some Key Algebra I Facts and Skills

6. What is the mode of the data in this histogram?

(1) 23
(2) 24
(3) 15
(4) 16

7. What is the median of the data in this box plot?

(1) 17
(2) 15.2
(3) 8
(4) 25

8. What is true about this data set: 1, 2, 10, 11, 11?
(1) The median is greater than the mean.
(2) The median is equal to the mean.
(3) The median is less than the mean.
(4) The median is greater than the mode.

Statistics

9. For which data set is the interquartile range equal to 0?
   (1) 2, 3, 6, 6, 6, 7, 8
   (2) 2, 6, 6, 6, 6, 6, 8
   (3) 1, 2, 3, 6, 7, 8, 9
   (4) 2, 6, 6, 6, 6, 7, 8

10. In a set of seven numbers, the largest number is increased by 10. Which measure of central tendency must increase because of this?
    (1) Mean
    (2) Mode
    (3) First quartile
    (4) Interquartile range

Solutions

1. To get the mean, add the 11 values and divide by 11. $(4 + 5 + 8 + 8 + 8 + 10 + 10 + 13 + 15 + 17 + 23)/11 = 121/11 = 11$.

   The correct choice is (**4**).

2. The median is the middle number after the numbers have been arranged from least to greatest. These numbers are already arranged from least to greatest so the middle number is the sixth number, which is 10.

   The correct choice is (**2**).

3. The interquartile range is the difference between the first quartile and the third quartile. The median is the sixth number so the first quartile is the median of the first five numbers. The first five numbers are 4, 5, 8, 8, 8 with a median of 8. The third quartile is the median of the last five numbers. The last five numbers are 10, 13, 15, 17, 23, with a median of 15. The first quartile is 8. The third quartile is 15. The interquartile range is $15 - 8 = 7$.

   The correct choice is (**1**).

4. The sum of a set of numbers is the product of the average and the number of numbers. If the first four days had an average of 80, the sum of all four temperatures was 320. If the first five days need to have an average of 82, the sum of all five temperatures needs to be $82 \cdot 5 = 410$. The fifth temperature needs to be $410 - 320 = 90$.

   The correct choice is (**3**).

5. The median for all four data sets is 10. The mean of the numbers in choice 4 is $(1 + 2 + 10 + 11 + 12)/5 = 7.2$, which is less than the median. All the other choices have the mean greater than or equal to the median.

   The correct choice is (**4**).

6. The mode is the most frequent number. In a histogram the most frequent number is the tallest bar, which is 16 in this example.

   The correct choice is (**4**).

7. The vertical line in the middle of the rectangle represents the median in a box plot. Since this is at 17, the median is 17.

   The correct choice is (**1**).

8. The median of the data set is 10. The mode is 11. The mean is 7. The median is greater than the mean.

   The correct choice is (**1**).

9. The interquartile range is zero if the first quartile is equal to the third quartile. This is the case for choice 2 where both the first quartile and the third quartile are 6.

   The correct choice is (**2**).

10. When the largest number is increased by 10, the sum of all the numbers is increased by 10 so the mean will increase. The other measures will not be affected by increasing the largest number.

    The correct choice is (**1**).

Glossary of Terms

Addition property of equality A property of algebra that states that when equal values are added to both sides of a true equation, the equation continues to be true. To solve the equation $x - 2 = 5$, add 2 to both sides of the equation by using the addition property of equality.

Arithmetic sequence A number sequence in which the difference between two consecutive terms is a constant. The sequence 2, 5, 8, 11, 14,...is an arithmetic sequence because the difference between consecutive terms is always 3.

Axis of symmetry An imaginary vertical line that passes through the vertex of a parabola. The equation for the axis of symmetry of a parabola is defined by $y = ax^2 + bx + c$ is $x = \frac{-b}{2a}$.

Base The number being raised to a power in an exponential expression. In the expression $2 \cdot 3^x$, the 3 is the base.

Binomial A polynomial with only two terms. $3x + 5$ is a binomial.

Box plot A graphical way to summarize data. The five numbers represented by the minimum, first quartile, median, third quartile, and maximum are graphed on a number line. A line segment connects the minimum to the first quartile. A rectangle is drawn around the first quartile and third quartile with a vertical line at the median. A line segment connects the third quartile to the maximum.

Glossary of Terms

Closed form defined sequence A formula that defines the nth term of a sequence. The formula $a_n = 3 + 2(n - 1)$ is a closed form definition of a sequence. To get the 50th term of the sequence, substitute 50 for n in the definition.

Coefficient A number multiplied by a variable expression. In the expression $5x + 2$, 5 is the coefficient of x.

Common difference In an arithmetic sequence, the difference between consecutive terms. The common difference in the sequence 2, 5, 8, 11, 14,... is 3.

Common ratio In a geometric sequence, the ratio between consecutive terms. The common ratio in the sequence 2, 6, 18, 54, 162,... is 3 since $162/54 = 54/18 = 18/6 = 6/2 = 3$.

Commutative property of addition or multiplication The law from arithmetic that states that the order in which two numbers are added or multiplied does not matter. Because of the commutative property of addition, $5 + 2 = 2 + 5$.

Completing the square A method of solving a quadratic equation that involves turning one side of the equation into a perfect square trinomial.

Constant A number that does not have a variable part. In the expression $5x + 2$, 2 is a constant.

Correlation coefficient A number represented by r that measures how well a curve of best fit matches the points in a scatter plot. When the correlation coefficient is very close to 1 or to -1, the curve is a very good fit.

Degree of a polynomial The exponent on the highest power of a polynomial. In the polynomial $x^3 - 2x^2 + 5x - 2$, the degree is 3 since the highest power is a 3.

Difference of perfect squares Factoring a quadratic binomial in the form $x^2 - a^2$ into $(x - a)(x + a)$. For example, $x^2 - 9 = (x - 3)(x + 3)$.

Distributive property of multiplication over addition The rule that allows expressions of the form $a(b + c)$ to become $a \cdot b + a \cdot c$. For example, $2(3x + 5) = 6x + 10$.

Division property of equality A property of algebra that states that when both sides of a true equation are divided by the same non-zero number, the equation continues to be true. To solve the equation $2x = 8$, divide both sides of the equation by 2 using the division property of equality.

Domain The numbers that can be input into a function. When the function is defined as a set of ordered pairs or as a graph, the domain is the set of x-coordinates.

Glossary of Terms

Dot plot A graphical way of representing a data set where each piece of data is represented with a dot.

Elimination method A way of solving a system of linear equations by combining the two equations in such a way as to eliminate one of the variables.
In the set of equations,

$$x + 2y = 12$$
$$3x - 2y = -4$$

the y variable is eliminated by adding the two equations together.

Equation Two mathematical expressions with an equals sign between them. $3x + 2 = 8$ is an equation.

Exponential equation An equation in which the variable is an exponent. $2 \cdot 3^x = 18$ is an exponential equation.

Exponential function A function in which the variable is an exponent. $f(x) = 2 \cdot 3^x$ is an exponential function.

Expression Numbers and variables that are combined with the operations from math.

Factoring a polynomial Finding two polynomials that can be multiplied to become another polynomial. The polynomial $x^2 + 5x + 6$ can be factored into $(x + 2)(x + 3)$.

Factors The polynomials that evenly divide into a polynomial. The factors of $x^2 + 5x + 6$ are $(x + 2)$ and $(x + 3)$.

First quartile The number in a data set that is bigger than just 25% of the numbers in the set.

FOIL A way of multiplying two binomials of the form $(a + b)(c + d)$ where (F)irst the a and c are multiplied, then the (O)uters a and d are multiplied, then the (I)nners b and c are multiplied, and finally the (L)asts b and d are multiplied. Then the four results are added together. The product of $(x + 2)$ and $(x + 3)$ is $x^2 + 3x + 2x + 6 = x^2 + 5x + 6$ by this method.

Function Something that takes numbers as inputs and outputs numbers. Functions are often labeled with the letters f or g. The notation $f(2) = 7$ means that when the number 2 is input into function f, it outputs the number 7.

Geometric sequence A number sequence in which the ratio between two consecutive terms (what you get when you divide one term by the term before it) is a constant. The sequence 2, 6, 18, 54, 162,... is a geometric sequence because the ratio between consecutive terms is always 3.

Graph A visual way to describe the solution set to an equation. Each solution to the equation corresponds to an ordered pair that is graphed as a point on the coordinate plane. Each of the ordered pairs that satisfy an equation produce a point on the coordinate plane and the collection of all the points is the graph of the equation.

Greatest common factor The largest expression that divides evenly into two or more monomials. The greatest common factor of $6x^2$ and $8x^3$ is $2x^2$.

Growth rate In an exponential expression $a \cdot b^x$, the b is the growth rate. For example, in the equation $y = 500 \cdot 1.05^x$, the growth rate is 1.05.

Histogram A way of representing data with repeated values. Each value is represented by a bar whose height corresponds to the number of times that value is repeated. There are no spaces between the bars.

Increasing A function is increasing on an interval if making the input value larger also makes the y output larger. On a graph, an increasing function "goes up" from left to right.

Inequality Like an equation, but there is a $<$, $>$, $\leq$, or $\geq$ sign between the two expressions. $x + 2 > 5$ in a one-variable inequality. $y \leq 2x + 6$ is a two-variable inequality.

Interquartile range The difference between the number that is the third quartile of a data set and the number that is the first quartile of a data set.

Isolating a variable A variable is isolated when it is by itself on one side of an equation. In the equation $x + 2 = 5$, the x is not yet isolated. Subtracting 2 from both sides of the equation transforms the original equation into $x = 3$ with the x now isolated.

Like terms Terms that have the same variable part. They can be combined by adding or subtracting. $2x^2$ and $3x^2$ are like terms. $2x^2$ and $3x^3$ are not like terms.

Line of best fit A line that comes closest to the set of points in a scatter plot.

Linear equation An equation in which the greatest exponent is a 1. The equation $2x + 3 = 7$ is a linear equation.

Linear function A function in which the greatest exponent is a 1. The function $f(x) = 2x + 3$ is a linear function.

Mean The average of the numbers in a data set. Calculate the mean by adding all the numbers and dividing the total by the number of numbers in the set.

Glossary of Terms

Median The middle number in a data set after it has been arranged from least to greatest. If there are an even number of numbers in the data set, the median is found by adding the two middle numbers and dividing by 2.

Mode The most frequent number in a data set.

Monomial A mathematical expression that has a coefficient and/or a variable part. $3x^2$ is a monomial. $3 + x^2$ is not a monomial.

Multiplication property of equality A property of algebra that states that when equal values are multiplied by both sides of a true equation, the equation continues to be true. To solve the equation $(1/2)x = 5$, multiply both sides of the equation by 2, using the multiplication property of equality.

Ordered pair Two numbers written in the form (x, y). An ordered pair can be a solution to a two-variable equation. For example, $(2, 5)$ is one solution to the equation $y = 2x + 1$. Ordered pairs can be graphed on the coordinate axes by locating the point with the x-coordinate equal to the x value and the y-coordinate equal to the y value.

Parabola A "U"-shaped curve that is the graph of the solution set of a quadratic equation.

Perfect square trinomial A quadratic trinomial of the form $x^2 + bx + (b/2)^2$, which can be factored into $(x + b/2)^2$. For example, $x^2 + 6x + 9 = (x + 3)^2$.

Piecewise function A function that has multiple rules for determining output values from input values, depending on what the input values are. If the function $f(x)$ is defined as

$$f(x) = 2x + 1 \quad \text{if } x < 0$$
$$ x^2 \quad\quad\ \ \text{if } x \geq 0$$

then $f(-3) = 2(-3) + 1 = -1$ and $f(5) = 5^2 = 25$.

Polynomial The sum of one or more monomials. Each term of the polynomial has the form ax^n. $3x^4 + 2x^3 - 3x^2 + 6x - 1$ is a polynomial.

Quadratic equation An equation in which the highest power on a variable is a 2. $x^2 + 5x + 6 = 0$ is a quadratic equation.

Quadratic formula The formula $x = \left(-b \pm \sqrt{(b^2 - 4ac)}\right)/2a$. This formula can be used to find the two solutions to the quadratic equation $ax^2 + bx + c = 0$.

Quadratic function A function in which the highest power on a variable is 2. $f(x) = x^2 + 5x + 6$ is a quadratic function.

Glossary of Terms

Quadratic polynomial A polynomial in which the highest power on a variable is 2. $x^2 + 5x + 6$ is a quadratic polynomial.

Range The set of values that can be output from a function is the range of that function.

Recursively defined sequence A way to define a sequence in which the first term or terms of the sequence are given and a formula is given for calculating the next term based on the previous term or terms.

$$a_1 = 5$$
$$a_n = 3 + a_{n-1} \text{ for } n > 1$$

is a recursive definition for the sequence 5, 8, 11, 14, ...

Regression Finding a curve that best fits a scatterplot. Three types of regression are linear, quadratic, and exponential.

Residual plot A set of points that represent how far points on a graph deviate from a curve of best fit.

Roots The roots of an equation are the values that solve that equation. The roots of $x^2 + 5x + 6 = 0$ are -3 and -2.

Sequence A list of numbers that usually have some kind of pattern.

Slope The slope of a line is a measure of the line's steepness. The equation for the slope of a line that passes through the two points (x_1, y_1) and (x_2, y_2) is $m = \dfrac{y_2 - y_1}{x_2 - x_1}$.

Slope-intercept form An equation in the form $y = mx + b$ where m and b are numbers in slope intercept form. $y = 2x - 1$ is in slope intercept form. When a two-variable equation is in slope intercept form, the graph of the equation has a y-intercept of $(0, b)$ and a slope of m.

Solution set The set of numbers or ordered pairs that satisfies an equation. The solution set of $x + 2 = 5$ is $\{3\}$. The solution set of $x + y = 10$ has an infinite number of ordered pairs in its solution set, including (2, 8), (3, 7), and (4, 6).

Substitution method A method for solving a system of equations in which one variable is isolated in one of the equations, and the expression equal to that variable is substituted for it in the other equation.

Subtraction property of equality A property of algebra that states that when equal values are subtracted from both sides of a true equation, the equation continues to be true. To solve the equation $x + 2 = 5$, subtract 2 from both sides of the equation by using the subtraction property of equality.

System of equations Two or more equations with two or more unknowns to solve for. An example of a system of equations with a solution of (8, 2) is

$$x + y = 10$$
$$x - y = 6$$

Third quartile The number in a data set that is greater than just 75% of the numbers in the set.

Translation A translation of a graph is when the points on it are each shifted the same amount in the same direction. Examples of translations are vertical translations, horizontal translations, and combinations of vertical and horizontal translations.

Trinomial A polynomial with three terms. The polynomial $x^2 + 5x + 6$ is a trinomial.

Variable A letter, often an x, y, or z, that represents a value in a mathematical expression. In an algebraic equation, the variable is often the unknown that needs to be solved for.

Vertex The turning point of a parabola is its vertex. If the parabola opens upward, the vertex is the minimum point. If the parabola opens downward, the vertex is the maximum point.

Vertical line test A way of testing to see if a graph can represent a function. If at least one vertical line can pass through at least two points on the graph, the graph fails the vertical line test and cannot represent a function. If there are no vertical lines that can pass through at least two points, then the graph can be the graph of a function.

x-intercept The location where a curve crosses the x-axis. The y-coordinate of the x-intercept is 0.

y-intercept The location where a curve crosses the y-axis. The x-coordinate of the y-intercept is 0. In slope-intercept form, $y = mx + b$, the y-intercept is located at $(0, b)$.

Zeros The zeros of a function f are the numbers that can be input into the function so that 0 is output from the function. For example, the function $f(x) = 2x - 6$ has the number 3 as its only zero since $f(3) = 2(3) - 6 = 6 - 6 = 0$.

Regents Examinations, Answers, and Self-Analysis Charts

Examination
June 2017
Algebra I

HIGH SCHOOL MATH REFERENCE SHEET

Conversions

1 inch = 2.54 centimeters

1 meter = 39.37 inches

1 mile = 5280 feet

1 mile = 1760 yards

1 mile = 1.609 kilometers

1 kilometer = 0.62 mile

1 pound = 16 ounces

1 pound = 0.454 kilogram

1 kilogram = 2.2 pounds

1 ton = 2000 pounds

1 cup = 8 fluid ounces

1 pint = 2 cups

1 quart = 2 pints

1 gallon = 4 quarts

1 gallon = 3.785 liters

1 liter = 0.264 gallon

1 liter = 1000 cubic centimeters

Formulas

Triangle	$A = \frac{1}{2}bh$
Parallelogram	$A = bh$
Circle	$A = \pi r^2$
Circle	$C = \pi d$ or $C = 2\pi r$

Formulas (continued)

General Prisms	$V = Bh$
Cylinder	$V = \pi r^2 h$
Sphere	$V = \dfrac{4}{3}\pi r^3$
Cone	$V = \dfrac{1}{3}\pi r^2 h$
Pyramid	$V = \dfrac{1}{3}Bh$
Pythagorean Theorem	$a^2 + b^2 = c^2$
Quadratic Formula	$x = \dfrac{-b \pm \sqrt{b^2 - 4ac}}{2a}$
Arithmetic Sequence	$a_n = a_1 + (n-1)d$
Geometric Sequence	$a_n = a_1 r^{n-1}$
Geometric Series	$S_n = \dfrac{a_1 - a_1 r^n}{1 - r}$ where $r \neq 1$
Radians	1 radian = $\dfrac{180}{\pi}$ degrees
Degrees	1 degree = $\dfrac{\pi}{180}$ radians
Exponential Growth/Decay	$A = A_0 e^{k(t - t_0)} + B_0$

EXAMINATION June 2017

PART I

Answer all 24 questions in this part. Each correct answer will receive 2 credits. No partial credit will be allowed. For each statement or question, write in the space provided the numeral preceding the word or expression that best completes the statement or answers the question. [48 credits]

1 To keep track of his profits, the owner of a carnival booth decided to model his ticket sales on a graph. He found that his profits only declined when he sold between 10 and 40 tickets. Which graph could represent his profits?

(1)

(2)

(3)

(4)

1 _____

2 The formula for the surface area of a right rectangular prism is $A = 2lw + 2hw + 2lh$, where l, w, and h represent the length, width, and height, respectively. Which term of this formula is *not* dependent on the height?
(1) A
(2) 2lw
(3) 2hw
(4) 2lh

3 Which graph represents $y = \sqrt{x-2}$?

(1)

(2)

(3)

(4)

4 A student plotted the data from a sleep study as shown in the graph below.

The student used the equation of the line

$$y = -0.09x + 9.24$$

to model the data. What does the rate of change represent in terms of these data?

(1) The average number of hours of sleep per day increases 0.09 hour per year of age.

(2) The average number of hours of sleep per day decreases 0.09 hour per year of age.

(3) The average number of hours of sleep per day increases 9.24 hours per year of age.

(4) The average number of hours of sleep per day decreases 9.24 hours per year of age.

5 Lynn, Jude, and Anne were given the function $f(x) = -2x^2 + 32$, and they were asked to find $f(3)$. Lynn's answer was 14, Jude's answer was 4, and Anne's answer was ±4. Who is correct?
   (1) Lynn, only
   (2) Jude, only
   (3) Anne, only
   (4) Both Lynn and Jude

6 Which expression is equivalent to $16x^4 - 64$?
   (1) $(4x^2 - 8)^2$
   (2) $(8x^2 - 32)^2$
   (3) $(4x^2 + 8)(4x^2 - 8)$
   (4) $(8x^2 + 32)(8x^2 - 32)$

7 Vinny collects population data, $P(h)$, about a specific strain of bacteria over time in hours, h, as shown in the graph below.

Which equation represents the graph of $P(h)$?
   (1) $P(h) = 4(2)^h$
   (2) $P(h) = \frac{46}{5}h + \frac{6}{5}$
   (3) $P(h) = 3h^2 + 0.2h + 4.2$
   (4) $P(h) = \frac{2}{3}h^3 - h^2 + 3h + 4$

8 What is the solution to the system of equations below?

$$y = 2x + 8$$
$$3(-2x + y) = 12$$

(1) no solution (3) $(-1, 6)$
(2) infinite solutions (4) $\left(\frac{1}{2}, 9\right)$

9 A mapping is shown in the diagram below.

This mapping is
(1) a function, because Feb has two outputs, 28 and 29
(2) a function, because two inputs, Jan and Mar, result in the output 31
(3) not a function, because Feb has two outputs, 28 and 29
(4) not a function, because two inputs, Jan and Mar, result in the output 31

10 Which polynomial function has zeros at -3, 0, and 4?
(1) $f(x) = (x + 3)(x^2 + 4)$
(2) $f(x) = (x^2 - 3)(x - 4)$
(3) $f(x) = x(x + 3)(x - 4)$
(4) $f(x) = x(x - 3)(x + 4)$

11 Jordan works for a landscape company during his summer vacation. He is paid $12 per hour for mowing lawns and $14 per hour for planting gardens. He can work a maximum of 40 hours per week, and would like to earn at least $250 this week. If m represents the number of hours mowing lawns and g represents the number of hours planting gardens, which system of inequalities could be used to represent the given conditions?

(1) $m + g \leq 40$
 $12m + 14g \geq 250$

(2) $m + g \geq 40$
 $12m + 14g \leq 250$

(3) $m + g \leq 40$
 $12m + 14g \leq 250$

(4) $m + g \geq 40$
 $12m + 14g \geq 250$

11 _____

12 Anne invested $1000 in an account with a 1.3% annual interest rate. She made no deposits or withdrawals on the account for 2 years. If interest was compounded annually, which equation represents the balance in the account after the 2 years?

(1) $A = 1000(1 - 0.013)^2$
(2) $A = 1000(1 + 0.013)^2$
(3) $A = 1000(1 - 1.3)^2$
(4) $A = 1000(1 + 1.3)^2$

12 _____

13 Which value would be a solution for x in the inequality $47 - 4x < 7$?

(1) -13
(2) -10
(3) 10
(4) 11

13 _____

14 Bella recorded data and used her graphing calculator to find the equation for the line of best fit. She then used the correlation coefficient to determine the strength of the linear fit.

Which correlation coefficient represents the strongest linear relationship?

(1) 0.9
(2) 0.5
(3) -0.3
(4) -0.8

14 _____

15 The heights, in inches, of 12 students are listed below.
   61, 67, 72, 62, 65, 59, 60, 79, 60, 61, 64, 63
Which statement best describes the spread of these data?

(1) The set of data is evenly spread.
(2) The median of the data is 59.5.
(3) The set of data is skewed because 59 is the only value below 60.
(4) 79 is an outlier, which would affect the standard deviation of these data.

15 _____

16 The graph of a quadratic function is shown below.

An equation that represents the function could be

(1) $q(x) = \frac{1}{2}(x + 15)^2 - 25$

(2) $q(x) = -\frac{1}{2}(x + 15)^2 - 25$

(3) $q(x) = \frac{1}{2}(x - 15)^2 + 25$

(4) $q(x) = -\frac{1}{2}(x - 15)^2 + 25$

16 _____

17 Which statement is true about the quadratic functions g(x), shown in the table below, and $f(x) = (x-3)^2 + 2$?

x	g(x)
0	4
1	-1
2	-4
3	-5
4	-4
5	-1
6	4

(1) They have the same vertex.
(2) They have the same zeros.
(3) They have the same axis of symmetry.
(4) They intersect at two points.

17 _____

18 Given the function $f(n)$ defined by the following:

$$f(1) = 2$$
$$f(n) = -5f(n-1) + 2$$

Which set could represent the range of the function?
(1) $\{2, 4, 6, 8,...\}$
(2) $\{2, -8, 42, -208,...\}$
(3) $\{-8, -42, -208, 1042,...\}$
(4) $\{-10, 50, -250, 1250,...\}$

18 _____

19 An equation is given below.

$$4(x-7) = 0.3(x+2) + 2.11$$

The solution to the equation is
(1) 8.3
(2) 8.7
(3) 3
(4) -3

19 _____

EXAMINATION June 2017

20 A construction worker needs to move 120 ft^3 of dirt by using a wheelbarrow. One wheelbarrow load holds 8 ft^3 of dirt and each load takes him 10 minutes to complete. One correct way to figure out the number of hours he would need to complete this job is

(1) $\dfrac{120\,\text{ft}^3}{1} \cdot \dfrac{10\,\text{min}}{1\,\text{load}} \cdot \dfrac{60\,\text{min}}{1\,\text{hr}} \cdot \dfrac{1\,\text{load}}{8\,\text{ft}^3}$

(2) $\dfrac{120\,\text{ft}^3}{1} \cdot \dfrac{60\,\text{min}}{1\,\text{hr}} \cdot \dfrac{8\,\text{ft}^3}{10\,\text{min}} \cdot \dfrac{1}{1\,\text{load}}$

(3) $\dfrac{120\,\text{ft}^3}{1} \cdot \dfrac{1\,\text{load}}{10\,\text{min}} \cdot \dfrac{8\,\text{ft}^3}{1\,\text{load}} \cdot \dfrac{1\,\text{hr}}{60\,\text{min}}$

(4) $\dfrac{120\,\text{ft}^3}{1} \cdot \dfrac{1\,\text{load}}{8\,\text{ft}^3} \cdot \dfrac{10\,\text{min}}{1\,\text{load}} \cdot \dfrac{1\,\text{hr}}{60\,\text{min}}$

20 _____

21 One characteristic of all linear functions is that they change by

(1) equal factors over equal intervals
(2) unequal factors over equal intervals
(3) equal differences over equal intervals
(4) unequal differences over equal intervals

21 _____

22 What are the solutions to the equation $x^2 - 8x = 10$?

(1) $4 \pm \sqrt{10}$
(2) $4 \pm \sqrt{26}$
(3) $-4 \pm \sqrt{10}$
(4) $-4 \pm \sqrt{26}$

22 _____

23 The formula for blood flow rate is given by $F = \dfrac{p_1 - p_2}{r}$, where F is the flow rate, p_1 the initial pressure, p_2 the final pressure, and r the resistance created by blood vessel size. Which formula can *not* be derived from the given formula?

(1) $p_1 = Fr + p_2$ (3) $r = F(p_2 - p_1)$

(2) $p_2 = p_1 - Fr$ (4) $r = \dfrac{p_1 - p_2}{F}$

24 Morgan throws a ball up into the air. The height of the ball above the ground, in feet, is modeled by the function $h(t) = -16t^2 + 24t$, where t represents the time, in seconds, since the ball was thrown. What is the appropriate domain for this situation?

(1) $0 \leq t \leq 1.5$ (3) $0 \leq h(t) \leq 1.5$

(2) $0 \leq t \leq 9$ (4) $0 \leq h(t) \leq 9$

EXAMINATION June 2017

PART II

Answer all 8 questions in this part. Each correct answer will receive 2 credits. Clearly indicate the necessary steps, including appropriate formula substitutions, diagrams, graphs, charts, etc. For all questions in this part, a correct numerical answer with no work shown will receive only 1 credit. [16 credits]

25 Express in simplest form: $(3x^2 + 4x - 8) - (-2x^2 + 4x + 2)$

$5x^2 - 10$

26 Graph the function $f(x) = -x^2 - 6x$ on the set of axes below.

x	f(x)
-6	0 ✓
-5	5 ✓
-4	8 ✓✓
-3	9 ✓
-2	8 ✓
-1	5 ✓
0	0 ✓
1	-7

State the coordinates of the vertex of the graph.

$(-3, 9)$

27 State whether $7 - \sqrt{2}$ is rational or irrational. Explain your answer.

Irrational because the number does not end

28 The value, $v(t)$, of a car depreciates according to the function $v(t) = P(.85)^t$, where P is the purchase price of the car and t is the time, in years, since the car was purchased. State the percent that the value of the car *decreases* by each year. Justify your answer.

15 %

29. A survey of 100 students was taken. It was found that 60 students watched sports, and 34 of these students did not like pop music. Of the students who did *not* watch sports, 70% liked pop music. Complete the two-way frequency table.

	Watch Sports	Don't Watch Sports	Total
Like Pop	26	28	54
Don't Like Pop	34	12	46
Total	60	40	100

30 Graph the inequality $y + 4 < -2(x - 4)$ on the set of axes below.

31 If $f(x) = x^2$ and $g(x) = x$, determine the value(s) of x that satisfy the equation $f(x) = g(x)$.

$x^2 = x$

$[0, 1)$

32 Describe the effect that each transformation below has on the function $f(x) = |x|$, where $a > 0$.

$g(x) = |x - a|$

g(x) is shifted a units to the right

$h(x) = |x| - a$

h(x) is shifted a units down

PART III

Answer all 4 questions in this part. Each correct answer will receive 4 credits. Clearly indicate the necessary steps, including appropriate formula substitutions, diagrams, graphs, charts, etc. For all questions in this part, a correct numerical answer with no work shown will receive only 1 credit. [16 credits]

33 The function $r(x)$ is defined by the expression $x^2 + 3x - 18$. Use factoring to determine the zeros of $r(x)$.

−6 and 3

Explain what the zeros represent on the graph of $r(x)$.

They represent the x intercept of the graph

EXAMINATION June 2017 187

34 The graph below models Craig's trip to visit his friend in another state. In the course of his travels, he encountered both highway and city driving.

Based on the graph, during which interval did Craig most likely drive in the city? Explain your reasoning.

D and E because the speed is 15 miles per hour which is reasonable for city driving

Question 34 is continued on the next page.

Question 34 continued

Explain what might have happened in the interval between *B* and *C*.

Interval B and C would stop.

Determine Craig's average speed, to the *nearest tenth of a mile per hour*, for his entire trip.

Average speed for the entire trip is 32.0 miles per hour

EXAMINATION June 2017

35 Given:
$$g(x) = 2x^2 + 3x + 10$$
$$k(x) = 2x + 16$$

Solve the equation $g(x) = 2k(x)$ algebraically for x, to the *nearest tenth*.

[3·8, −3·1]

Explain why you chose the method you used to solve this quadratic equation.

The quadratic formula is the easiest

EXAMINATION June 2017

36 Michael has $10 in his savings account. Option 1 will add $100 to his account each week. Option 2 will double the amount in his account at the end of each week.

Write a function in terms of x to model each option of saving.

option 1: $y = 100x + 10$

option 2: $y = 10 \cdot 2^x$

Michael wants to have at least $700 in his account at the end of 7 weeks to buy a mountain bike. Determine which option(s) will enable him to reach his goal. Justify your answer.

both are over $700 at the end of the week

PART IV

Answer the question in this part. A correct answer will receive 6 credits. Clearly indicate the necessary steps, including appropriate formula substitutions, diagrams, graphs, charts, etc. A correct numerical answer with no work shown will receive only 1 credit. [6 credits]

37 Central High School had five members on their swim team in 2010. Over the next several years, the team increased by an average of 10 members per year. The same school had 35 members in their chorus in 2010. The chorus saw an increase of 5 members per year.

Write a system of equations to model this situation, where x represents the number of years since 2010.

Swim team $y = 10x + 5$

Chorus $y = 5x + 35$

The

Question 37 is continued on the next page.

Question 37 continued

Graph this system of equations on the set of axes below.

[blank coordinate grid with y-axis and x-axis labeled]

Explain in detail what each coordinate of the point of intersection of these equations means in the context of this problem.

(6, 65)

Answers June 2017
Algebra I

Answer Key

PART I

1. (3)
2. (2)
3. (4)
4. (2)
5. (1)
6. (3)
7. (1)
8. (1)
9. (3)
10. (3)
11. (1)
12. (2)
13. (4)
14. (1)
15. (4)
16. (4)
17. (3)
18. (2)
19. (1)
20. (4)
21. (3)
22. (2)
23. (3)
24. (1)

PART II

25. $5x^2 - 10$
26. Vertex: $(-3, 9)$
27. Irrational
28. 15%
29.

	Watch Sports	Don't Watch Sports	Total
Like Pop	26	28	54
Don't Like Pop	34	12	46
Total	60	40	100

30. See graph
31. $\{0, 1\}$
32. $g(x)$ is shifted a units to the right. $h(x)$ is shifted a units down.

PART III

33. The zeros are -6 and 3. They represent the x-intercepts of the graph.
34. Between D and E, the speed is 15 miles per hour, which is reasonable for city driving. Between B and C, he was stopped. The average speed for the entire trip is 32.9 miles per hour.
35. $\{3.6, -3.1\}$ The quadratic formula is simplest.
36. Option 1: $y = 100x + 10$. Option 2: $y = 10 \cdot 2^x$. Both are over $700 after 7 weeks.

PART IV

37. Swim team: $y = 10x + 5$. Chorus: $y = 5x + 35$. The intersection point is $(6, 65)$. In 2016, they both have 65 members.

In **Parts II–IV**, you are required to show how you arrived at your answers. For sample methods of solutions, see the *Answers Explained* section.

Answers and Explanations

PART I

1. If the profits are declining only when he sells between 10 and 40 tickets, the graph should show a curve that is going down as it goes from left to right between 10 and 40.

   The graph in choice (1) is flat between 10 and 40. The graph in choice (2) is going up between 10 and 40. The graph in choice (4) is going up from 10 to 25 but then down from 25 to 40. Only the graph in choice (3) is going down for the entire interval between 10 and 40.

   The correct choice is (**3**).

2. A term is dependent on a variable if a change in that variable's value would also change the value of that entire term. When a term contains a variable, the term is dependent on that variable.

   Choices (3) and (4) have the variable h in them, so those can be eliminated. Even though choice (1) doesn't look like it has an h in it, it does. If you look at the definition of A, its equation has two hs in it. So A is also dependent on h. Choice (2) has no h in it or in the definition of any of the variables involved in that choice.

   The correct choice is (**2**).

3. Graphing the relation $y = \sqrt{x-2}$ on the graphing calculator makes a graph that resembles choice (**4**).

   For the TI-84:

For the TI-Nspire:

This graph can also be made without the graphing calculator by knowing that the graph of the function $y = \sqrt{x}$ has the following shape.

The graph of $y = \sqrt{x-2}$ is a transformation of the graph of $y = \sqrt{x}$ shifted 2 units to the right.

The correct choice is (**4**).

4. The rate of change is equivalent to the slope of the line. Since m represents the slope when the equation is of the form $y = mx + b$, the rate of change for this model is -0.09. For every 1 unit that a point on the line goes to the right, the line will go down by 0.09 units.

For this model, a change by 1 unit to the right corresponds to an increase in age by 1 year. A change of 0.09 units down corresponds to a 0.09 decrease in the average number of hours of sleep per day.

The correct choice is (**2**).

5. To calculate $f(3)$, substitute 3 for x in the equation:

$$f(3) = -2 \cdot 3^2 + 32$$
$$= -2 \cdot 9 + 32$$
$$= -18 + 32$$
$$= 14$$

Lynn got 14 for her answer. The other two people seem to have tried to solve the equation $0 = -2x^2 + 32$.

The correct choice is (**1**).

6. Since both $16x^4$ and 64 are perfect squares, this expression can be written as $(4x^2)^2 - 8^2$. This can be factored as the difference of perfect squares:

$$a^2 - b^2 = (a-b)(a+b)$$
$$\left(4x^2\right)^2 - 8^2 = \left(4x^2 - 8\right)\left(4x^2 + 8\right)$$

The correct choice is (**3**).

7. Each time the x-coordinate increases by 1 in this graph, the y-coordinate doubles. This is what happens when the graph is of an exponential equation, and only choice (1) is exponential.

Another way to solve this question is to calculate to see if $P(0) = 4$ and $P(1) = 8$ for any of the answer choices. Test choice (1):

$$P(0) = 4(2)^0 = 4 \cdot 1 = 4$$
$$P(1) = 4(2)^1 = 4 \cdot 2 = 8$$

The correct choice is (**1**).

8. Since the first equation has the y-variable isolated on one side of the equation, the simplest way to solve this system of equations is with the substitution method. Substitute $2x + 8$ for the y in the second equation:

$$3(-2x + (2x + 8)) = 12$$
$$3(-2x + 2x + 8) = 12$$
$$3(8) = 12$$
$$24 = 12$$

Since 24 can never equal 12, this system of equations has no solution.

The correct choice is (**1**).

9. In a function, each input value can be mapped to only one output value. When that happens, you can make statements like $f(\text{Jan}) = 31$ and $f(\text{Apr}) = 30$. In this mapping diagram though, Feb is mapped to two different numbers. So this mapping is not considered a function.

The correct choice is (**3**).

10. A zero of a polynomial is a number that when put into the function makes an output of zero. For -3 to be a zero, $(x + 3)$ must be a factor since $-3 + 3 = 0$. For 4 to be a zero, $(x - 4)$ must be a factor. For 0 to be a zero, x must be a factor.

So the function is $f(x) = x(x + 3)(x - 4)$.

The correct choice is (**3**).

11. Since Jordan can work at most 40 hours, the sum of his hours mowing and planting gardens must be less than or equal to 40. If he mows for m hours at $12 per hour, the amount he makes mowing is $12m$. If he plants gardens for g hours at $14 per hour, the amount he makes planting gardens is $14g$. Together he wants these amounts to be greater than or equal to 250.

So the system of inequalities is

$$m + g \leq 40$$
$$12m + 14g \geq 250$$

The correct choice is (**1**).

12. After the first deposit, Anne has $1000 in the account. After 1 year, she earns 1.3% interest. Her new balance can be calculated by multiplying $1000 by 0.013 to find the amount of interest. Then add the interest to the original $1000 to get $1000 + $13 = $1013. To find the amount Anne has after 2 years, multiply $1013 by 0.013 to find the amount of interest. Then add the interest to the $1013 to get $1013 + $13.169 = $1026.169. If you test each of the choices, only choice (2) gets this same answer.

A shorter way is to know that interest compounded annually can be calculated with the formula $A = P(1 + r)^n$, where P is the initial deposit, r is the percent interest as a decimal, and n is the number of years. In this example, $P = 1000$, $r = 0.013$, and $n = 2$.

The correct choice is (**2**).

13. Start solving the inequality the same way you would solve an equation that contained an equal sign instead of a < sign:

$$47 - 4x < 7$$
$$-47 = -47$$
$$-4x < -40$$

When you divide both sides of an inequality by a negative, you must flip the direction of the inequality sign:

$$\frac{-4x}{-4} > \frac{-40}{-4}$$
$$x > 10$$

Of the choices, only the number 11 is greater than 10.

You could answer this question by just testing each of the answer choices to see which of them make the inequality true. Test choice (4):

$$47 - 4(11) < 7$$
$$47 - 44 < 7$$
$$3 < 7$$

This is true. Testing the other choices would result in an inequality that is false.

The correct choice is (**4**).

14. The closer the correlation coefficient is to either 1 or -1, the stronger the linear relationship is. Since 0.9 is just 0.1 away from 1 while -0.8 is 0.2 away from -1, 0.9 is the closest.

The correct choice is **(1)**.

15. Arrange the numbers from smallest to largest:

$$59, 60, 60, 61, 61, 62, 63, 64, 65, 67, 72, 79$$

Analyze each of the answer choices.
- Choice (1): Evenly spaced numbers would be something like 59, 61, 63, 65, 67, where each number is the same amount more than the number before. That is not what is happening with this set of numbers.
- Choice (2): The median is the middle number. Since there are an even amount of numbers, there is no one middle number. For an even amount of numbers, the median is the average of the two middle numbers, in this case the 6th and 7th numbers:

$$\text{median} = \frac{62 + 63}{2} = 62.5$$

The median is 62.5, not 59.5. Note that 59.5 is what you would get if you did not arrange the numbers in order before taking the average of the 6th and 7th numbers.
- Choice (3): Most data would be considered skewed. In this case, the data are not skewed because 59 is the only number less than 60.
- Choice (4): The standard deviation is a measure of how close together the numbers are. If all the numbers were the same, the standard deviation would be zero. Without the 79, the standard deviation would be much lower since all the other numbers are pretty close to 65.

For the TI-84:

For the TI-Nspire:

The correct choice is (**4**).

16. The graph shows a parabola with a vertex at (15, 25) and that opens downward. The vertex form for the equation of a parabola is $y = a(x - h)^2 + k$, where h is the x-coordinate of the vertex, k is the y-coordinate of the vertex, a will be positive if the parabola opens upward, and a will be negative if the parabola opens downward. The graphs for choices (3) and (4) are parabolas with a vertex at (15, 25). Since choice (4) has a negative a-value, the parabola will open downward like in the question.

On the graphing calculator, the four choices can be graphed. Only choice (4) will resemble the graph in the question.

For the TI-84:

For the TI-Nspire:

The correct choice is (**4**).

17. The graphs of $f(x)$ and $g(x)$ look like this.

Analyze the answer choices:

- Choice (1): The vertex of $g(x)$ is $(3, -5)$, while the vertex of $f(x)$ is $(3, 2)$.
- Choice (2): Zeros can be seen on a graph as x-intercepts. Function $g(x)$ has two zeros. One is between 0 and 1, and the other is between 5 and 6. Function $f(x)$ does not have any zeros.
- Choice (3): The axis of symmetry of $g(x)$ is $x = 3$, and the axis of symmetry of $f(x)$ is also $x = 3$.
- Choice (4): The graphs do not intersect.

The correct choice is (**3**).

18. Since $f(1) = 2$, the first number in the sequence must be 2. Eliminate choices (3) and (4). To calculate the second number in the sequence, substitute 2 for n in the recursive part of the definition:

$$f(2) = -5f(2-1) + 2$$
$$= -5f(1) + 2$$
$$= -5 \cdot 2 + 2$$
$$= -10 + 2 = -8$$
$$= -8$$

The correct choice is (**2**).

19. First use the distributive property:

$$4(x-7) = 0.3(x+2) + 2.11$$
$$4x - 28 = 0.3x + 0.6 + 2.11$$

Then combine like terms:

$$4x - 28 = 0.3x + 2.71$$

Complete the problem by getting the variables on one side of the equation and the constants on the other side of the equation:

$$4x - 28 = 0.3x + 2.71$$
$$-0.3x = -0.3x$$
$$3.7x - 28 = 2.71$$
$$+28 = +28$$
$$\frac{3.7x}{3.7} = \frac{30.71}{3.7}$$
$$x = 8.3$$

The correct choice is (**1**).

20. The simplest way to solve this is to calculate the number of hours to complete the job. To calculate the number of wheelbarrow loads needed, divide 120 by 8 to get 15 wheelbarrows. To calculate the number of minutes needed to move the dirt, multiply the number of wheelbarrow loads by 10, which is $15 \times 10 = 150$ minutes. Finally, convert 150 minutes to hours by dividing 150 by 60 to get 2.5 hours.

    Test the four answer choices to see which one equals 2.5 hours.

    The correct choice is (**4**).

21. A linear function is something like $f(x) = 2x^1 - 3 = 2x - 3$. In other words, the exponent on the variable is 1. The graph of a linear function is a line.

$y = 2x - 3$

Start from any point on the line, such as (2, 1), and pick an interval, such as 3. The point 3 to the right has the coordinates (5, 7). The interval in this case is 3. The difference in the y-coordinates is $7 - 1 = 6$. If you start from some different point, such as (4, 5), and choose the same interval, 3, the other point will be (7, 11). The difference in the y-coordinates is 6. The differences are the same.

The correct choice is (3).

22. There are three ways to solve this question.

One way is to complete the square by adding $\left(\frac{-8}{2}\right)^2 = (-4)^2 = 16$ to both sides of the equation:

$$x^2 - 8x = 10$$
$$+16 = +16$$
$$x^2 - 8x + 16 = 26$$
$$(x - 4)^2 = 26$$
$$\sqrt{(x - 4)^2} = \pm\sqrt{26}$$
$$x - 4 = \pm\sqrt{26}$$

Another way to answer this is to substitute each of the answers for x into the expression $x^2 - 8x$ to see which one makes the expression evaluate to 16.

For the TI-84:

```
4+√26→X
        9.099019514
X²-8X
             10
```

For the TI-Nspire:

```
4+√26 →x        9.09902
x²-8·x             10.
```

The third method is to use the quadratic formula:

$$x = \frac{-b \pm \sqrt{b^2 - 4ac}}{2a}$$

Remember to start by rearranging the given equation into the format $ax^2 + bx + c = 0$:

$$x^2 - 8x = 10$$
$$x^2 - 8x - 10 = 0$$

$$x = \frac{-(-8) \pm \sqrt{(-8)^2 - 4(1)(-10)}}{(2)(1)}$$

$$x = \frac{8 \pm \sqrt{64 + 40}}{2}$$

$$x = \frac{8 \pm \sqrt{104}}{2}$$

$$x = \frac{8 \pm \sqrt{4 \cdot 26}}{2}$$

$$x = \frac{8 \pm 2\sqrt{26}}{2}$$

$$x = 4 \pm \sqrt{26}$$

The correct choice is (**2**).

23. To isolate the other variables, start by multiplying both sides of the equation by r:

$$Fr = r \cdot \frac{p_1 - p_2}{r}$$
$$Fr = p_1 - p_2$$

The isolate r:

$$\frac{Fr}{F} = \frac{p_1 - p_2}{F}$$

This is choice (4).

Then isolate p_1:

$$Fr = p_1 - p_2$$
$$+p_2 = +p_2$$
$$Fr + p_2 = p_1$$

This is choice (1).

Then isolate p_2:

$$Fr = p_1 - p_2$$
$$+p_2 = +p_2$$
$$Fr + p_2 = p_1$$
$$-Fr = -Fr$$
$$p_2 = p_1 - Fr$$

This is choice (2).

Only choice (3) cannot be derived from the given formula.

The correct choice is (**3**).

24. Since the ball never goes underground, the appropriate domain is the values of t of which $h(t) \geq 0$. By graphing the function on a graphing calculator, it can be seen that the ball starts on the ground at $t = 0$ and lands when $t = 1.5$.

For the TI-84:

For the TI-Nspire:

This can also be solved using algebra by calculating the values of t that make $h(t) = 0$:

$$-16t^2 + 24t = 0$$
$$8t(-2t + 3) = 0$$
$$8t = 0 \quad \text{or} \quad -2t + 3 = 0$$
$$t = 0 \quad \text{or} \quad -2t = -3$$
$$t = \frac{3}{2}$$

If the ball is on the ground at $t = 0$ and at $t = \frac{3}{2} = 1.5$, then it is above the ground between those two values.

The correct choice is (**1**).

PART II

25. Eliminate the parentheses by distributing the negative sign through the second expression. Then combine like terms:

$$(3x^2 + 4x - 8) - (2x^2 + 4x + 2)$$
$$3x^2 + 4x - 8 - 2x^2 + 4x - 2$$
$$5x^2 - 10$$

26. The graph can be created with a table.

x	$f(x) = -x^2 - 6x$
-6	$-(-6)^2 - 6(-6) = -36 + 36 = 0$
-5	$-(-5)^2 - 6(-5) = -25 + 30 = 5$
-4	$-(-4)^2 - 6(-4) = -16 + 24 = 8$
-3	$-(-3)^2 - 6(-3) = -9 + 18 = 9$
-2	$-(-2)^2 - 6(-2) = -4 + 12 = 8$
-1	$-(-1)^2 - 6(-1) = -1 + 6 = 5$
0	$-(0)^2 - 6(0) = 0 + 0 = 0$

This table can also be created with the graphing calculator.

For the TI-84:

For the TI-Nspire:

The graph looks like this:

The vertex is the highest point of the parabola (or the lowest if it is a "smiling" parabola). The coordinates of the vertex, according to the graph, are $(-3, 9)$.

You can also calculate the x-coordinate of the vertex by using the following formula:

$$x = \frac{-b}{2a}$$
$$= \frac{-(-6)}{2(-1)}$$
$$= \frac{6}{-2}$$
$$= -3$$

The y-coordinate of the vertex is what you get when you substitute the x-coordinate for x in the function:

$$y = f(-3)$$
$$= -(-3)^2 - 6(-3)$$
$$= -9 + 18$$
$$= 9$$

So the coordinates of the vertex are $(-3, 9)$.

27. When an irrational number is added to or subtracted from a rational number, the result will be irrational.

    If you calculate this difference on a calculator it becomes 5.585786..., which appears to be irrational since it does not terminate and does not seem to have a repeating pattern.

28. One way to solve this is to substitute a value for the initial price of the car, like $P = 10000$.

For $t = 0$:

$$v(0) = 10000(.85)^0$$
$$= 10000 \cdot 1$$
$$= 10000$$

One year later for $t = 1$:

$$v(1) = 10000(.85)^1$$
$$= 10000 \cdot .85$$
$$= 8500$$

Calculate the percent change from $t = 0$ to $t = 1$:

$$\% \text{ change} = \frac{\text{old price} - \text{new price}}{\text{old price}}$$
$$= \frac{10000 - 8500}{10000}$$
$$= \frac{15000}{1000}$$
$$= 0.15$$
$$= 15\%$$

A quicker way to solve this is to write the equation in the form $v(t) = P(1 + r)^t$, where r is the percent change. Since $.85 = (1 - .15)$, the percent change is 15%. Since r is negative, this is a percent decrease.

29. From the given information, three values from the chart can be quickly filled in.

	Watch Sports	Don't Watch Sports	Total
Like Pop			
Don't Like Pop	34		
Total	60		100

From these three values, the number of students who do not watch sports can be calculated by subtracting the number of students who watch sports from the total number of students:

$$100 - 60 = 40$$

	Watch Sports	Don't Watch Sports	Total
Like Pop			
Don't Like Pop	34		
Total	60	40	100

Since 70% of the students who do not watch sports do like pop music, calculate 70% of 40 and put that number into the table:

$$40 \cdot 0.7 = 28$$

	Watch Sports	Don't Watch Sports	Total
Like Pop		28	
Don't Like Pop	34		
Total	60	40	100

Now there is enough information to calculate the remaining four cells in the table. Each row and each column must add up to the total:

$$60 - 34 = 26$$
$$40 - 28 = 12$$
$$34 + 12 = 46$$
$$26 + 28 = 54$$

	Watch Sports	Don't Watch Sports	Total
Like Pop	26	28	54
Don't Like Pop	34	12	46
Total	60	40	100

30. First rewrite the inequality in slope-intercept form:

$$y + 4 < -2(x - 4)$$
$$y + 4 < -2x + 8$$
$$-4 = -4$$
$$y < -2x + 4$$

Graph the boundary line $y = -2x + 4$. Make it a dotted line because the inequality sign is $<$ and not $\leq$. It has a slope of -2 and a y-intercept of 4.

To decide which side of the line to shade, test to see if the point $(0, 0)$ makes the inequality true:

$$0 < -2(0) + 4$$
$$0 < 0 + 4$$
$$0 < 4$$

Since this is true, shade the side of the line containing (0, 0).

This can also be done on the graphing calculator.

For the TI-84:

For the TI-Nspire:

31. If $f(x) = g(x)$, then $x^2 = x$. To solve this equation, the first step is to subtract x from both sides of the equal sign. If you instead start by dividing both sides by x, you will lose one of the correct answers. Then factor and solve for x.

$$x^2 = x$$
$$-x = -x$$
$$x^2 - x = 0$$
$$x(x-1) = 0$$
$$x = 0 \quad \text{or} \quad x - 1 = 0$$
$$+1 = +1$$
$$x = 1$$

The solution set is $\{0, 1\}$.

32. The graph of a function $f(x - a)$ will look like the graph of $f(x)$ shifted to the right by a units. The graph of a function $f(x) - a$ will look like the graph of $f(x)$ shifted down by a units.

By choosing a value for a, like $a = 2$, and graphing $f(x)$, $g(x)$, and $h(x)$ on a graphing calculator it is easier to see how the graphs of $g(x)$ and $h(x)$ compare to the graph of $f(x)$.

For the TI-84:

For the TI-Nspire:

So $g(x)$ is shifted a units to the right, and $h(x)$ is shifted a units down.

PART III

33. The zeros of a function are the input values that make the function equal to 0. To find zeros, solve the equation:

$$r(x) = 0$$
$$x^2 + 3x - 18 = 0$$

Factor a polynomial in the form $x^2 + bx + c$ into $(x + p)(x + q)$ by finding, if possible, two numbers that have a sum of b and a product of c. In this case since b is 3 and c is -18, the two numbers are 6 and -3 since $6 + -3 = 3$ and $6 \cdot -3 = 18$:

$$x^2 + 3x - 18 = 0$$
$$(x + 6)(x - 3) = 0$$

$$
\begin{array}{ccc}
x + 6 = 0 & \text{or} & x - 3 = 0 \\
-6 = -6 & & +3 = +3 \\
x = -6 & & x = 3
\end{array}
$$

On the graph of the function, the zeros are the x-intercepts. The x-intercepts of the graph of $r(x)$ are $(-6, 0)$ and $(3, 0)$. This can be seen on the graphing calculator.

For the TI-84:

For the TI-Nspire:

34. In a graph where the x-axis is time and the y-axis is distance, the slope of a line segment is the speed at which the object is moving. There are four line segments in this graph. Find the slope of each segment.

- Calculate the slope from A to B.

    From hours $= 0$ to hours $= 2$, the distance changed from 0 miles to 110 miles:

    $$\begin{aligned} \text{speed} &= \text{slope} \\ &= \frac{110 - 0}{2 - 0} \\ &= \frac{110}{2} \\ &= 55 \text{ miles per hour} \end{aligned}$$

- Calculate the slope from B to C.

    From hours $= 2$ to hours $= 3.5$, the distance stayed at 110 for the whole time:

    $$\begin{aligned} \text{speed} &= \text{slope} \\ &= \frac{110 - 110}{3.5 - 2} \\ &= \frac{0}{1.5} \\ &= 0 \text{ miles per hour (stopped)} \end{aligned}$$

- Calculate the slope from C to D.

    From hours $= 3.5$ to hours $= 5$, the distance changed from 110 miles to 200 miles:

    $$\begin{aligned} \text{speed} &= \text{slope} \\ &= \frac{200 - 110}{5 - 3.5} \\ &= \frac{90}{1.5} \\ &= 60 \text{ miles per hour} \end{aligned}$$

- Calculate the slope from D to E.

    From hours = 5 to hours = 7, the distance changed from 200 miles to 230 miles:

    $$\begin{aligned} \text{speed} &= \text{slope} \\ &= \frac{230 - 200}{7 - 5} \\ &= \frac{30}{2} \\ &= 15 \text{ miles per hour} \end{aligned}$$

Craig most likely drove in the city between D and E because city driving is most likely to be 15 miles per hour.

In the interval between B and C, his speed was zero. So Craig did not move at all. He may have stopped to eat.

Since Craig travels a total of 230 miles in 7 hours, his average speed is $\frac{230}{7} \approx 32.9$ miles per hour.

35. Start by setting $g(x)$ and $2k(x)$ equal to each other:

$$\begin{aligned} g(x) &= 2k(x) \\ 2x^2 + 3x + 10 &= 2(2x + 16) \\ 2x^2 + 3x + 10 &= 4x + 32 \\ -4x \quad &= -4x \\ 2x^2 - x + 10 &= 32 \\ -32 &= -32 \\ 2x^2 - x - 22 &= 0 \end{aligned}$$

Then use the quadratic formula:

$$x = \frac{-b \pm \sqrt{b^2 - 4ac}}{2a}$$

$$x = \frac{-(-1) \pm \sqrt{(-1)^2 - 4 \cdot 2 \cdot (-22)}}{2 \cdot 2}$$

$$x = \frac{1 \pm \sqrt{1 + 176}}{4} = \frac{1 \pm \sqrt{177}}{4}$$

$$x = \frac{1 + \sqrt{177}}{4} \approx 3.6 \quad \text{or} \quad x = \frac{1 - \sqrt{177}}{4} \approx -3.1$$

Of the three options: factoring, completing the square, and the quadratic formula, the quadratic formula was simplest for this example. The coefficient of 2 in front of the x^2 makes using the other two options complicated.

36. A function to model Option 1 is $f(x) = 10 + 100x$.

A function to model Option 2 is $g(x) = 10 \cdot 2^x$.

After 7 weeks with Option 1, he will have:

$$f(7) = 10 + 100 \cdot 7$$
$$= 10 + 700$$
$$= 710$$

After 7 weeks with Option 2, he will have:

$$g(7) = 10 \cdot 2^7$$
$$= 10 \cdot 128$$
$$= 1280$$

Both Option 1 and Option 2 will enable Michael to have more than $700. Either option will let him reach his goal.

PART IV

37. The equation for the swim team can be written as $y = 5 + 10x$ or $y = 10x + 5$.

The equation for the chorus can be written as $y = 35 + 5x$ or $y = 5x + 35$.

Before graphing, choose an appropriate scale. Since the point of intersection must be on the graph, you must find that point of intersection before deciding on the scale.

Using a chart is the simplest way to determine the approximate intersection point.

x	swim	chorus
0	5	35
1	15	40
2	25	45
3	35	50
4	45	55
5	55	60
6	65	65
7	75	70

Since the grid is 20 by 20 and the point (6, 65) must be on it, one possible scale is to have one box equal 0.5 units on the x-axis and 5 units on the y-axis.

The graph looks like this:

The point of intersection is (6, 65). It means that in the year 2016, both the swim team and the chorus had 65 members.

SELF-ANALYSIS CHART June 2017

Topic	Question Numbers	Number of Points	Your Points	Your Percentage
1. Polynomials	5, 10, 25, 31, 33	2 + 2 + 2 + 2 + 4 = 12		
2. Properties of Algebra	2, 19, 23	2 + 2 + 2 = 6		
3. Functions	9, 17, 20, 21, 24, 32, 35	2 + 2 + 2 + 2 + 2 + 2 + 4 = 16		
4. Creating and Interpreting Equations	4, 28	2 + 2 = 4		
5. Inequalities	11, 13, 30	2 + 2 + 2 = 6		
6. Sequences and Series	18	2		
7. Systems of Equations	8, 37	2 + 6 = 8		
8. Quadratic Equations and Factoring	6, 22	2 + 2 = 4		
9. Regression	14	2		
10. Exponential Equations	7, 12, 36	2 + 2 + 4 = 8		
11. Graphing	1, 16, 26, 34	2 + 2 + 2 + 4 = 10		
12. Statistics	15, 29	2 + 2 = 4		
13. Number Properties	20, 27	2 + 2 = 4		

HOW TO CONVERT YOUR RAW SCORE TO YOUR ALGEBRA I REGENTS EXAMINATION SCORE

The accompanying conversion chart must be used to determine your final score on the June 2017 Regents Examination in Algebra I. To find your final exam score, locate in the column labeled "Raw Score" the total number of points you scored out of a possible 86 points. Since partial credit is allowed in Parts II, III, and IV of the test, you may need to approximate the credit you would receive for a solution that is not completely correct. Then locate in the adjacent column to the right the scale score that corresponds to your raw score. The scale score is your final Algebra I Regents Examination score.

SELF-ANALYSIS CHART June 2017

Regents Examination in Algebra I—June 2017
Chart for Converting Total Test Raw Scores to Final Examination Scores (Scaled Scores)

Raw Score	Scale Score	Performance Level	Raw Score	Scale Score	Performance Level	Raw Score	Scale Score	Performance Level
86	100	5	57	81	4	28	66	3
85	99	5	56	81	4	27	65	3
84	97	5	55	80	4	26	64	2
83	96	5	54	80	4	25	62	2
82	95	5	53	80	4	24	61	2
81	94	5	52	80	4	23	60	2
80	93	5	51	79	3	22	58	2
79	92	5	50	79	3	21	57	2
78	91	5	49	79	3	20	55	2
77	90	5	48	78	3	19	53	1
76	90	5	47	78	3	18	52	1
75	89	5	46	78	3	17	50	1
74	88	5	45	77	3	16	48	1
73	88	5	44	77	3	15	46	1
72	87	5	43	77	3	14	44	1
71	86	5	42	76	3	13	42	1
70	86	5	41	76	3	12	39	1
69	86	5	40	75	3	11	37	1
68	85	5	39	75	3	10	34	1
67	84	4	38	74	3	9	32	1
66	84	4	37	73	3	8	29	1
65	83	4	36	73	3	7	26	1
64	83	4	35	72	3	6	23	1
63	83	4	34	71	3	5	19	1
62	82	4	33	70	3	4	16	1
61	82	4	32	70	3	3	12	1
60	82	4	31	69	3	2	8	1
59	82	4	30	68	3	1	4	1
58	81	4	29	67	3	0	0	1

Examination August 2017
Algebra I

HIGH SCHOOL MATH REFERENCE SHEET

Conversions

1 inch = 2.54 centimeters
1 meter = 39.37 inches
1 mile = 5280 feet
1 mile = 1760 yards
1 mile = 1.609 kilometers

1 cup = 8 fluid ounces
1 pint = 2 cups
1 quart = 2 pints
1 gallon = 4 quarts
1 gallon = 3.785 liters
1 liter = 0.264 gallon
1 liter = 1000 cubic centimeters

1 kilometer = 0.62 mile
1 pound = 16 ounces
1 pound = 0.454 kilogram
1 kilogram = 2.2 pounds
1 ton = 2000 pounds

Formulas

Triangle	$A = \frac{1}{2}bh$
Parallelogram	$A = bh$
Circle	$A = \pi r^2$
Circle	$C = \pi d$ or $C = 2\pi r$

Formulas (continued)

General Prisms	$V = Bh$
Cylinder	$V = \pi r^2 h$
Sphere	$V = \frac{4}{3}\pi r^3$
Cone	$V = \frac{1}{3}\pi r^2 h$
Pyramid	$V = \frac{1}{3}Bh$
Pythagorean Theorem	$a^2 + b^2 = c^2$
Quadratic Formula	$x = \frac{-b \pm \sqrt{b^2 - 4ac}}{2a}$
Arithmetic Sequence	$a_n = a_1 + (n-1)d$
Geometric Sequence	$a_n = a_1 r^{n-1}$
Geometric Series	$S_n = \frac{a_1 - a_1 r^n}{1 - r}$ where $r \neq 1$
Radians	1 radian = $\frac{180}{\pi}$ degrees
Degrees	1 degree = $\frac{\pi}{180}$ radians
Exponential Growth/Decay	$A = A_0 e^{k(t - t_0)} + B_0$

PART I

Answer all 24 questions in this part. Each correct answer will receive 2 credits. No partial credit will be allowed. For each statement or question, write in the space provided the numeral preceding the word or expression that best completes the statement or answers the question. [48 credits]

1 A part of Jennifer's work to solve the equation $2(6x^2 - 3) = 11x^2 - x$ is shown below.

Given: $2(6x^2 - 3) = 11x^2 - x$
Step 1: $12x^2 - 6 = 11x^2 - x$

Which property justifies her first step?

(1) identity property of multiplication
(2) multiplication property of equality
(3) commutative property of multiplication
(4) distributive property of multiplication over subtraction

2 Which value of x results in equal outputs for $j(x) = 3x - 2$ and $b(x) = |x + 2|$?

(1) -2 (3) $\frac{2}{3}$
(2) 2 (4) 4

3 The expression $49x^2 - 36$ is equivalent to

(1) $(7x - 6)^2$
(2) $(24.5x - 18)^2$
(3) $(7x - 6)(7x + 6)$
(4) $(24.5x - 18)(24.5x + 18)$

4 If $f(x) = \frac{1}{2}x^2 - \left(\frac{1}{4}x + 3\right)$, what is the value of $f(8)$?

(1) 11 (3) 27
(2) 17 (4) 33

5 The graph below models the height of a remote-control helicopter over 20 seconds during flight.

Over which interval does the helicopter have the *slowest* average rate of change?

(1) 0 to 5 seconds (3) 10 to 15 seconds
(2) 5 to 10 seconds (4) 15 to 20 seconds

6 In the functions $f(x) = kx^2$ and $g(x) = |kx|$, k is a positive integer. If k is replaced by $\frac{1}{2}$, which statement about these new functions is true?

(1) The graphs of both $f(x)$ and $g(x)$ become wider.
(2) The graph of $f(x)$ becomes narrower and the graph of $g(x)$ shifts left.
(3) The graphs of both $f(x)$ and $g(x)$ shift vertically.
(4) The graph of $f(x)$ shifts left and the graph of $g(x)$ becomes wider.

7 Wenona sketched the polynomial $P(x)$ as shown on the axes below.

Which equation could represent $P(x)$?
(1) $P(x) = (x + 1)(x - 2)^2$
(2) $P(x) = (x - 1)(x + 2)^2$
(3) $P(x) = (x + 1)(x - 2)$
(4) $P(x) = (x - 1)(x + 2)$

8 Which situation does *not* describe a causal relationship?
(1) The higher the volume on a radio, the louder the sound will be.
(2) The faster a student types a research paper, the more pages the research paper will have.
(3) The shorter the time a car remains running, the less gasoline it will use.
(4) The slower the pace of a runner, the longer it will take the runner to finish the race.

9 A plumber has a set fee for a house call and charges by the hour for repairs. The total cost of her services can be modeled by $c(t) = 125t + 95$.

Which statements about this function are true?

   I. A house call fee costs $95.
   II. The plumber charges $125 per hour.
   III. The number of hours the job takes is represented by t.

(1) I and II, only
(2) I and III, only
(3) II and III, only
(4) I, II, and III

10 What is the domain of the relation shown below?

$$\{(4, 2), (1, 1), (0, 0), (1, -1), (4, -2)\}$$

(1) $\{0, 1, 4\}$
(2) $\{-2, -1, 0, 1, 2\}$
(3) $\{-2, -1, 0, 1, 2, 4\}$
(4) $\{-2, -1, 0, 0, 1, 1, 1, 2, 4, 4\}$

11 What is the solution to the inequality $2 + \frac{4}{9}x \geq 4 + x$?

(1) $x \leq -\frac{18}{5}$
(2) $x \geq -\frac{18}{5}$
(3) $x \leq \frac{54}{5}$
(4) $x \geq \frac{54}{5}$

12 Konnor wants to burn 250 Calories while exercising for 45 minutes at the gym. On the treadmill, he can burn 6 Cal/min. On the stationary bike, he can burn 5 Cal/min.

If t represents the number of minutes on the treadmill and b represents the number of minutes on the stationary bike, which expression represents the number of Calories that Konnor can burn on the stationary bike?

(1) b
(2) $5b$
(3) $45 - b$
(4) $250 - 5b$

13 Which value of x satisfies the equation $\frac{5}{6}\left(\frac{3}{8} - x\right) = 16$?

(1) -19.575
(2) -18.825
(3) -16.3125
(4) -15.6875

14 If a population of 100 cells triples every hour, which function represents $p(t)$, the population after t hours?

(1) $p(t) = 3(100)t$
(2) $p(t) = 100(3)t$
(3) $p(t) = 3t + 100$
(4) $p(t) = 100t + 3$

15 A sequence of blocks is shown in the diagram below.

This sequence can be defined by the recursive function $a_1 = 1$ and $a_n = a_{n-1} + n$. Assuming the pattern continues, how many blocks will there be when $n = 7$?

(1) 13
(2) 21
(3) 28
(4) 36

16 Mario's $15,000 car depreciates in value at a rate of 19% per year. The value, V, after t years can be modeled by the function $V = 15{,}000(0.81)^t$. Which function is equivalent to the original function?

(1) $V = 15{,}000(0.9)^{9t}$
(2) $V = 15{,}000(0.9)^{2t}$
(3) $V = 15{,}000(0.9)^{\frac{t}{9}}$
(4) $V = 15{,}000(0.9)^{\frac{t}{2}}$

17 The highest possible grade for a book report is 100. The teacher deducts 10 points for each day the report is late.

Which kind of function describes this situation?

(1) linear
(2) quadratic
(3) exponential growth
(4) exponential decay

18 The function $h(x)$, which is graphed below, and the function $g(x) = 2|x+4| - 3$ are given.

Which statements about these functions are true?

I. $g(x)$ has a lower minimum value than $h(x)$.
II. For all values of x, $h(x) < g(x)$.
III. For any value of x, $g(x) \neq h(x)$.

(1) I and II, only
(2) I and III, only
(3) II and III, only
(4) I, II, and III

19 The zeros of the function $f(x) = 2x^3 + 12x - 10x^2$ are
(1) {2, 3}
(2) {−1, 6}
(3) {0, 2, 3}
(4) {0, −1, 6}

20 How many of the equations listed below represent the line passing through the points (2, 3) and (4, −7)?

$$5x + y = 13$$
$$y + 7 = -5(x - 4)$$
$$y = -5x + 13$$
$$y - 7 = 5(x - 4)$$

(1) 1
(2) 2
(3) 3
(4) 4

20 ____

21 The Ebola virus has an infection rate of 11% per day as compared to the SARS virus, which has a rate of 4% per day.

If there were one case of Ebola and 30 cases of SARS initially reported to authorities and cases are reported each day, which statement is true?

(1) At day 10 and day 53 there are more Ebola cases.
(2) At day 10 and day 53 there are more SARS cases.
(3) At day 10 there are more SARS cases, but at day 53 there are more Ebola cases.
(4) At day 10 there are more Ebola cases, but at day 53 there are more SARS cases.

21 ____

22 The results of a linear regression are shown below.

$$y = ax + b$$
$$a = -1.15785$$
$$b = 139.3171772$$
$$r = -0.896557832$$
$$r^2 = 0.8038159461$$

Which phrase best describes the relationship between x and y?

(1) strong negative correlation
(2) strong positive correlation
(3) weak negative correlation
(4) weak positive correlation

22 ____

23 Abigail's and Gina's ages are consecutive integers. Abigail is younger than Gina and Gina's age is represented by x. If the difference of the square of Gina's age and eight times Abigail's age is 17, which equation could be used to find Gina's age?

(1) $(x+1)^2 - 8x = 17$
(2) $(x-1)^2 - 8x = 17$
(3) $x^2 - 8(x+1) = 17$
(4) $x^2 - 8(x-1) = 17$

23 _____

24 Which system of equations does *not* have the same solution as the system below?

$$4x + 3y = 10$$
$$-6x - 5y = 16$$

(1) $-12x - 9y = -30$
    $12x + 10y = 32$
(2) $20x + 15y = 50$
    $-18x - 15y = -48$
(3) $24x + 18y = 60$
    $-24x - 20y = -64$
(4) $40x + 30y = 100$
    $36x - 30y = -96$

24 _____

PART II

Answer all 8 questions in this part. Each correct answer will receive 2 credits. Clearly indicate the necessary steps, including appropriate formula substitutions, diagrams, graphs, charts, etc. For all questions in this part, a correct numerical answer with no work shown will receive only 1 credit. [16 credits]

25 A teacher wrote the following set of numbers on the board:

$$a = \sqrt{20} \qquad b = 2.5 \qquad c = \sqrt{225}$$

Explain why $a + b$ is irrational, but $b + c$ is rational.

A + b is irrational because the numbers do not end or repeat.

B + C is rational because it ends.

26 Determine and state whether the sequence 1, 3, 9, 27, . . . displays exponential behavior. Explain how you arrived at your decision.

Yes because it goes up by the same amount each time

27 Using the formula for the volume of a cone, express r in terms of V, h, and π.

$$r = \sqrt{\frac{3V}{\pi h}}$$

28 The graph below models the cost of renting video games with a membership in Plan A and Plan B.

Explain why Plan B is the better choice for Dylan if he only has $50 to spend on video games, including a membership fee.

For $50
Dylan can get 14 games

Bobby wants to spend $65 on video games, including a membership fee. Which plan should he choose? Explain your answer.

For $65 either plan Bobby gets 20 games

29 Samantha purchases a package of sugar cookies. The nutrition label states that each serving size of 3 cookies contains 160 Calories. Samantha creates the graph below showing the number of cookies eaten and the number of Calories consumed.

Explain why it is appropriate for Samantha to draw a line through the points on the graph.

so she does not have to eat cookies 3 at a time.

She can also eat a fraction of a cookie

30 A two-inch-long grasshopper can jump a horizontal distance of 40 inches. An athlete, who is five feet nine, wants to cover a distance of one mile by jumping. If this person could jump at the same ratio of body-length to jump-length as the grasshopper, determine, to the *nearest jump*, how many jumps it would take this athlete to jump one mile.

48 jumps

31 Write the expression $5x + 4x^2(2x + 7) - 6x^2 - 9x$ as a polynomial in standard form.

$5x + 8x^3 + 28x^2 - 6x^2 - 9x$

$8x^3 + 22x^2 - 4x$

32 Solve the equation $x^2 - 6x = 15$ by completing the square.

PART III

Answer all 4 questions in this part. Each correct answer will receive 4 credits. Clearly indicate the necessary steps, including appropriate formula substitutions, diagrams, graphs, charts, etc. For all questions in this part, a correct numerical answer with no work shown will receive only 1 credit. [16 credits]

33 Loretta and her family are going on vacation. Their destination is 610 miles from their home. Loretta is going to share some of the driving with her dad. Her average speed while driving is 55 mph and her dad's average speed while driving is 65 mph.

The plan is for Loretta to drive for the first 4 hours of the trip and her dad to drive for the remainder of the trip. Determine the number of hours it will take her family to reach their destination.

10 hours

After Loretta has been driving for 2 hours, she gets tired and asks her dad to take over. Determine, to the *nearest tenth of an hour*, how much time the family will save by having Loretta's dad drive for the remainder of the trip.

0.3 hours

34 The heights, in feet, of former New York Knicks basketball players are listed below.

6.4	6.9	6.3	6.2	6.3	6.0	6.1	6.3	6.8	6.2
6.5	7.1	6.4	6.3	6.5	6.5	6.4	7.0	6.4	6.3
6.2	6.3	7.0	6.4	6.5	6.5	6.5	6.0	6.2	

Using the heights given, complete the frequency table below.

Interval	Frequency
6.0–6.1	
6.2–6.3	
6.4–6.5	
6.6–6.7	
6.8–6.9	
7.0–7.1	

Question 34 is continued on the next page.

Question 34 continued

Based on the frequency table created, draw and label a frequency histogram on the grid below.

Determine and state which interval contains the upper quartile. Justify your response.

The quartile is in the 6-4, 6-5 intervals

246 EXAMINATION August 2017

35 Solve the following system of inequalities graphically on the grid below and label the solution S.

$$3x + 4y > 20$$
$$x < 3y - 18$$

Is the point (3, 7) in the solution set? Explain your answer.

36 An Air Force pilot is flying at a cruising altitude of 9000 feet and is forced to eject from her aircraft. The function $h(t) = -16t^2 + 128t + 9000$ models the height, in feet, of the pilot above the ground, where t is the time, in seconds, after she is ejected from the aircraft.

Determine and state the vertex of $h(t)$. Explain what the second coordinate of the vertex represents in the context of the problem.

(4, 9,256)

After the pilot was ejected, what is the maximum number of feet she was above the aircraft's cruising altitude? Justify your answer.

256 feet

PART IV

Answer the question in this part. A correct answer will receive 6 credits. Clearly indicate the necessary steps, including appropriate formula substitutions, diagrams, graphs, charts, etc. A correct numerical answer with no work shown will receive only 1 credit. [6 credits]

37 Zeke and six of his friends are going to a baseball game. Their combined money totals $28.50. At the game, hot dogs cost $1.25 each, hamburgers cost $2.50 each, and sodas cost $0.50 each. Each person buys one soda. They spend all $28.50 on food and soda.

Write an equation that can determine the number of hot dogs, x, and hamburgers, y, Zeke and his friends can buy.

$1.25x + 2.50y = 25$

$2.50y = 25 - 1.25x$

$y = 10 - 0.5x$

Question 37 is continued on the next page.

Question 37 continued

Graph your equation on the grid below.

Determine how many different combinations, including those combinations containing zero, of hot dogs and hamburgers Zeke and his friends can buy, spending all $28.50. Explain your answer.

Answers August 2017
Algebra I

Answer Key

PART I
1. (4)
2. (2)
3. (3)
4. (3)
5. (2)
6. (1)
7. (1)
8. (2)
9. (4)
10. (1)
11. (1)
12. (2)
13. (2)
14. (2)
15. (3)
16. (2)
17. (1)
18. (2)
19. (3)
20. (3)
21. (3)
22. (1)
23. (4)
24. (4)

PART II
25. Rational + Irrational = Irrational; Rational + Rational = Rational
26. Yes, the common ratio is 3.
27. $r = \sqrt{\dfrac{3V}{\pi h}}$
28. For $50, Dylan can get 14 games with B compared to 12 games with A. For $65, either plan gets Bobby 20 games.
29. She does not have to eat cookies 3 at a time. She can also eat a fraction of a cookie.
30. 46 jumps
31. $8x^3 + 22x^2 - 4x$
32. $x = 3 \pm \sqrt{24}$ or $x = 3 \pm 2\sqrt{6}$

PART III
33. 10 hours, 0.3 hours
34. See diagram, the top quartile is in the 6.4–6.5 interval
35. See diagram, (3, 7) is not in the solution set
36. (4, 9,256), 256 feet

PART IV
37. $1.25x + 2.50y = 25$, 11 different combinations

In **Parts II–IV**, you are required to show how you arrived at your answers. For sample methods of solutions, see the *Answers Explained* section.

Answers and Explanations

PART I

1. The distributive property of multiplication over subtraction says that an expression in the form $a \cdot (b - c)$ is equivalent to the expression $a \cdot b - a \cdot c$. In the solution, this property is applied to change $2(6x^3 - 3)$ to $2 \cdot 6x^3 - 2 \cdot 3 = 12x^3 - 6$.

   The correct choice is (**4**).

2. Test each of the answer choices to see which one makes $j(x) = b(x)$.

   - Choice (1):

   $$j(-2) = 3 \cdot (-2) - 2$$
   $$= -6 - 2$$
   $$= -8$$

   $$b(-2) = |-2 + 2|$$
   $$= |0|$$
   $$= 0$$

   - Choice (2):

   $$j(2) = 3 \cdot 2 - 2$$
   $$= 6 - 2$$
   $$= 4$$

   $$b(2) = |2 + 2|$$
   $$= |4|$$
   $$= 4$$

   The correct choice is (**2**).

3. The difference of perfect squares factoring pattern says that an expression of the form $a^2 - b^2$ is equivalent to the expression $(a-b)(a+b)$. The expression $49x^2 - 36$ can be rewritten as $(7x)^2 - 6^2$. Since this is a difference of perfect squares, it is equal to $(7x - 6)(7x + 6)$.

The correct choice is (**3**).

4. Replace each x with the number 8 and simplify:

$$f(8) = \frac{1}{2} \cdot 8^2 - \left(\frac{1}{4} \cdot 8 + 3\right)$$
$$= \frac{1}{2} \cdot 64 - (2 + 3)$$
$$= 32 - 5$$
$$= 27$$

The correct choice is (**3**).

5. The average rate of change over an interval is the slope of the line segment that joins the endpoints of that interval. The slowest rate of change corresponds to the interval whose slope has the smallest absolute value. Even without doing any calculations, it can be seen that the interval from 5 to 10 seconds has the slope with the smallest absolute value since a line segment joining (5, 19) and (10, 20) is closest to being a horizontal line segment.

The average rate of change can also be calculated by using the slope formula $m = \frac{y_2 - y_1}{x_2 - x_1}$ for each of the intervals.

- Choice (1):

$$m = \frac{19 - 46}{5 - 0}$$
$$= \frac{-27}{5}$$
$$= -5\frac{2}{5}$$

The absolute value is $5\frac{2}{5}$.

- Choice (2):
$$m = \frac{20 - 19}{10 - 5}$$
$$= \frac{1}{5}$$

The absolute value is $\frac{1}{5}$.

- Choice (3):
$$m = \frac{14 - 20}{15 - 10}$$
$$= \frac{-6}{5}$$
$$= -1\frac{1}{5}$$

The absolute value is $1\frac{1}{5}$.

- Choice (4):
$$m = \frac{17 - 14}{20 - 15}$$
$$= \frac{3}{5}$$

The absolute value is $\frac{3}{5}$.

The correct choice is (**2**).

6. If you choose an initial integer value for k, such as $k = 2$, you can use a graphing calculator to compare the graphs of $f(x) = 2x^2$ to $f(x) = \frac{1}{2}x^2$ and the graphs of $g(x) = |2x|$ to $g(x) = \left|\frac{1}{2}x\right|$.

For the TI-84:

For the TI-Nspire:

When k is replaced with $\frac{1}{2}$, each of the graphs becomes wider.

The correct choice is (**1**).

7. Each of the answer choices can be graphed on the graphing calculator. When choice (1) is graphed, it looks like the graph from the question.

For the TI-84:

For the TI-Nspire:

Even without the graphing calculator, it can be seen that the graph has an x-intercept at -1. So $(x + 1)$ would be one factor. The graph also has a double root at $x = 2$ since the graph bounces off the axis rather than passing through the axis. So there will be two $(x - 2)$ factors.

The correct choice is (**1**).

8. A causal relationship is one where one event causes the other to happen.
   - Choice (1): Raising the volume does cause the radio to become louder.
   - Choice (2): How fast the student types does not matter. If the paper is five pages long, the student will complete typing the paper more quickly if he or she types faster, but the paper will not be longer.
   - Choice (3): Assuming that the car uses the same amount of gasoline each minute it runs, the shorter it remains running, the less gasoline the car will use.
   - Choice (4): Running slower does cause a runner to take more time to complete the race.

   There is an argument to be made that choice (2) could be considered causal if, for example, the student decides to spend exactly two hours on the paper and to type for that amount of time. In that case, the faster typist will have completed more pages after two hours. Compared to the other choices, though, choice (2) is still the best choice.

   The correct choice is (**2**).

9. One way to determine what the components of this equation represent is to calculate the values of $c(0)$, $c(1)$, and $c(2)$:

$$c(0) = 125 \cdot 0 + 95$$
$$= 0 + 95$$
$$= 95$$
$$c(1) = 125 \cdot 1 + 95$$
$$= 125 + 95$$
$$= 220$$
$$c(2) = 125 \cdot 2 + 95$$
$$= 250 + 95$$
$$= 345$$

   So when $t = 0$, the function evaluates to 95. This means that $95 must be the cost to get the plumber to come to the house even without spending any time doing the repairs. When $t = 1$, the fee increases by $125. So 125 must be the price per hour.

   In general, when a real-world scenario has the form $y = mx + b$, the b represents the fixed costs and the m represents the cost increase for every time x increases by 1.

   The correct choice is (**4**).

10. The domain of a set of ordered pairs is a set containing each unique x-coordinate from each of the ordered pairs. The x-coordinates of the five ordered pairs are 4, 1, 0, 1, and 4. Since a set does not contain repeated numbers, this domain can be described as {0, 1, 4}.

The correct choice is (**1**).

11. Start by solving the inequality the way you would solve an equality:

$$2 + \frac{4}{9}x \geq 4 + x$$
$$-x = -x$$
$$2 + \frac{4}{9}x - \frac{9}{9}x \geq 4$$
$$2 - \frac{5}{9}x \geq 4$$
$$-2 = -2$$
$$-\frac{5}{9}x \geq 2$$

For the last step, multiply both sides of the equation by $-\frac{9}{5}$. With inequalities, you have to switch the direction of the inequality sign anytime you multiply or divide both sides of the inequality by a negative:

$$\left(-\frac{9}{5}\right)\left(-\frac{5}{9}\right)x \leq \left(-\frac{9}{5}\right)2$$
$$x \leq -\frac{18}{5}$$

The correct choice is (**1**).

12. If you were trying to solve for the number of minutes on each of the exercise machines, you would need to solve a system of equations:

$$t + b = 45$$
$$6t + 5b = 250$$

However, this question does not ask you to solve for t and b. It just asks how many calories Konnor burns on the bike. If he spends b hours on the bike and burns 5 Cal/min each minute on the bike, he burns $5b$ calories on the bike. The other information in the question is irrelevant.

The correct choice is (**2**).

13. Solve the equation for x:

$$\frac{5}{6}\left(\frac{3}{8} - x\right) = 16$$

$$\left(\frac{5}{6}\right)\left(\frac{3}{8}\right) - \frac{5}{6}x = 16$$

$$\frac{15}{48} - \frac{5}{6}x = 16$$

$$-\frac{15}{48} = -\frac{15}{48}$$

$$-\frac{5}{6}x = 16 - \frac{15}{48}$$

$$= 15.6875$$

$$\left(-\frac{6}{5}\right)\left(-\frac{6}{5}x\right) = \left(-\frac{6}{5}\right) \cdot 15.6875$$

$$x = -18.825$$

The correct choice is (**2**).

14. The population of cells forms the sequence 100, 300, 900, 2,700, ...

If you substitute 0, 1, and 2 into the functions given in the choices, choice (2) becomes

$$p(0) = 100(3)^0$$
$$= 100 \cdot 1$$
$$= 100$$

$$p(1) = 100(3)^1$$
$$= 100 \cdot 3$$
$$= 300$$

$$p(0) = 100(3)^2$$
$$= 100 \cdot 9$$
$$= 900$$

An alternate way of answering this is to notice that since the numbers increase exponentially, the function must be of the form $p(t) = a \cdot b^t$. In a function of this form, a represents the initial value and b represents the growth factor. The initial value is 100, and the growth factor is 3. So the function is $p(t) = 100(3)^t$.

The correct choice is (**2**).

15. The given values of the sequence are $a_1 = 1$, $a_2 = 3$, $a_3 = 6$, and $a_4 = 10$. To get the next three values, substitute $n = 5, 6$, and 7 into the recursive function definition:

$$a_5 = a_{5-1} + 5$$
$$= a_4 + 5$$
$$= 10 + 5$$
$$= 15$$
$$a_6 = a_{6-1} + 6$$
$$= a_5 + 6$$
$$= 15 + 6$$
$$= 21$$
$$a_7 = a_{7-1} + 7$$
$$= a_6 + 7$$
$$= 21 + 7$$
$$= 28$$

The correct choice is (**3**).

16. The simplest way to solve this question is to substitute $t = 1$ into each of the answer choices to see which gives the same answer as $V = 15{,}000(0.81)^1 = 15{,}000 \cdot 0.81 = 12{,}150$.

When you substitute $t = 1$ into choice (2), it becomes:

$$V = 15{,}000(0.9)^{2 \cdot 1}$$
$$= 15{,}000(0.9)^2$$
$$= 15{,}000 \cdot 0.81$$
$$= 12{,}150$$

Another way to see that choice (2) is equivalent to the original equation is to use the property of exponents that $x^{ab} = (x^a)^b$:

$$15{,}000 \cdot (0.9)^{2t} = 15{,}000\left(0.9^2\right)^t$$
$$= 15{,}000(0.81)^t$$

The correct choice is (**2**).

17. If you make a graph where for each point the x-coordinate is the number of days late and the y-coordinate is the highest possible grade, it will look like this:

Since the points all fall on a line, this is a linear relationship.

The correct choice is (**1**).

18. When you graph $g(x)$ on the same set of axes as $h(x)$, they look like this:

Using the graphing calculator, plot function g. It becomes the same as the graph of h but shifted down 5 units.

For the TI-84:

For the TI-Nspire:

- Statement I: The minimum value of $g(x)$ is -3, while the minimum value of $h(x)$ is 2. So the minimum value of $g(x)$ is lower than the minimum value of $h(x)$.
- Statement II: For every x-value, the y-coordinate of the graph of $h(x)$ is above the y-coordinate of the graph of $g(x)$. So it is not true that for all (or any) values of x that $h(x) < g(x)$.
- Statement III: Since $g(x)$ is the same as $h(x)$ but just shifted down 5 units, there is no x-value where $g(x) = h(x)$. So it is true that for any value of x, $g(x) \neq h(x)$.

Statements I and III are both true.

The correct choice is (**2**).

19. The zeros of a function are the values that make the function evaluate to zero. To calculate the zeros of f, find all the solutions to the equation $0 = 2x^3 + 12x - 10x^2$:

$$0 = 2x^3 + 12x - 10x^2$$
$$0 = 2x(x^2 + 6 - 5x)$$
$$= 2x(x^2 - 5x + 6)$$
$$= 2x(x - 2)(x - 3)$$

For the product to be equal to zero, $2x$, $x - 2$, or $x - 3$ must equal zero:

$$2x = 0$$
$$\frac{2x}{2} = \frac{0}{2}$$
$$x = 0$$

or

$$x - 2 = 0$$
$$+2 = +2$$
$$x = 2$$

or

$$x - 3 = 0$$
$$+3 = +3$$
$$x = 3$$

The solution set is $\{0, 2, 3\}$.

If you divided both sides of the equation by x, you would have "lost" the $x = 0$ solution and found only $\{2, 3\}$ as your solution, which is incorrect.

The correct choice is (**3**).

20. To find the equation of a line through two given points, first find the slope of the line. In this case, the slope is

$$m = \frac{-7 - 3}{4 - 2}$$
$$= \frac{-10}{2}$$
$$= -5$$

To get the equation of the line in slope-intercept form, pick one of the two points. Insert its coordinates and the slope you just calculated into the equation $y = mx + b$. Then solve for b:

$$y = mx + b$$
$$3 = -5 \cdot 2 + b$$
$$3 = -10 + b$$
$$+10 = +10$$
$$13 = b$$

In slope-intercept form, the equation of the line is $y = -5x + 13$.

Of the four equations listed, this exactly matches the third. However, you must rearrange the others and then compare them to $y = -5x + 13$:

- First equation:

$$5x + y = 13$$
$$-5x = -5x$$
$$y = -5x + 13$$
✓

- Second equation:

$$y + 7 = -5(x - 4)$$
$$y + 7 = -5x + 20$$
$$-7 = -7$$
$$y = -5x + 13$$
✓

- Fourth equation:

$$y - 7 = 5(x - 4)$$
$$y - 7 = 5x - 20$$
$$+7 = +7$$
$$y = 5x - 13$$
NO

Of the four equations, the first three are equivalent to $y = -5x + 13$ while the fourth is not. Three of the four equations, then, represent the line passing through the two points.

The correct choice is (**3**).

21. The formula for exponential growth is $y = a(1 + r)^x$, where a is the starting value and r is the growth rate.

    For the Ebola virus the formula is

    $$y = 1(1 + 0.11)^x$$
    $$= 1(1.11)^x$$

    For the SARS virus the formula is:

    $$y = 30(1 + .04)^x$$
    $$= 30(1.04)^x$$

    At day 10, the number of Ebola cases is

    $$y = 1(1.11)^{10}$$
    $$= 2.84$$
    $$\approx 3$$

At day 10, the number of SARS cases is

$$y = 30(1.04)^{10}$$
$$= 44.4$$
$$\approx 44$$

At day 53, the number of Ebola cases is

$$y = 1(1.11)^{53}$$
$$= 252.4$$
$$\approx 252$$

At day 53, the number of SARS cases is

$$y = 30(1.04)^{53}$$
$$= 239.8$$
$$\approx 240$$

At day 10, there are more SARS cases. At day 53, there are more Ebola cases.

The correct choice is (**3**).

22. This information tells you that the equation for the line of best fit for a scatter plot is $y = -1.15785x + 139.3171772$. Since the slope of the line is negative, the relationship between x and y is negative. So eliminate choices (3) and (4).

The strength of the correlation is based on the r^2-value. The closer that is to 1, the stronger the correlation. Since 0.8038159461 is pretty close to 1, this can be considered a strong negative correlation.

The correct choice is (**1**).

23. Gina's age is represented by x. If Abigail is younger than Gina and their ages are consecutive integers, Gina must be one year younger than Abigail. So Gina's age can be represented by $x - 1$.

The square of Gina's age is x^2.
Eight times Abigail's age is $8(x - 1)$.

If the difference between these is 17, an equation that could be used to find Gina's age is $x^2 - 8(x - 1) = 17$.

The correct choice is **(4)**.

24. In a system of equations, if you multiply both sides of one (or both) of the equations by a constant, the new system will have the same solution as the original system. Look at each of the four choices.

- Choice (1): This is what you would get if you multiplied both sides of the top equation by -3 and both sides of the bottom equation by -2.
- Choice (2): This is what you would get if you multiplied both sides of the top equation by 5 and both sides of the bottom equation by 3.
- Choice (3): This is what you would get if you multiplied both sides of the top equation by 6 and both sides of the bottom equation by 4.
- Choice (4): Although you could get this top equation by multiplying both sides of the original top equation by 10, there is no way to get the bottom equation by multiplying both sides of the original bottom equation by the same number. If you multiply the left side of the bottom equation by -6 and the right side by $+6$, the new system of equations will have a different solution than the original system.

The correct choice is **(4)**.

PART II

25. a is irrational because 20 is not a perfect square. b is rational because it is a terminating decimal. Even though c is a square root, it is rational because 225 is a perfect square and $\sqrt{225} = 15$.

$a + b$ is irrational because an irrational plus a rational is always irrational.
$b + c$ is rational because a rational plus a rational is always rational.

Note: Sometimes an irrational plus an irrational is irrational while other times an irrational plus an irrational is a rational number.

26. A sequence in which each term is equal to the previous term multiplied by some constant exhibits exponential behavior. In this case, the geometric sequence 1, 3, 9, 27, . . . has an initial term of 1 and a constant ratio of 3.

27. The formula for the volume of a cone is given in the reference sheet: $V = \frac{1}{3}\pi r^2 h$. To express r in terms of the other variables, treat the V and h as numbers and "solve" for r:

$$V = \frac{1}{3}\pi r^2 h$$
$$3 \cdot V = 3 \cdot \frac{1}{3}\pi r^2 h$$
$$3V = \pi r^2 h$$
$$\frac{3V}{\pi h} = \frac{\pi r^2 h}{\pi h}$$
$$\frac{3V}{\pi h} = r^2$$
$$\sqrt{\frac{3V}{\pi h}} = \sqrt{r^2}$$
$$\sqrt{\frac{3V}{\pi h}} = r$$

You should not include a $\pm$ symbol in front of the radical because the radius of a cone must be a positive number.

28. If you draw a horizontal line at the $50 mark on the vertical axis, you will see that the line for Plan *A* contains a point at approximately (12.25, 50) while the line for Plan *B* contains a point at (14, 50). This means that with Plan *A*, you can get only 12 games for $50 while with Plan *B* you can get 14. Plan *B* is a better choice for Dylan since he can get two additional games.

Because both lines contain the point (20, 65), it doesn't matter which plan Bobby chooses. With either plan, he can get 20 games for $65.

29. It is appropriate for Samantha to draw a line through the points on the graph because it is possible to eat a number of cookies that is not a multiple of 3. You can have 1 cookie or 2 cookies or even a half a cookie. The number of calories is proportional to the number of cookies. So one cookie has one-third the number of calories as 3 cookies, and half of a cookie has half as many calories as one cookie.

In mathematical terms, we say that the domain of the graph is all non-negative real numbers.

30. First set up a proportion to find the number of inches that the athlete can jump. Convert the athlete's height to inches so everything is expressed in the same unit:

$$5 \text{ feet } 9 \text{ inches} = (12 \times 5) + 9$$
$$= 60 + 9$$
$$= 69 \text{ inches}$$

$$\frac{2 \text{ in.}}{40 \text{ in.}} = \frac{69 \text{ in.}}{x \text{ in.}}$$
$$2x = 40 \cdot 69 = 2760$$
$$\frac{2x}{2} = \frac{2760}{2}$$
$$x = 1380$$

The athlete can jump 1,380 inches each jump.

To calculate the number of jumps the athlete needs to accomplish to make a mile, divide the number of inches in a mile by 1,380:

$$\text{Feet in a mile} = 5,280$$
$$\text{Inches in a mile} = 12 \times 5,280$$
$$= 63,360$$

$$63,360 \div 1,380 = 45.9$$
$$\approx 46 \text{ jumps}$$

31. First use the distributive property, and then combine the like terms:

$$5x + 4x^2(2x + 7) - 6x^2 - 9x$$
$$5x + 8x^3 + 28x^2 - 6x^2 - 9x$$
$$8x^3 + 28x^2 - 6x^2 + 5x - 9x$$
$$8x^3 + 22x^2 - 4x$$

32. To complete the square with an equation like $x^2 + bx = c$, add $\left(\frac{b}{2}\right)^2$ to both sides of the equation:

$$x^2 - 6x = 15$$
$$+\left(\frac{-6}{2}\right)^2 = +\left(\frac{-6}{2}\right)^2$$
$$x^2 - 6x + 9 = 15 + 9$$
$$x^2 - 6x + 9 = 24$$

The left side of this equation is now a perfect square and can be factored:

$$x^2 - 6x + 9 = 24$$
$$(x - 3)^2 = 24$$

Take the square root of both sides of the equation, and isolate the x-term:

$$(x - 3)^2 = 24$$
$$\sqrt{(x-3)^2} = \pm\sqrt{24}$$
$$x - 3 = \pm\sqrt{24}$$
$$+3 = +3$$
$$x = 3 \pm \sqrt{24}$$

This can be simplified to $x = 3 \pm 2\sqrt{6}$.

PART III

33. In 4 hours, Loretta drives $4 \times 55 = 220$ miles. This leaves $610 - 220 = 390$ miles. At 65 mph, 390 miles can be traveled in $390 \div 65 = 6$ hours. This, together with Loretta's 4 hours of driving, is a total of 10 hours.

    In 2 hours, Loretta drives $2 \times 55 = 110$ miles. This leaves $610 - 110 = 500$ miles. At 65 mph, 500 miles can be traveled in $500 \div 65 = 7.7$ hours. This, together with Loretta's 2 hours of driving, is a total of approximately 9.7 hours. This is 0.3 hours faster than the original plan.

34.

Interval	Frequency
6.0–6.1	3
6.2–6.3	10
6.4–6.5	11
6.6–6.7	0
6.8–6.9	2
7.0–7.1	3

A histogram is like a bar graph with no spaces between the bars. In a histogram you can choose the size of the intervals. Using an interval of 0.2, you get a histogram like this

With 29 numbers, the 15th is the median. To get the top quartile, find the median of the top 14 numbers. Since the 7th and 8th largest numbers are both 6.5, the top quartile is 6.5, which is contained in the 6.4–6.5 interval.

35. Start by changing the inequalites into slope-intercept form.

    First inequality:

    $$3x + 4y > 20$$
    $$-3x = -3x$$
    $$4y > -3x + 20$$
    $$\frac{4y}{4} > \frac{-3x + 20}{4}$$
    $$y > -\frac{3}{4}x + 5$$

Second inequality:

$$x < 3y - 18$$
$$+18 = +18$$
$$x + 18 < 3y$$
$$\frac{x + 18}{3} < \frac{3y}{3}$$
$$\frac{1}{3}x + 6 < y$$
$$y > \frac{1}{3}x + 6$$

Graph the boundary line for the first inequality like $y = -\frac{3}{4}x + 5$ but as a dotted line because of the $>$ sign. If there was a $\geq$ sign, the graph would need to be a solid line.

To decide which side of the line to shade, check if the ordered pair $(0, 0)$ makes the inequality true:

$$0 > -\frac{3}{4} \cdot 0 + 5$$
$$0 > 0 + 5$$
$$0 > 5$$

NO

Since it is not true that $0 > 5$, shade the side of the line that does not contain $(0, 0)$.

Graph the boundary line for the second inequality like $y = \frac{1}{3}x + 6$ but as a dotted line because of the $>$ sign.

To decide which side of the line to shade, check if the ordered pair $(0, 0)$ makes the inequality true:

$$0 > \frac{1}{3} \cdot 0 + 6$$
$$0 > 0 + 6$$
$$0 > 6$$
NO

Since it is not true that $0 > 6$, shade the side of the line that does not contain $(0, 0)$.

This can also be done with the graphing calculator.

For the TI-84:

For the TI-Nspire:

The point (3, 7) is not in the solution set because it is on the dotted boundary line. You can also verify this by checking if (3, 7) makes both inequalities true. Since $3 < 3 \cdot 7 - 18$ is not true (they are equal), the ordered pair (3, 7) does not satisfy the second inequality. So (3, 7) cannot be part of the solution set.

36. One way to find the vertex of $h(x)$ is to graph the function on the graphing calculator and to use the maximum command to find the coordinates of the vertex.

    For the TI-84:

    For the TI-Nspire:

The x-coordinate of the vertex of a quadratic function (a parabola) can also be found by using the formula $x = \frac{-b}{2a}$. For this function, $x = \frac{-b}{2a} = \frac{-128}{2(-16)} = 4$. The y-coordinate of the vertex will be

$$h(4) = -16 \cdot 4^2 + 128 \cdot 4 + 9{,}000$$
$$= 9{,}256$$

The vertex of $h(t)$ is $(4, 9{,}256)$.

Since cruising altitude is 9,000 feet and the maximum height achieved is 9,256 feet, the answer is $9{,}256 - 9{,}000 = 256$ feet above the cruising altitude.

PART IV

37. Since there are 7 people and they each buy one soda, they spend 7 × 0.50 = $3.50 on sodas. This leaves $28.50 − $3.50 = $25.00 for the hot dogs and hamburgers.

Since each hot dog costs $1.25 and there are x hot dogs, the total cost of the hotdogs is 1.25x. Since each hamburger costs $2.50 and there are y hamburgers, the total cost of the hamburgers is 2.50y. For the total cost of all the hot dogs and hamburgers to be $25.00, the equation is

$$1.25x + 2.50y = 25$$

To graph this equation, convert to slope-intercept form:

$$1.25x + 2.50y = 25$$
$$-1.25x = -1.25x$$
$$2.50y = -1.25x + 25$$
$$\frac{2.50y}{2.50} = \frac{-1.25x + 25}{2.50}$$
$$y = -\frac{1}{2}x + 10$$

Graph this line on the grid.

Since you cannot buy a fraction of a hot dog or a hamburger, find the coordinates of the points that have integers as both the x-coordinate and y-coordinate. There are 11 different combinations.

$$y = -\frac{1}{2}x + 10$$

Hot Dogs	Hamburgers
20	0
18	1
16	2
14	3
12	4
10	5
8	6
6	7
4	8
2	9
0	10

SELF-ANALYSIS CHART August 2017

Topic	Question Numbers	Number of Points	Your Points	Your Percentage
1. Polynomials	7, 31	$2+2=4$		
2. Properties of Algebra	1, 13, 27, 30	$2+2+2+2=8$		
3. Functions	2, 4, 5, 10, 17, 20, 19	$2+2+2+2+2+2+2=14$		
4. Creating and Interpreting Equations	9, 23, 33, 37	$2+2+4+6=14$		
5. Inequalities	11, 35	$2+4=6$		
6. Sequences and Series	15, 26	$2+2=4$		
7. Systems of Equations	24, 28	$2+2=4$		
8. Quadratic Equations and Factoring	3, 19, 32, 36	$2+2+2+4=10$		
9. Regression	22	2		
10. Exponential Equations	14, 16, 21	$2+2+2=6$		
11. Graphing	6, 18	$2+2=4$		
12. Statistics	8, 12, 24, 28, 34	$2+2+2+2+4=12$		
13. Number Properties	25	2		

HOW TO CONVERT YOUR RAW SCORE TO YOUR ALGEBRA I REGENTS EXAMINATION SCORE

The accompanying conversion chart must be used to determine your final score on the August 2017 Regents Examination in Algebra I. To find your final exam score, locate in the column labeled "Raw Score" the total number of points you scored out of a possible 86 points. Since partial credit is allowed in Parts II, III, and IV of the test, you may need to approximate the credit you would receive for a solution that is not completely correct. Then locate in the adjacent column to the right the scale score that corresponds to your raw score. The scale score is your final Algebra I Regents Examination score.

Regents Examination in Algebra I—August 2017
Chart for Converting Total Test Raw Scores to Final Examination Scores (Scaled Scores)

Raw Score	Scale Score	Performance Level	Raw Score	Scale Score	Performance Level	Raw Score	Scale Score	Performance Level
86	100	5	57	81	4	28	66	3
85	98	5	56	81	4	27	65	3
84	97	5	55	81	4	26	64	2
83	96	5	54	80	4	25	63	2
82	94	5	53	80	4	24	62	2
81	93	5	52	80	4	23	60	2
80	92	5	51	80	3	22	59	2
79	91	5	50	79	3	21	57	2
78	91	5	49	79	3	20	56	2
77	90	5	48	79	3	19	55	1
76	89	5	47	78	3	18	52	1
75	88	5	46	78	3	17	50	1
74	88	5	45	78	3	16	49	1
73	87	5	44	77	3	15	47	1
72	87	5	43	77	3	14	44	1
71	86	5	42	77	3	13	42	1
70	86	5	41	76	3	12	40	1
69	86	5	40	76	3	11	37	1
68	85	5	39	75	3	10	35	1
67	84	4	38	74	3	9	32	1
66	84	4	37	74	3	8	29	1
65	84	4	36	73	3	7	26	1
64	83	4	35	73	3	6	23	1
63	83	4	34	72	3	5	19	1
62	83	4	33	71	3	4	16	1
61	82	4	32	70	3	3	12	1
60	82	4	31	69	3	2	8	1
59	82	4	30	68	3	1	4	1
58	82	4	29	67	3	0	0	1

Examination
June 2018
Algebra I

HIGH SCHOOL MATH REFERENCE SHEET

Conversions

1 inch = 2.54 centimeters
1 meter = 39.37 inches
1 mile = 5280 feet
1 mile = 1760 yards
1 mile = 1.609 kilometers

1 kilometer = 0.62 mile
1 pound = 16 ounces
1 pound = 0.454 kilogram
1 kilogram = 2.2 pounds
1 ton = 2000 pounds

1 cup = 8 fluid ounces
1 pint = 2 cups
1 quart = 2 pints
1 gallon = 4 quarts
1 gallon = 3.785 liters
1 liter = 0.264 gallon
1 liter = 1000 cubic centimeters

Formulas

Triangle	$A = \frac{1}{2}bh$
Parallelogram	$A = bh$
Circle	$A = \pi r^2$
Circle	$C = \pi d$ or $C = 2\pi r$

Formulas (continued)

General Prisms	$V = Bh$
Cylinder	$V = \pi r^2 h$
Sphere	$V = \frac{4}{3}\pi r^3$
Cone	$V = \frac{1}{3}\pi r^2 h$
Pyramid	$V = \frac{1}{3}Bh$
Pythagorean Theorem	$a^2 + b^2 = c^2$
Quadratic Formula	$x = \frac{-b \pm \sqrt{b^2 - 4ac}}{2a}$
Arithmetic Sequence	$a_n = a_1 + (n-1)d$
Geometric Sequence	$a_n = a_1 r^{n-1}$
Geometric Series	$S_n = \frac{a_1 - a_1 r^n}{1 - r}$ where $r \neq 1$
Radians	1 radian = $\frac{180}{\pi}$ degrees
Degrees	1 degree = $\frac{\pi}{180}$ radians
Exponential Growth/Decay	$A = A_0 e^{k(t-t_0)} + B_0$

EXAMINATION June 2018

PART I

Answer all 24 questions in this part. Each correct answer will receive 2 credits. No partial credit will be allowed. For each statement or question, write in the space provided the numeral preceding the word or expression that best completes the statement or answers the question. [48 credits]

1. The solution to $4p + 2 < 2(p + 5)$ is
    (1) $p > -6$ (3) $p > 4$
    (2) $p < -6$ (4) $p < 4$

2. If $k(x) = 2x^2 - 3\sqrt{x}$, then $k(9)$ is
    (1) 315   (3) 159
    (2) 307   (4) 153

3. The expression $3(x^2 + 2x - 3) - 4(4x^2 - 7x + 5)$ is equivalent to
    (1) $-13x - 22x + 11$ (3) $19x^2 - 22x + 11$
    (2) $-13x^2 + 34x - 29$ (4) $19x^2 + 34x - 29$

4. The zeros of the function $p(x) = x^2 - 2x - 24$ are
    (1) -8 and 3 (3) -4 and 6
    (2) -6 and 4 (4) -3 and 8

5. The box plot below summarizes the data for the average monthly high temperatures in degrees Fahrenheit for Orlando, Florida.

The third quartile is
    (1) 92   (3) 83
    (2) 90   (4) 71

EXAMINATION June 2018

6 Joy wants to buy strawberries and raspberries to bring to a party. Strawberries cost $1.60 per pound and raspberries cost $1.75 per pound. If she only has $10 to spend on berries, which inequality represents the situation where she buys x pounds of strawberries and y pounds of raspberries?

(1) $1.60x + 1.75y \leq 10$ (3) $1.75x + 1.60y \leq 10$
(2) $1.60x + 1.75y \geq 10$ (4) $1.75x + 1.60y \geq 10$ 6 ____

7 On the main floor of the Kodak Hall at the Eastman Theater, the number of seats per row increases at a constant rate. Steven counts 31 seats in row 3 and 37 seats in row 6. How many seats are there in row 20?

(1) 65   (3) 69
(2) 67   (4) 71   7 ____

8 Which ordered pair below is *not* a solution to $f(x) = x^2 - 3x + 4$?

(1) $(0, 4)$ (3) $(5, 14)$
(2) $(1.5, 1.75)$ (4) $(-1, 6)$ 8 ____

9 Students were asked to name their favorite sport from a list of basketball, soccer, or tennis. The results are shown in the table below.

	Basketball	Soccer	Tennis
Girls	42	58	20
Boys	84	41	5

What percentage of the students chose soccer as their favorite sport?

(1) 39.6%   (3) 50.4%
(2) 41.4%   (4) 58.6%   9 ____

10 The trinomial $x^2 - 14x + 49$ can be expressed as

(1) $(x - 7)^2$ (3) $(x - 7)(x + 7)$
(2) $(x + 7)^2$ (4) $(x - 7)(x + 2)$ 10 ____

EXAMINATION June 2018

11 A function is defined as {(0, 1), (2, 3), (5, 8), (7, 2)}. Isaac is asked to create one more ordered pair for the function. Which ordered pair can he add to the set to keep it a function?
(1) (0, 2)
(2) (5, 3)
(3) (7, 0)
(4) (1, 3)

12 The quadratic equation $x^2 - 6x = 12$ is rewritten in the form $(x + p)^2 = q$, where q is a constant. What is the value of p?
(1) −12
(2) −9
(3) −3
(4) 9

13 Which of the quadratic functions below has the *smallest* minimum value?
(1) $h(x) = x^2 + 2x - 6$
(3) $k(x) = (x + 5)(x + 2)$
(2) [graph of g(x)]
(4)
x	$f(x)$
−1	−2
0	−5
1	−6
2	−5
3	−2

14 Which situation is *not* a linear function?
(1) A gym charges a membership fee of $10.00 down and $10.00 per month.
(2) A cab company charges $2.50 initially and $3.00 per mile.
(3) A restaurant employee earns $12.50 per hour.
(4) A $12,000 car depreciates 15% per year.

15 The Utica Boilermaker is a 15-kilometer road race. Sara is signed up to run this race and has done the following training runs:

I. 10 miles

II. 44,880 feet

III. 15,560 yards

Which run(s) are at least 15 kilometers?

(1) I, only
(2) II, only
(3) I and III
(4) II and III

16 If $f(x) = x^2 + 2$, which interval describes the range of this function?

(1) $(-\infty, \infty)$
(2) $[0, \infty)$
(3) $[2, \infty)$
(4) $(-\infty, 2]$

17 The amount Mike gets paid weekly can be represented by the expression $2.50a + 290$, where a is the number of cell phone accessories he sells that week. What is the constant term in this expression and what does it represent?

(1) $2.50a$, the amount he is guaranteed to be paid each week
(2) $2.50a$, the amount he earns when he sells a accessories
(3) 290, the amount he is guaranteed to be paid each week
(4) 290, the amount he earns when he sells a accessories

18 A cubic function is graphed on the set of axes below.

Which function could represent this graph?

(1) $f(x) = (x - 3)(x - 1)(x + 1)$
(2) $g(x) = (x + 3)(x + 1)(x - 1)$
(3) $h(x) = (x - 3)(x - 1)(x + 3)$
(4) $k(x) = (x + 3)(x + 1)(x - 3)$

18 _____

19 Mrs. Allard asked her students to identify which of the polynomials below are in standard form and explain why.

I. $15x^4 - 6x + 3x^2 - 1$

II. $12x^3 + 8x + 4$

III. $2x^5 + 8x^2 + 10x$

Which student's response is correct?

(1) Tyler said I and II because the coefficients are decreasing.
(2) Susan said only II because all the numbers are decreasing.
(3) Fred said II and III because the exponents are decreasing.
(4) Alyssa said II and III because they each have three terms.

19 _____

20 Which graph does *not* represent a function that is always increasing over the entire interval $-2 < x < 2$?

(1)

(2)

(3)

(4)

21 At an ice cream shop, the profit, $P(c)$, is modeled by the function $P(c) = 0.87c$, where c represents the number of ice cream cones sold. An appropriate domain for this function is

(1) an integer ≤ 0
(2) an integer ≥ 0
(3) a rational number ≤ 0
(4) a rational number ≥ 0

22 How many real-number solutions does $4x^2 + 2x + 5 = 0$ have?

(1) one
(2) two
(3) zero
(4) infinitely many

23 Students were asked to write a formula for the length of a rectangle by using the formula for its perimeter, $p = 2\ell + 2w$. Three of their responses are shown below.

   I. $\ell = \frac{1}{2}p - w$

   II. $\ell = \frac{1}{2}(p - 2w)$

   III. $\ell = \frac{p - 2w}{2}$

Which responses are correct?

(1) I and II, only
(2) II and III, only
(3) I and III, only
(4) I, II, and III

23 _____

24 If $a_n = n(a_{n-1})$ and $a_1 = 1$, what is the value of a_5?

(1) 5
(2) 20
(3) 120
(4) 720

24 _____

PART II

Answer all 8 questions in this part. Each correct answer will receive 2 credits. Clearly indicate the necessary steps, including appropriate formula substitutions, diagrams, graphs, charts, etc. For all questions in this part, a correct numerical answer with no work shown will receive only 1 credit. [16 credits]

25 Graph $f(x) = \sqrt{x+2}$ over the domain $-2 \leq x \leq 7$.

26 Caleb claims that the ordered pairs shown in the table below are from a nonlinear function.

x	$f(x)$
0	2
1	4
2	8
3	16

State if Caleb is correct. Explain your reasoning.

No Caleb is not correct because every input has one output.

27 Solve for x to the nearest tenth: $x^2 + x - 5 = 0$.

$x = 1.9$

and $x = -2.9$

EXAMINATION June 2018

28 The graph of the function $p(x)$ is represented below. On the same set of axes, sketch the function $p(x+2)$.

29 When an apple is dropped from a tower 256 feet high, the function $h(t) = -16t^2 + 256$ models the height of the apple, in feet, after t seconds. Determine, algebraically, the number of seconds it takes the apple to hit the ground.

4 seconds

30 Solve the equation below algebraically for the exact value of x.

$$6 - \frac{2}{3}(x+5) = 4x$$

31 Is the product of $\sqrt{16}$ and $\frac{4}{7}$ rational or irrational? Explain your reasoning.

rational

32 On the set of axes below, graph the piecewise function:

$$f(x) = \begin{cases} -\frac{1}{2}x, & x < 2 \\ x, & x \geq 2 \end{cases}$$

PART III

Answer all 4 questions in this part. Each correct answer will receive 4 credits. Clearly indicate the necessary steps, including appropriate formula substitutions, diagrams, graphs, charts, etc. For all questions in this part, a correct numerical answer with no work shown will receive only 1 credit.   [16 credits]

33 A population of rabbits in a lab, $p(x)$, can be modeled by the function $p(x) = 20(1.014)^x$, where x represents the number of days since the population was first counted.

Explain what 20 and 1.014 represent in the context of the problem.

20 is the starting population 1.014 is the growth rate

Determine, to the *nearest tenth*, the average rate of change from day 50 to day 100.

0.2

34 There are two parking garages in Beacon Falls. Garage A charges $7.00 to park for the first 2 hours, and each additional hour costs $3.00. Garage B charges $3.25 per hour to park.

When a person parks for at least 2 hours, write equations to model the cost of parking for a total of x hours in Garage A and Garage B.

Garage A

$A = 7 + 3(x - 2)$

Garage B

$B = 3.25x$

Determine algebraically the number of hours when the cost of parking at both garages will be the same.

both garages cost the same when you park for 4 hours

EXAMINATION June 2018

35 On the set of axes below, graph the following system of inequalities:

$$2y + 3x \leq 14$$
$$4x - y < 2$$

Determine if the point (1, 2) is in the solution set. Explain your answer.

No it is not in the solution set

36 The percentage of students scoring 85 or better on a mathematics final exam and an English final exam during a recent school year for seven schools is shown in the table below.

Percentage of Students Scoring 85 or Better	
Mathematics, x	English, y
27	46
12	28
13	45
10	34
30	56
45	67
20	42

Write the linear regression equation for these data, rounding all values to the *nearest hundredth*.

$y = 0.93x + 23.95$

State the correlation coefficient of the linear regression equation, to the *nearest hundredth*. Explain the meaning of this value in the context of these data.

0.92 which is a strong positive correlation

PART IV

Answer the question in this part. A correct answer will receive 6 credits. Clearly indicate the necessary steps, including appropriate formula substitutions, diagrams, graphs, charts, etc. A correct numerical answer with no work shown will receive only 1 credit. [6 credits]

37 Dylan has a bank that sorts coins as they are dropped into it. A panel on the front displays the total number of coins inside as well as the total value of these coins. The panel shows 90 coins with a value of $17.55 inside of the bank.

If Dylan only collects dimes and quarters, write a system of equations in two variables or an equation in one variable that could be used to model this situation.

$0.10d + 0.25q = 17.55$

Using your equation or system of equations, algebraically determine the number of quarters Dylan has in his bank.

57 quarters

Question 37 is continued on the next page.

Question 37 continued

Dylan's mom told him that she would replace each one of his dimes with a quarter. If he uses all of his coins, determine if Dylan would then have enough money to buy a game priced at $20.98 if he must also pay an 8% sales tax. Justify your answer.

No he would not

Answers June 2018
Algebra I

Answer Key

PART I
1. (4)
2. (4)
3. (2)
4. (3)
5. (2)
6. (1)
7. (1)
8. (4)
9. (1)
10. (1)
11. (4)
12. (3)
13. (2)
14. (4)
15. (1)
16. (3)
17. (3)
18. (2)
19. (3)
20. (3)
21. (2)
22. (3)
23. (4)
24. (3)

PART II
25.

26. Caleb is correct. It is a nonlinear function.
27. $x = 1.8$ and $x = -2.8$
28. See graph on page 327.
29. 4 seconds
30. $\frac{4}{7}$
31. The product is rational.
32. See graphs on pages 329 and 330.

PART III
33. 20 is the starting population. 1.014 is the growth factor. The average rate of change is 0.8.
34. For Garage A, $A = 7 + 3(x - 2)$. For Garage B, $B = 3.25x$. Both garages cost the same when you park for 4 hours.
35. (1, 2) is not in the solution set.
36. $y = 0.96x + 23.95$ The correlation coefficient is 0.92, which is a strong positive correlation.

PART IV
37. $0.10d + 0.25q = 17.55$
    $d + q = 90$

Dylan has 57 quarters. No, he would not be able to buy the game.

In **Parts II–IV**, you are required to show how you arrived at your answers. For sample methods of solutions, see the *Answers Explained* section.

Answers and Explanations

PART I

1. Solving an inequality is very similar to solving an equality. Use the properties of algebra to isolate the variable.

   First distribute the 2 on the right-hand side:
   $$4p + 2 < 2(p + 5)$$
   $$4p + 2 < 2p + 10$$

   Then eliminate the constant term from the left-hand side:
   $$4p + 2 < 2p + 10$$
   $$-2 = -2$$
   $$4p < 2p + 8$$

   Next eliminate the variables from the right-hand side:
   $$4p < 2p + 8$$
   $$-2p = -2p$$
   $$2p < 8$$

   Finally, eliminate the coefficient by dividing both sides of the inequality by 2. If the coefficient is negative, the direction of the inequality would need to be reversed. In this case, the coefficient is positive. So do not reverse the inequality:
   $$\frac{2p}{2} < \frac{8}{2}$$
   $$p < 4$$

   The correct choice is **(4)**.

2. Substitute 9 for all occurrences of x in the function:

$$k(9) = 2(9)^2 - 3\sqrt{9}$$
$$= 2(81) - 3(3)$$
$$= 162 - 9$$
$$= 153$$

The correct choice is **(4)**.

3. Distribute the 3 to the terms in the first parentheses:

$$3x^2 + 6x - 9 - 4(4x^2 - 7x + 5)$$

Then distribute the -4 to the terms in the second parentheses. Be sure to change the signs since you are multiplying by a negative:

$$3x^2 + 6x - 9 - 16x^2 + 28x - 20$$

Now combine like terms:

$$3x^2 - 16x^2 + 6x + 28x - 9 - 20 = -13x^2 + 34x - 29$$

The correct choice is **(2)**.

4. The zeros of a function are the numbers that evaluate to zero when you plug them into the function. To find the zeros, solve the equation $x^2 - 2x - 24 = 0$. There are several ways to solve this equation. The simplest is by factoring:

$$x^2 - 2x - 24 = 0$$
$$(x - 6)(x + 4) = 0$$
$$x - 6 = 0 \text{ or } x + 4 = 0$$
$$x = 6 \text{ or } x = -4$$

Another way to find the zeros of a function is to graph it on a graphing calculator and find the x-intercepts of the graph. This can be done with the zero function on the calculator.

For the TI-84:

For the TI-Nspire:

The correct choice is **(3)**.

5. In a box plot, the five values marked with vertical lines are the minimum, the first quartile, the median, the third quartile, and the maximum. For this box plot, the minimum is about 71, the first quartile is 75, the median is about 83, the third quartile is 90, and the maximum is about 92.

The correct choice is **(2)**.

6. Since 1 pound of strawberries costs $1.60, x pounds of strawberries cost $1.60x$. Since 1 pound of raspberries costs $1.75, y pounds of raspberries cost $1.75y$. Together, the cost of x pounds of strawberries and y pounds of raspberries is $1.60x + 1.75y$. Since Joy has only $10 to spend, the total cost must be less than or equal to 10. So the inequality that represents the situation is $1.60x + 1.75y \leq 10$.

The correct choice is **(1)**.

7. One way to solve this is to make a list with the given information:

Row 1.

Row 2.

Row 3. 31 seats

Row 4.

Row 5.

Row 6. 37 seats

The seats increase at a constant rate. Since the number of seats increases by 6 when the number of rows increases by 3, the increase in seats for each new row must be 6 ÷ 3 = 2. Create a table to show the number of seats in each row:

1. 27	5. 35	9. 43	13. 51	17. 59
2. 29	6. 37	10. 45	14. 53	18. 61
3. 31	7. 39	11. 47	15. 55	19. 63
4. 33	8. 41	12. 49	16. 57	20. 65

A quicker way to solve this problem is to use the arithmetic sequence formula from the reference sheet, $a_n = a_1 + (n-1)d$. Remember that a_1 is the first term of the sequence, n is the term you want to find, and d is the difference from one term to the next.

For this question, $a_1 = 27$, $d = 2$, and $n = 20$. Plug the values into the formula:

$$a_{20} = 27 + (20-1)2$$
$$= 27 + 19 \cdot 2$$
$$= 27 + 38$$
$$= 65$$

A third way to solve this problem is to treat x as the row number and y as the number of seats in row x. Then find the equation of the line in slope-intercept form that passes through (3, 31) and (6, 37):

$$m = \frac{37 - 31}{6 - 3}$$
$$= \frac{6}{3}$$
$$= 2$$
$$y = mx + b$$
$$37 = 2 \cdot 6 + b$$
$$37 = 12 + b$$
$$-12 = -12$$
$$25 = b$$
$$y = 2x + 25$$

Substitute 20 for x:

$$y = 2 \cdot 20 + 25$$
$$= 40 + 25$$
$$= 65$$

The correct choice is **(1)**.

8. To check if an ordered pair is a solution, plug in the x-coordinate of the ordered pair and see if the function evaluates to the corresponding y-coordinate.

- Testing choice (1):

$$f(0) = 0^2 - 3 \cdot 0 + 4 \overset{\checkmark}{=} 4$$

- Testing choice (2):

$$f(1.5) = 1.5^2 - 3 \cdot 1.5 + 4 \overset{\checkmark}{=} 1.75$$

- Testing choice (3):

$$f(5) = 5^2 - 3 \cdot 5 + 4 \stackrel{\checkmark}{=} 14$$

- Testing choice (4):

$$f(-1) = (-1)^2 - 3 \cdot (-1) + 4 = 1 + 3 + 4 = 8 \neq 6$$

The correct choice is **(4)**.

9. To calculate the percentage of students who chose soccer as their favorite sport, divide the number of students who chose soccer by the total number of students. There were 58 boys who chose soccer and 41 girls who chose soccer. So $58 + 41 = 99$ students chose soccer. The total number of students who answered the question was

$$42 + 58 + 20 + 84 + 41 + 5 = 250$$

The percentage of students who chose soccer was

$$\frac{99}{250} \approx 39.6\%$$

The correct choice is **(1)**.

10. You can factor an expression of the form $x^2 + bx + c$ into the product $(x + r_1)(x + r_2)$ by finding two numbers whose sum is b and whose product is c. In this problem, the two numbers whose sum is -14 and whose product is 49 are -7 and -7. So the expression factors into $(x - 7)(x - 7) = (x - 7)^2$.

Another way to solve this is to recognize that in an expression like $x^2 + bx + c$ when the c-term is equal to the square of half the b-term, the expression is a perfect square trinomial. That expression can be factored into $\left(x + \frac{b}{2}\right)^2$. In this example, $\left(\frac{-14}{2}\right)^2 = 49$. So the expression factors into $\left(x + \frac{-14}{2}\right)^2 = (x - 7)^2$.

Since this is a multiple-choice question, a third way this can be solved is to test each of the four choices to see which one becomes $x^2 - 14x + 49$ when multiplied.

- Testing choice (1):
$$(x - 7)^2 = (x - 7)(x - 7)$$
$$= x^2 - 7x - 7x + 49$$
$$= x^2 - 14x + 49$$

- Testing choice (2):
$$(x + 7)^2 = (x + 7)(x + 7)$$
$$= x^2 + 7x + 7x + 49$$
$$= x^2 + 14x + 49$$

- Testing choice (3):
$$(x - 7)(x + 7) = x^2 + 7x - 7x - 49$$
$$= x^2 - 49$$

- Testing choice (4):
$$(x - 7)(x + 2) = x^2 + 2x - 7x - 14$$
$$= x^2 - 5x - 14$$

The correct choice is **(1)**.

11. When a function is described as a set of ordered pairs, none of the x-coordinates can repeat.
    - Testing choice (1): (0, 2) cannot be part of the function since the ordered pair (0, 1) already has 0 as an x-coordinate.
    - Testing choice (2): (5, 3) cannot be part of the function since the ordered pair (5, 8) already has 5 as an x-coordinate.
    - Testing choice (3): (7, 0) cannot be part of the function since the ordered pair (7, 2) already has 7 as an x-coordinate.
    - Testing choice (4): (1, 3) can be part of the function since no other ordered pairs in the set have an x-coordinate of 1. It does not matter that the ordered pair (2, 3) already has a y-coordinate of 3.

    The correct choice is **(4)**.

12. This question requires that you do the first two steps of solving a quadratic equation by completing the square process.

Step 1: The first step of solving an equation of the form $x^2 + bx = c$ by completing the square is to add $\left(\dfrac{b}{2}\right)^2$ to both sides of the equation.

For this example:

$$x^2 - 6x + \left(\dfrac{-6}{2}\right)^2 = 12 + \left(\dfrac{-6}{2}\right)^2$$
$$x^2 - 6x + 9 = 12 + 9$$
$$x^2 - 6x + 9 = 21$$

Step 2: Factor the left side of the equation:

$$(x-3)(x-3) = 21$$
$$(x-3)^2 = 21$$

The completing the square process is not finished. Since this question just asks for the value of p, though, there is no more work to do. If $(x+p) = (x-3)$, then p must be -3.

The correct choice is **(3)**.

13. The quadratic function with the smallest minimum value is the one whose vertex has the smallest y-coordinate.

- Testing choice (1): Use the graphing calculator to find the vertex of the parabola. The minimum is at $(-1, -7)$.

For the TI-84:

For the TI-Nspire:

- Testing choice (2): As shown on the graph, the vertex of the parabola is at $(2, -10)$.
- Testing choice (3): Use the graphing calculator to find the vertex of the parabola. The minimum is at $(-3.5, -2.25)$.
- Testing choice (4): When you graph the five given points, they resemble a parabola with a vertex at $(1, -6)$.

Since -10 is the minimum of all the y-coordinates of the four vertices, it is the smallest minimum value.

The correct choice is **(2)**.

14. One way to answer this question is to create an equation to model each scenario and determine which equation is not in the form $y = mx + b$.

- Testing choice (1): The equation is $y = 10x + 10$.
- Testing choice (2): The equation is $y = 3x + 2.5$.
- Testing choice (3): The equation is $y = 12.5x + 0$.
- Testing choice (4): The equation is $y = 12{,}000 \cdot 0.85^x$.

Choices (1), (2), and (3) are linear functions, while choice (4) is an exponential function.

Another way to answer this question is to make a sketch of each scenario and see which of them does not form a line.

- Testing choice (1):

- Testing choice (2):

- Testing choice (3):

- Testing choice (4):

Only choice (4) does not form a line.

The correct choice is **(4)**.

15. Use the reference sheet to convert the measurements in each of the three runs into kilometers.

- Run I:

$$1 \text{ mile} = 1.609 \text{ kilometers}$$

$$10 \text{ miles} \left(\frac{1.609 \text{ kilometers}}{1 \text{ mile}} \right) = 16.09 \text{ kilometers}$$

$$16.09 \text{ kilometers} \geq 15 \text{ kilometers}$$

- Run II:

$$5280 \text{ feet} = 1 \text{ mile}$$

$$44,880 \text{ feet} \left(\frac{1 \text{ mile}}{5280 \text{ feet}} \right) = 8.5 \text{ miles}$$

$$1 \text{ mile} = 1.609 \text{ kilometers}$$

$$8.5 \text{ miles} \left(\frac{1.609 \text{ kilometers}}{1 \text{ mile}} \right) = 13.6765 \text{ kilometers} \approx 13.7 \text{ kilometers}$$

$$13.7 \text{ kilometers} < 15 \text{ kilometers}$$

- Run III:

$$1760 \text{ yards} = 1 \text{ mile}$$

$$15,560 \text{ yards} \left(\frac{1 \text{ mile}}{1760 \text{ yards}} \right) = 8.8409 \text{ miles} \approx 8.84 \text{ miles}$$

$$1 \text{ mile} = 1.609 \text{ kilometers}$$

$$8.84 \text{ miles} \left(\frac{1.609 \text{ kilometers}}{1 \text{ mile}} \right) = 14.22356 \text{ kilometers} \approx 14.2 \text{ kilometers}$$

$$14.2 \text{ kilometers} < 15 \text{ kilometers}$$

The correct choice is **(1)**.

16. The range of a function is the set of all the different y-coordinates of the graph of that function. You can graph this function on the graphing calculator.

For the TI-84:

For the TI-Nspire:

The minimum point has a y-coordinate of 2, and there is no maximum point. So the range is $y \geq 2$. In interval notation, this is written as $[2, \infty)$. The square bracket indicates that the interval includes the endpoint of 2. For the ∞, you always use a rounded parenthesis.

The correct choice is **(3)**.

17. In an expression in the form $mx + b$, the b is the constant term, the m is the coefficient, the x is the variable, and the mx is the variable part. In the expression $2.50a + 290$, the number 290 is the constant term.

When an expression like this is used to model a real-world scenario, you can interpret the meaning of the coefficient and of the constant by plugging in 0 and 1 for the variable.

If you plug in 0 for a, you get $2.50 \cdot 0 + 290 = 290$. So 290 is the amount Mike is guaranteed to be paid even if he sells 0 cell phone accessories.

If you plug in 1 for a, you get $2.50 \cdot 1 + 290 = 292.50$. So 2.50 must be the amount Mike earns for each accessory he sells. Therefore $2.50a$ is the amount Mike earns when he sells a accessories.

The correct choice is **(3)**.

18. If the graph of a polynomial function has an x-intercept at $(a, 0)$, the function will have a factor of $(x - a)$. Since the graph of this function has x-intercepts at $(-3, 0), (-1, 0)$, and $(1, 0)$, the function must have factors of $(x - (-3)) = (x + 3), (x - (-1)) = (x + 1)$, and $(x - (+1)) = (x - 1)$.

Another way to answer this is to graph the four choices to see which most resembles the given graph. On the graphing calculator, choice (2) looks most like the given graph.

For the TI-84:

For the TI-Nspire:

The correct choice is **(2)**.

19. A polynomial is in standard form if the exponents are decreasing from left to right. Whether the coefficients are decreasing, increasing, or fluctuating up and down makes no difference.

    - For option I, the exponents go from 4 to 1 to 2 to 0, which is not always decreasing.
    - For option II, the exponents decrease from 3 to 1 to 0.
    - For option III, the exponents decrease from 5 to 2 to 1.

    Both options II and III are in standard form because their exponents decrease in value. Both choices (3) and (4) identify which of the Roman numerals are in standard form. However, choice (4) is incorrect since the number of terms has nothing to do with whether or not a polynomial is in standard form.

    The correct choice is **(3)**.

20. A function is increasing over an interval if the graph is always going "up" as you move from left to right.

- Testing choice (1): Between $x = -2$ and $x = 2$, the function is always increasing.
- Testing choice (2): Between $x = -2$ and $x = 2$, the function is always increasing. There is a "jump" at $x = -2$. So it isn't accurate to say that this function is always increasing on all intervals. However between $x = -2$ and $x = 2$, it is increasing.
- Testing choice (3): Between $x = -2$ and approximately $x = 0$, the function is increasing. From approximately $x = 0$ to $x = 2$, the function is decreasing.
- Testing choice (4): Even though this function starts to decrease at $x = 3$, it is increasing from $x = -2$ to $x = 2$.

The correct choice is **(3)**.

21. Since you cannot generally buy a fraction of an ice cream cone, the domain must be integers. Since you cannot buy a negative number of ice cream cones, the domain must be a nonnegative integer (including 0). If you were permitted to buy fractions of an ice cream cone, the answer would be choice (4).

The correct choice is **(2)**.

22. The quickest way to solve this question is to graph the parabola $y = 4x^2 + 2x + 5$ and count how many x-intercepts it has.

For the TI-84:

For the TI-Nspire:

If you want to use algebra instead, calculate $b^2 - 4ac$. If $b^2 - 4ac = 0$, the equation has one real solution. If $b^2 - 4ac > 0$, the equation has two real solutions. If $b^2 - 4ac < 0$, the equation has no real solutions.

In this example, $b^2 - 4ac = 2^2 - 4 \cdot 4 \cdot 5 = 4 - 80 = -76 < 0$. So the equation has no real solutions.

The correct choice is **(3)**.

23. Isolate the l in the equation.

Step 1: Subtract $2w$ from both sides of the equation:

$$p = 2l + 2w$$
$$-2w = -2w$$
$$p - 2w = 2l$$

Step 2: Divide both sides by 2:

$$\frac{p-2w}{2} = \frac{2l}{2}$$
$$\frac{p-2w}{2} = l$$
$$l = \frac{p-2w}{2}$$

This is option III.

Since dividing by 2 is the same thing as multiplying by $\frac{1}{2}$, the equation can also be written as:

$$\tfrac{1}{2}(p - 2w) = l$$
$$l = \tfrac{1}{2}(p - 2w)$$

This is option II.

If you distribute the $\frac{1}{2}$ through the right side of this equation you get:

$$l = \tfrac{1}{2}p - \tfrac{1}{2} \cdot 2w$$
$$l = \tfrac{1}{2}p - w$$

This is option I.

The correct choice is (4).

24. To calculate a_5 with this recursive definition of the function, you first have to calculate a_2, a_3, and a_4.

For $n = 2$:

$$\begin{aligned} a_2 &= 2(a_{2-1}) \\ &= 2(a_1) \\ &= 2 \cdot 1 \\ &= 2 \end{aligned}$$

For $n = 3$:

$$\begin{aligned} a_3 &= 3(a_{3-1}) \\ &= 3(a_2) \\ &= 3 \cdot 2 \\ &= 6 \end{aligned}$$

For $n = 4$:

$$\begin{aligned} a_4 &= 4(a_{4-1}) \\ &= 4(a_3) \\ &= 4 \cdot 6 \\ &= 24 \end{aligned}$$

For $n = 5$:

$$\begin{aligned} a_5 &= 5(a_{5-1}) \\ &= 5(a_4) \\ &= 5 \cdot 24 \\ &= 120 \end{aligned}$$

The correct choice is **(3)**.

PART II

25. First make a chart for all integer values from -2 to 7:

x	$f(x)$
-2	$\sqrt{-2+2} = \sqrt{0} = 0$
-1	$\sqrt{-1+2} = \sqrt{1} = 1$
0	$\sqrt{0+2} = \sqrt{2} \approx 1.4$
1	$\sqrt{1+2} = \sqrt{3} \approx 1.7$
2	$\sqrt{2+2} = \sqrt{4} = 2$
3	$\sqrt{3+2} = \sqrt{5} \approx 2.2$
4	$\sqrt{4+2} = \sqrt{6} \approx 2.4$
5	$\sqrt{5+2} = \sqrt{7} \approx 2.6$
6	$\sqrt{6+2} = \sqrt{8} \approx 2.8$
7	$\sqrt{7+2} = \sqrt{9} = 3$

Plot the 10 points on the graph, and join the points with a curve.

26. The graph of a nonlinear function is not a line. When the four ordered pairs are plotted, the graph looks like this:

Since there is no line that passes through all four of these points, Caleb is correct. The ordered pairs are from a nonlinear function.

Even without graphing, there is a way to tell that the ordered pairs are from a nonlinear function. In a linear function every time x increases by 1, $f(x)$ increases by a specific amount. In this example when x increases from 0 to 1, $f(x)$ increases by 2. When x increases from 1 to 2, though, $f(x)$ increases by 4. This can happen only in a nonlinear function.

Caleb is correct. It is a nonlinear function.

27. Use the quadratic formula from the reference sheet, $x = \dfrac{-b \pm \sqrt{b^2 - 4ac}}{2a}$. For this question, $a = 1$, $b = 1$, and $c = -5$.

$$x = \frac{-1 \pm \sqrt{1^2 - 4 \cdot 1 \cdot (-5)}}{2 \cdot 1}$$
$$= \frac{-1 \pm \sqrt{1 + 20}}{2}$$
$$= \frac{-1 \pm \sqrt{21}}{2}$$

$\dfrac{-1 + \sqrt{21}}{2} \approx 1.79 \approx 1.8$

$\dfrac{-1 - \sqrt{21}}{2} \approx -2.79 \approx -2.8$

To the *nearest tenth*, $x = 1.8$ and $x = -2.8$.

28. The graph of $p(x+2)$ is the graph of $p(x)$ shifted to the *left* by 2 units.

29. When the apple hits the ground, the height of the apple will be 0. To determine algebraically when this will happen, solve the equation $0 = -16t^2 + 256$. The answer must be positive since the number of seconds for the apple to fall cannot be negative:

$$0 = -16t^2 + 256$$
$$+16t^2 = +16t^2$$
$$\frac{16t^2}{16} = \frac{256}{16}$$
$$t^2 = 16$$
$$t = \sqrt{16} = 4$$

The apple takes 4 seconds to hit the ground.

30. Start with the distributive property. Then use the properties of algebra to get the constants on one side of the equals sign and the variables on the other side of the equals sign. To finish solving the equation, isolate the x variable:

$$6 - \frac{2}{3}(x+5) = 4x$$

$$6 - \frac{2}{3}x - \frac{2}{3} \cdot 5 = 4x$$

$$6 - \frac{2}{3}x - \frac{10}{3} = 4x$$

$$+\frac{2}{3}x = +\frac{2}{3}x$$

$$6 - \frac{10}{3} = 4x + \frac{2}{3}x$$

$$\frac{18}{3} - \frac{10}{3} = \frac{12}{3}x + \frac{2}{3}x$$

$$\frac{8}{3} = \frac{14}{3}x$$

$$\frac{3}{14} \cdot \frac{8}{3} = \frac{3}{14} \cdot \frac{14}{3}x$$

$$\frac{8}{14} = x$$

$$\frac{4}{7} = x$$

The exact value of x is $\frac{4}{7}$.

31. A rational number is one that can be written as a fraction that has an integer in both the numerator and the denominator. Since 16 is a perfect square, $\sqrt{16} = 4$ is a rational number. The product of 4 and $\frac{4}{7}$ is $4 \cdot \frac{4}{7} = \frac{16}{7}$. Since this fraction has an integer in both the numerator and the denominator, it is rational. The product is rational.

32. Since the boundary between the two pieces is $x = 2$, make a chart with some x-values less than 2 and some x-values greater than 2. For $x = -1, 0,$ and 1, use the first equation in the function. For $x = 2, 3, 4,$ and 5, use the second equation in the function:

x	$f(x)$
-1	$-\frac{1}{2} \cdot (-1) = \frac{1}{2}$
0	$-\frac{1}{2} \cdot 0 = 0$
1	$-\frac{1}{2} \cdot 1 = -\frac{1}{2}$
2	2
3	3
4	4
5	5

This is what the 7 points look like on the graph:

To plot the first piece of the graph, find the boundary point by plugging 2 into the first equation $-\frac{1}{2} \cdot 2 = -1$. This corresponds to the point $(2, -1)$.

Even though $(2, -1)$ is not an actual point on the graph, it is important to plot it on the graph as a small open circle. Make a ray that starts at this open circle and that passes through $\left(1, -\frac{1}{2}\right)$.

To plot the second piece of the the graph, make a ray that starts at $(2, 2)$ and that goes through $(3, 3)$.

The finished graph looks like this:

PART III

33. In an exponential equation of the form $y = a \cdot b^x$, a represents the starting amount and b represents the growth factor. In this question, the 20 represents the number of rabbits when the population was first counted. The 1.014 is the growth factor. The growth factor is also 1 plus the growth rate, so this population is increasing by 1.4% a year.

The average rate of change of a function $p(x)$ between $x = a$ and $x = b$ can be calculated with the formula $\dfrac{p(b) - p(a)}{b - a}$. In this question, $a = 50$ and $b = 100$. Calculate the average rate of change:

$$\frac{p(100) - p(50)}{100 - 50} = \frac{20(1.014)^{100} - 20(1.014)^{50}}{100 - 50}$$

$$= \frac{80.3 - 40.1}{100 - 50}$$

$$= \frac{40.2}{50}$$

$$= 0.804 \approx 0.8$$

To the *nearest tenth*, the average rate of change from day 50 to day 100 is 0.8.

34. For Garage A, the cost of parking for x hours (where x is greater than 2) is $A = 7 + 3(x - 2)$. For example, if you park for 6 hours in Garage A, the cost is $7 for the first 2 hours and then $3 for the additional 4 (not 6) hours. A common mistake is to write $A = 7 + 3x$ for the cost. However, that equation would make you pay an extra $3 for each of the first 2 hours.

For Garage B, the cost of parking for x hours is $B = 3.25x$.

The cost of parking at both garages will be the same when $7 + 3(x - 2) = 3.25x$.

Algebraically, this can be solved as follows:

$$7 + 3(x - 2) = 3.25x$$
$$7 + 3x - 6 = 3.25x$$
$$-3x = -3x$$
$$7 - 6 = 3.25x - 3x$$
$$\frac{1}{0.25} = \frac{0.25x}{0.25}$$
$$4 = x$$

Both garages cost the same when you park for 4 hours.

35. Linear inequalities are graphed as a boundary line and all the points on one side of the line are shaded. To graph the inequality $2y + 3x \leq 14$, first graph the boundary line. Start by changing the inequality sign into an equals sign. Then rewrite the equation in slope-intercept form $y = mx + b$:

$$2y + 3x \leq 14$$
$$2y + 3x = 14$$
$$-3x = -3x$$
$$\frac{2y}{2} = \frac{-3x + 14}{2}$$
$$y = -\frac{3}{2}x + 7$$

Since the inequality has a $\leq$ instead of a $<$, graph a solid line.

To determine which side of the line to shade, test if the point $(0, 0)$ makes the inequality true. Substitute $(0, 0)$ into the original inequality $2y + 3x \leq 14$:

$$2 \cdot 0 + 3 \cdot 0 \leq 14$$
$$0 + 0 \leq 14$$
$$0 \leq 14$$

Since it is true that 0 is less than or equal to 14, shade the side of the boundary line that contains $(0, 0)$.

To find the boundary line, change the inequality $4x - y < 2$ into an equation by substituting an equals sign for the inequality sign. Then write the equation in slope-intercept form $y = mx + b$:

$$4x - y < 2$$

$$4x - y = 2$$
$$-4x = -4x$$
$$\frac{-y}{-1} = \frac{-4x + 2}{-1}$$
$$y = 4x - 2$$

Since the inequality has a $<$ instead of a $\leq$, graph a dotted line.

Test to see if $(0, 0)$ is a solution to the second inequality:

$$4 \cdot 0 - 0 < 2$$
$$0 - 0 < 2$$
$$0 < 2$$

Since this is true, shade the side of the line that contains $(0, 0)$.

Inequalities can also be graphed using the graphing calculator.

For the TI-84:

For the TI-Nspire:

A point is in the solution set if it is in the double-shaded area. The point (1, 2) is on the dotted line, so it is not part of the double-shaded area and, therefore, is not part of the solution set. If it was in the double-shaded area or even on the solid boundary line, it would be part of the solution set.

36. Enter the data into the graphing calculator, and use the linear regression function.

For the TI-84:

If your TI-84 calculator is not showing the value of r, which is the correlation coefficient, you have to turn on the diagnostics by pressing [2nd] [0] and scrolling down to DiagnosticOn.

For the TI-Nspire:

To the *nearest hundredth*, the equation is $y = 0.96x + 23.95$.

The correlation coefficient is 0.92, which is very close to $+1$. This means that there is a strong positive correlation. If the correlation coefficient was very close to -1, there would be a strong negative correlation. If the correlation coefficient was close to 0, there would be a weak correlation.

PART IV

37. If d is the number of dimes and q is the number of quarters, the following system of equations can model this situation:

$$0.10d + 0.25q = 17.55$$
$$d + q = 90$$

To solve this system of equations, start by isolating the d in the second equation. Then use the substitution method so the only remaining variable is q, which shows the number of quarters Dylan has in his bank:

$$d = 90 - q$$
$$0.10d + 0.25q = 17.55$$
$$0.10(90 - q) + 0.25q = 17.55$$
$$9 - 0.10q + 0.25q = 17.55$$
$$-9 = -9$$
$$-0.10q + 0.25q = 17.55 - 9$$
$$\frac{0.15q}{0.15} = \frac{8.55}{0.15}$$
$$q = 57$$

Dylan has 57 quarters in his bank.

If each of Dylan's dimes is replaced with a quarter, he would have 90 quarters. They would be worth:

$$90 \times \$0.25 = \$22.50$$

The game costs $20.98 plus 8% of $20.98 for sales tax:

$$\$20.98 + (0.08 \times \$20.98) = \$22.66$$

Including sales tax, the game costs $22.66, which is more than $22.50.

No, Dylan would not be able to buy the game if all his dimes are replaced with quarters.

SELF-ANALYSIS CHART June 2018

Topic	Question Numbers	Number of Points	Your Points	Your Percentage
1. Polynomials	3, 12, 19	$2 + 2 + 2 = 6$		
2. Properties of Algebra	30	2		
3. Functions	2, 8, 11, 20	$2 + 2 + 2 + 2 = 8$		
4. Creating and Interpreting Equations	6, 14, 17, 21, 23	$2 + 2 + 2 + 2 + 2 = 10$		
5. Inequalities	1, 35	$2 + 4 = 6$		
6. Sequences and Series	7, 24	$2 + 2 = 4$		
7. Systems of Equations	34, 37	$4 + 6 = 10$		
8. Quadratic Equations and Factoring	4, 10, 22, 27, 29	$2 + 2 + 2 + 2 + 2 = 10$		
9. Regression	36	4		
10. Exponential Equations	26, 33	$2 + 4 = 6$		
11. Graphing	13, 16, 18, 25, 28, 32	$2 + 2 + 2 + 2 + 2 + 2 = 12$		
12. Statistics	5, 9	$2 + 2 = 4$		
13. Number Properties	31	2		
13. Unit Conversions	15	2		

SELF-ANALYSIS CHART June 2018

HOW TO CONVERT YOUR RAW SCORE TO YOUR ALGEBRA I REGENTS EXAMINATION SCORE

The accompanying conversion chart must be used to determine your final score on the June 2018 Regents Examination in Algebra I. To find your final exam score, locate in the column labeled "Raw Score" the total number of points you scored out of a possible 86 points. Since partial credit is allowed in Parts II, III, and IV of the test, you may need to approximate the credit you would receive for a solution that is not completely correct. Then locate in the adjacent column to the right the scale score that corresponds to your raw score. The scale score is your final Algebra I Regents Examination score.

SELF-ANALYSIS CHART June 2018

Regents Examination in Algebra I—June 2018
Chart for Converting Total Test Raw Scores to Final Examination Scores (Scaled Scores)

Raw Score	Scale Score	Performance Level	Raw Score	Scale Score	Performance Level	Raw Score	Scale Score	Performance Level
86	100	5	57	81	4	28	67	3
85	99	5	56	81	4	27	66	3
84	98	5	55	81	4	26	65	4
83	97	5	54	81	4	25	63	2
82	96	5	53	80	4	24	62	2
81	94	5	52	80	4	23	60	2
80	93	5	51	80	4	22	59	2
79	93	5	50	80	4	21	57	2
78	92	5	49	79	3	20	56	2
77	91	5	48	79	3	19	55	2
76	90	5	47	79	3	18	52	1
75	89	5	46	78	3	17	51	1
74	89	5	45	78	3	16	49	1
73	88	5	44	78	3	15	47	1
72	87	5	43	77	3	14	44	1
71	87	5	42	77	3	13	42	1
70	86	5	41	76	3	12	40	1
69	86	5	40	76	3	11	37	1
68	86	5	39	75	3	10	35	1
67	85	5	38	75	3	9	32	1
66	84	4	37	74	3	8	29	1
65	84	4	36	74	3	7	26	1
64	84	4	35	73	3	6	23	1
63	83	4	34	72	3	5	20	1
62	83	4	33	71	3	4	16	1
61	83	4	32	71	3	3	13	1
60	82	4	31	70	3	2	9	1
59	82	4	30	69	3	1	5	1
58	82	4	29	68	3	0	0	1

Examination
August 2018
Algebra I

HIGH SCHOOL MATH REFERENCE SHEET

Conversions

1 inch = 2.54 centimeters
1 meter = 39.37 inches
1 mile = 5280 feet
1 mile = 1760 yards
1 mile = 1.609 kilometers

1 cup = 8 fluid ounces
1 pint = 2 cups
1 quart = 2 pints
1 gallon = 4 quarts
1 gallon = 3.785 liters
1 liter = 0.264 gallon
1 liter = 1000 cubic centimeters

1 kilometer = 0.62 mile
1 pound = 16 ounces
1 pound = 0.454 kilogram
1 kilogram = 2.2 pounds
1 ton = 2000 pounds

Formulas

Triangle	$A = \frac{1}{2}bh$
Parallelogram	$A = bh$
Circle	$A = \pi r^2$
Circle	$C = \pi d$ or $C = 2\pi r$

Formulas (continued)

General Prisms	$V = Bh$
Cylinder	$V = \pi r^2 h$
Sphere	$V = \dfrac{4}{3} \pi r^3$
Cone	$V = \dfrac{1}{3} \pi r^2 h$
Pyramid	$V = \dfrac{1}{3} Bh$
Pythagorean Theorem	$a^2 + b^2 = c^2$
Quadratic Formula	$x = \dfrac{-b \pm \sqrt{b^2 - 4ac}}{2a}$
Arithmetic Sequence	$a_n = a_1 + (n-1)d$
Geometric Sequence	$a_n = a_1 r^{n-1}$
Geometric Series	$S_n = \dfrac{a_1 - a_1 r^n}{1 - r}$ where $r \neq 1$
Radians	1 radian $= \dfrac{180}{\pi}$ degrees
Degrees	1 degree $= \dfrac{\pi}{180}$ radians
Exponential Growth/Decay	$A = A_0 e^{k(t - t_0)} + B_0$

EXAMINATION August 2018

PART I

Answer all 24 questions in this part. Each correct answer will receive 2 credits. No partial credit will be allowed. For each statement or question, write in the space provided the numeral preceding the word or expression that best completes the statement or answers the question. [48 credits]

1 The number of bacteria grown in a lab can be modeled by $P(t) = 300 \cdot 2^{4t}$, where t is the number of hours. Which expression is equivalent to $P(t)$?

(1) $300 \cdot 8^t$
(2) $300 \cdot 16^t$
(3) $300^t \cdot 2^4$
(4) $300^{2t} \cdot 2^{2t}$

1 _____

2 During physical education class, Andrew recorded the exercise times in minutes and heart rates in beats per minute (bpm) of four of his classmates. Which table best represents a linear model of exercise time and heart rate?

Student 1

Exercise Time (in minutes)	Heart Rate (bpm)
0	60
1	65
2	70
3	75
4	80

(1)

Student 2

Exercise Time (in minutes)	Heart Rate (bpm)
0	62
1	70
2	83
3	88
4	90

(2)

Student 3

Exercise Time (in minutes)	Heart Rate (bpm)
0	58
1	65
2	70
3	75
4	79

(3)

Student 4

Exercise Time (in minutes)	Heart Rate (bpm)
0	62
1	65
2	66
3	73
4	75

(4)

2 _____

3 David correctly factored the expression $m^2 - 12m - 64$. Which expression did he write?

(1) $(m-8)(m-8)$
(2) $(m-8)(m+8)$
(3) $(m-16)(m+4)$
(4) $(m+16)(m-4)$

3 _____

4 The solution to $-2(1-4x) = 3x + 8$ is

(1) $\dfrac{6}{11}$
(2) 2
(3) $-\dfrac{10}{7}$
(4) -2

4 _____

5 The graph of $f(x)$ is shown below.

What is the value of $f(-3)$?

(1) 6
(2) 2
(3) -2
(4) -4

5 _____

6 If the function $f(x) = x^2$ has the domain $\{0, 1, 4, 9\}$, what is its range?

(1) $\{0, 1, 2, 3\}$
(2) $\{0, 1, 16, 81\}$
(3) $\{0, -1, 1, -2, 2, -3, 3\}$
(4) $\{0, -1, 1, -16, 16, -81, 81\}$

6 _____

7 The expression $4x^2 - 25$ is equivalent to
  (1) $(4x - 5)(x + 5)$
  (2) $(4x + 5)(x - 5)$
  (3) $(2x + 5)(2x - 5)$
  (4) $(2x - 5)(2x - 5)$

  7 _____

8 Compared to the graph of $f(x) = x^2$, the graph of $g(x) = (x - 2)^2 + 3$ is the result of translating $f(x)$
  (1) 2 units up and 3 units right
  (2) 2 units down and 3 units up
  (3) 2 units right and 3 units up
  (4) 2 units left and 3 units right

  8 _____

9 Lizzy has 30 coins that total $4.80. All of her coins are dimes, D, and quarters, Q. Which system of equations models this situation?
  (1) $D + Q = 4.80$
      $.10D + .25Q = 30$
  (2) $D + Q = 30$
      $.10D + .25Q = 4.80$
  (3) $D + Q = 30$
      $.25D + .10Q = 4.80$
  (4) $D + Q = 4.80$
      $.25D + .10Q = 30$

  9 _____

10 Gretchen has $50 that she can spend at the fair. Ride tickets cost $1.25 each and game tickets cost $2 each. She wants to go on a minimum of 10 rides and play at least 12 games.

Which system of inequalities represents this situation when r is the number of ride tickets purchased and g is the number of game tickets purchased?
  (1) $1.25r + 2g < 50$
      $r \le 10$
      $g > 12$
  (2) $1.25r + 2g \le 50$
      $r \ge 10$
      $g \ge 12$
  (3) $1.25r + 2g \le 50$
      $r \ge 10$
      $g > 12$
  (4) $1.25r + 2g < 50$
      $r \le 10$
      $g \ge 12$

  10 _____

11 Three functions are shown below.

$g(x) = 3^x + 2$

x	h(x)
−5	30
−4	14
−3	6
−2	2
−1	0
0	−1
1	−1.5
2	−1.75

Which statement is true?

(1) The y-intercept for $h(x)$ is greater than the y-intercept for $f(x)$.
(2) The y-intercept for $f(x)$ is greater than the y-intercept for $g(x)$.
(3) The y-intercept for $h(x)$ is greater than the y-intercept for both $g(x)$ and $f(x)$.
(4) The y-intercept for $g(x)$ is greater than the y-intercept for both $f(x)$ and $h(x)$.

11 _____

12 Olivia entered a baking contest. As part of the contest, she needs to demonstrate how to measure a gallon of milk if she only has a teaspoon measure. She converts the measurement using the ratios below:

$$\frac{4 \text{ quarts}}{1 \text{ gallon}} \cdot \frac{2 \text{ pints}}{1 \text{ quart}} \cdot \frac{2 \text{ cups}}{1 \text{ pint}} \cdot \frac{\frac{1}{4} \text{ cup}}{4 \text{ tablespoons}} \cdot \frac{3 \text{ teaspoons}}{1 \text{ tablespoon}}$$

Which ratio is *incorrectly* written in Olivia's conversion?

(1) $\dfrac{4 \text{ quarts}}{1 \text{ gallon}}$

(2) $\dfrac{2 \text{ pints}}{1 \text{ quart}}$

(3) $\dfrac{\frac{1}{4} \text{ cup}}{4 \text{ tablespoons}}$

(4) $\dfrac{3 \text{ teaspoons}}{1 \text{ tablespoon}}$

12 _____

EXAMINATION August 2018   **347**

13 If $y = 3x^3 + x^2 - 5$ and $z = x^2 - 12$, which polynomial is equivalent to $2(y + z)$?

  (1) $6x^3 + 4x^2 - 34$ (3) $6x^3 + 3x^2 - 22$
  (2) $6x^3 + 3x^2 - 17$ (4) $6x^3 + 2x^2 - 17$ 13 _____

14 An outdoor club conducted a survey of its members. The members were asked to state their preference between skiing and snowboarding. Each member had to pick one. Of the 60 males, 45 stated they preferred to snowboard. Twenty-two of the 60 females preferred to ski. What is the relative frequency that a male prefers to ski?

  (1) 0.125   (3) $0.\overline{333}$
  (2) 0.25   (4) $0.\overline{405}$ 14 _____

15 When the function $g(x) = \begin{cases} 5x, & x \leq 3 \\ x^2 + 4, & x > 3 \end{cases}$ is graphed correctly, how should the points be drawn on the graph for an x-value of 3?

  (1) open circles at (3, 15) and (3, 13)
  (2) closed circles at (3, 15) and (3, 13)
  (3) an open circle at (3, 15) and a closed circle at (3, 13)
  (4) a closed circle at (3, 15) and an open circle at (3, 13)   15 _____

16 If $f(x) = 2x^2 + x - 3$, which equation can be used to determine the zeros of the function?

  (1) $0 = (2x - 3)(x + 1)$ (3) $0 = 2x(x + 1) - 3$
  (2) $0 = (2x + 3)(x - 1)$ (4) $0 = 2x(x - 1) - 3(x + 1)$ 16 _____

17 Each day, a local dog shelter spends an average of $2.40 on food per dog. The manager estimates the shelter's daily expenses, assuming there is at least one dog in the shelter, using the function $E(x) = 30 + 2.40x$.

Which statements regarding the function $E(x)$ are correct?
 I. x represents the number of dogs at the shelter per day.
 II. x represents the number of volunteers at the shelter per day.
 III. 30 represents the shelter's total expenses per day.
 IV. 30 represents the shelter's nonfood expenses per day.

(1) I and III (3) II and III
(2) I and IV (4) II and IV

18 Which point is not in the solution set of the equation $3y + 2 = x^2 - 5x + 17$?

(1) $(-2, 10)$ (3) $(2, 3)$
(2) $(-1, 7)$ (4) $(5, 5)$

19 The functions $f(x)$ and $g(x)$ are graphed below.

Based on the graph, the solutions to the equation $f(x) = g(x)$ are

(1) the x-intercepts
(2) the y-intercepts
(3) the x-values of the points of intersection
(4) the y-values of the points of intersection

20 For the sequence $-27, -12, 3, 18, \ldots$, the expression that defines the nth term where $a_1 = -27$ is

(1) $15 - 27n$
(2) $15 - 27(n-1)$
(3) $-27 + 15n$
(4) $-27 + 15(n-1)$

20 _____

21 The data obtained from a random sample of track athletes showed that as the foot size of the athlete decreased, the average running speed decreased. Which statement is best supported by the data?

(1) Smaller foot sizes cause track athletes to run slower.
(2) The sample of track athletes shows a causal relationship between foot size and running speed.
(3) The sample of track athletes shows a correlation between foot size and running speed.
(4) There is no correlation between foot size and running speed in track athletes.

21 _____

22 Which system of equations will yield the same solution as the system below?

$$x - y = 3$$
$$2x - 3y = -1$$

(1) $-2x - 2y = -6$
    $2x - 3y = -1$
(2) $-2x + 2y = 3$
    $2x - 3y = -1$
(3) $2x - 2y = 6$
    $2x - 3y = -1$
(4) $3x + 3y = 9$
    $2x - 3y = -1$

22 _____

23 Which of the three situations given below is best modeled by an exponential function?

   I. A bacteria culture doubles in size every day.
   II. A plant grows by 1 inch every 4 days.
   III. The population of a town declines by 5% every 3 years.

(1) I, only
(2) II, only
(3) I and II
(4) I and III

23 _____

24 The length, width, and height of a rectangular box are represented by $2x$, $3x + 1$, and $5x - 6$, respectively. When the volume is expressed as a polynomial in standard form, what is the coefficient of the 2nd term?

(1) −13

(2) 13

(3) −26

(4) 26

24 _____

EXAMINATION August 2018

PART II

Answer all 8 questions in this part. Each correct answer will receive 2 credits. Clearly indicate the necessary steps, including appropriate formula substitutions, diagrams, graphs, charts, etc. For all questions in this part, a correct numerical answer with no work shown will receive only 1 credit. [16 credits]

25 Explain how to determine the zeros of
$f(x) = (x+3)(x-1)(x-8)$.

put into
y= and look
at the table for
o's

State the zeros of the function.

$[-3, 1, 8]$

26 Four relations are shown below.

[I: graph] ~function~

[III: table] ~Function~

x	y
−4	1
0	3
4	5
6	6

I

$\{(1, 2), (2, 5), (3, 8), (2, -5), (1, -2)\}$

II

$y = x^2$

IV

State which relation(s) are functions.

IV III

Explain why the other relation(s) are *not* functions.

I is not a function because it fails the VLT ans

II is not a function because the x values of 1 and 2 repeat

27 The table below represents the height of a bird above the ground during flight, with $P(t)$ representing height in feet and t representing time in seconds.

t	P(t)
0	6.71
3	6.26
4	6
9	3.41

Calculate the average rate of change from 3 to 9 seconds, in feet per second.

$$\frac{y_2 - y_1}{x_2 - x_1}$$

$$\frac{3.41 - 4.75}{9 - 3}$$

28. Is the solution to the quadratic equation written below rational or irrational? Justify your answer.

$$0 = 2x^2 + 3x - 10$$

$a = 2$

$b = 3$

$c = -10$

$$\frac{-(3) \pm \sqrt{(3)^2 - 4(2)(-10)}}{2(2)}$$

$$\frac{-3 \pm \sqrt{89}}{4}$$

Irrational

29 The formula for converting degrees Fahrenheit (F) to degrees Kelvin (K) is:
$$K = \frac{5}{9}(F + 459.67)$$

Solve for F, in terms of K.

30. Solve the following equation by completing the square:
$$x^2 + 4x = 2$$

$a = 1$

$b = 4$

$c = -2$

$x^2 + 4x - 2 = 0$

$$\frac{-(4) \pm \sqrt{(4)^2 - 4(1)(-2)}}{2(1)}$$

$$\frac{-4 \pm \sqrt{24}}{2}$$

$$-2 \pm \sqrt{6}$$

31 The students in Mrs. Lankford's 4th and 6th period Algebra classes took the same test. The results of the scores are shown in the following table:

	$\bar{x}$	σ_x	n	min	Q_1	med	Q_3	max
4th Period	77.75	10.79	20	58	69	76.5	87.5	96
6th Period	78.4	9.83	20	59	71.5	78	88	96

Based on these data, which class has the largest spread of test scores? Explain how you arrived at your answer.

period 4 because the numbers are higher

32 Write the first five terms of the recursive sequence defined below.

$$a_1 = 0$$
$$a_n = 2(a_{n-1})^2 - 1, \text{ for } n > 1$$

0, -1, 1, 1, 1

PART III

Answer all 4 questions in this part. Each correct answer will receive 4 credits. Clearly indicate the necessary steps, including appropriate formula substitutions, diagrams, graphs, charts, etc. For all questions in this part, a correct numerical answer with no work shown will receive only 1 credit.   [16 credits]

33 Sarah wants to buy a snowboard that has a total cost of $580, including tax. She has already saved $135 for it. At the end of each week, she is paid $96 for babysitting and is going to save three-quarters of that for the snowboard.

Write an inequality that can be used to determine the *minimum* number of weeks Sarah needs to babysit to have enough money to purchase the snowboard.

$$72w + 135 \geq 580$$

Determine and state the *minimum* number of full weeks Sarah needs to babysit to have enough money to purchase this snowboard.

7 weeks

34 A car was purchased for $25,000. Research shows that the car has an average yearly depreciation rate of 18.5%.

Create a function that will determine the value, $V(t)$, of the car t years after purchase.

$V(t) = 25000(0.815)^t$

Determine, to the *nearest cent*, how much the car will depreciate from year 3 to year 4.

2503.71

EXAMINATION August 2018

35 Graph the following system of inequalities on the set of axes below:

$$2y \geq 3x - 16$$
$$y + 2x > -2$$

Based upon your graph, explain why (6, 1) is a solution to this system and why (−6, 7) is *not* a solution to this system.

(6, 1) is a solution because it is the shaded region.

(−6, 7) is not a solution because it is not in the shaded region.

36 Paul plans to have a rectangular garden adjacent to his garage. He will use 36 feet of fence to enclose three sides of the garden. The area of the garden, in square feet, can be modeled by $f(w) = w(36 - 2w)$, where w is the width in feet.

On the set of axes below, sketch the graph of $f(w)$.

Explain the meaning of the vertex in the context of the problem.

w = 9

the area is 9.

max of 162

PART IV

Answer the question in this part. A correct answer will receive 6 credits. Clearly indicate the necessary steps, including appropriate formula substitutions, diagrams, graphs, charts, etc. A correct numerical answer with no work shown will receive only 1 credit.   [6 credits]

37 At the present time, Mrs. Bee's age is six years more than four times her son's age. Three years ago, she was seven times as old as her son was then.

If b represents Mrs. Bee's age now and s represents her son's age now, write a system of equations that could be used to model this scenario.

$$b = 4s + 6$$

Use this system of equations to determine, algebraically, the ages of both Mrs. Bee and her son now.

$$b - 3 = 7(s - 3)$$

Determine how many years from now Mrs. Bee will be three times as old as her son will be then.

$$b = 38$$

$$s = 8, 7 \text{ years old}$$

Answers August 2018
Algebra I

Answer Key

PART I

1. (2)	5. (1)	9. (2)	13. (1)	17. (2)	21. (3)
2. (1)	6. (2)	10. (2)	14. (2)	18. (1)	22. (3)
3. (3)	7. (3)	11. (4)	15. (4)	19. (3)	23. (4)
4. (2)	8. (3)	12. (3)	16. (2)	20. (4)	24. (3)

PART II

25. $\{-3, 1, 8\}$
26. III, IV
27. -0.475
28. Irrational
29. $F = \frac{9}{5}K - 459.67$
30. $x = -2 \pm \sqrt{6}$
31. Period 4
32. $0, -1, 1, 1, 1$

PART III

33. $72w + 135 \geq 580$, 7 weeks
34. $V(t) = 25,000(0.815)^t$, 2503.71

35. $(6, 2)$ is a solution but $(-6, 7)$ is not.

36. When $w = 9$, the area is a maximum of 162.

PART IV

37. $b = 4s + 6$
$b - 3 = 7(s - 3)$
$b = 38, s = 8$, 7 years

In **Parts II–IV,** you are required to show how you arrived at your answers. For sample methods of solutions, see the *Answers Explained* section.

Answers Explained

PART I

1. There is a property of exponents that $x^{ab} = \left(x^a\right)^b$. Using this property, the expression

$$P(t) = 300 \cdot 2^{4t} = 300 \cdot \left(2^4\right)^t = 300 \cdot 16^t$$

The correct choice is **(2)**.

2. Make a scatter plot for each of the four choices and find the one that looks most like a straight line.

   - Choice (1):

- Choice (2):

- Choice (3):

- Choice (4):

The correct choice is (**1**).

3. To factor an expression of the form $x^2 + bx + c$ into the form $(x + p)(x + q)$, find two numbers that have a sum of b and a product of c. In this example, $b = -12$ and $c = -64$.

Since $-16 \cdot 4 = -64$ and $-16 + 4 = -12$, the two numbers are -16 and 4.

$$m^2 - 12m - 64 = (m - 16)(m + 4)$$

This question could also be answered by simplifying each of the answer choices to see which one becomes $m^2 - 12m - 64$.

- Testing choice (1):

$$(m - 8)(m - 8) = m^2 - 8m - 8m + 64 = m^2 - 16m + 64$$

- Testing choice (2):

$$(m - 8)(m + 8) = m^2 + 8m - 8m - 64 = m^2 - 64$$

- Testing choice (3):

$$(m - 16)(m + 4) = m^2 + 4m - 16m - 64 = m^2 - 12m - 64$$

- Testing choice (4):

$$(m + 16)(m - 4) = m^2 - 4m + 16m - 64 = m^2 + 12m - 64$$

The correct choice is **(3)**.

4. Use the properties of algebra to isolate the variable.

$$-2(1 - 4x) = 3x + 8$$
$$-2 + 8x = 3x + 8$$
$$-3x = -3x$$
$$-2 + 5x = 8$$
$$+2 = +2$$
$$\frac{5x}{5} = \frac{10}{5}$$
$$x = 2$$

The correct choice is **(2)**.

5. Find the point on the graph that has an x-coordinate of −3. The value of $f(-3)$ will be the y-coordinate of that point.

The correct choice is **(1)**.

6. Plug each of the numbers from the domain into the function. The values that the function evaluates to will be the range.

$$f(0) = 0^2 = 0$$
$$f(1) = 1^2 = 1$$
$$f(4) = 4^2 = 16$$
$$f(9) = 9^2 = 81$$

The range is the set of numbers {0, 1, 16, 81}.

The correct choice is **(2)**.

7. An expression of the form $a^2 - b^2$ factors into $(a-b)(a+b)$. The expression $4x^2 - 25$ can be written as $(2x)^2 - 5^2$. Using the factoring pattern, this becomes $(2x)^2 - 5^2 = (2x-5)(2x+5) = (2x+5)(2x-5)$.

This question could also be answered by simplifying each of the answer choices to see which one becomes $4x^2 - 25$.

- Testing choice (1):

$$(4x - 5)(x + 5) = 4x^2 + 20x - 5x - 25 = 4x^2 + 15x - 25$$

- Testing choice (2):

  $(4x+5)(x-5) = 4x^2 - 20x + 5x - 25 = 4x^2 - 15x - 25$

- Testing choice (3):

  $(2x+5)(2x-5) = 4x^2 - 10x + 10x - 25 = 4x^2 - 25$

- Testing choice (4):

  $(2x-5)(2x-5) = 4x^2 - 10x - 10x + 25 = 4x^2 - 20x + 25$

The correct choice is **(3)**.

8. On your graphing calculator, graph both functions.

The graph of $g(x)$ is a translation of the graph of $f(x)$ two units to the right and three units up. This can also be answered without a graphing calculator.

In general:

The graph of $f(x + a)$ is a translation of the graph of $f(x)$ by a units to the left.

The graph of $f(x - a)$ is a translation of the graph of $f(x)$ by a units to the right.

The graph of $f(x) + a$ is a translation of the graph of $f(x)$ by a units up.

The graph of $f(x) - a$ is a translation of the graph of $f(x)$ by a units down.

In this example, $g(x) = f(x - 2) + 3$, so the graph is a translation of the graph $f(x)$ two units to the right and three units up.

The correct choice is **(3)**.

9. Since D is the number of dimes and Q is the number of quarters, the total number of coins is $D + Q$. Since there are 30 coins, the first equation is $D + Q = 30$.

Since each dime is worth 10 cents, the value of D dimes is $.10D$. Since each quarter is worth 25 cents, the value of Q quarters is $.25Q$. The total value of the coins is $.10D + .25Q$. Since the coins are worth \$4.80, the second equation is $.10D + .25Q = 4.80$.

The correct equation is **(2)**.

10. Since rides cost \$1.25 each, the cost of r rides is $1.25r$. Since games cost \$2 each, the cost of g games is $2g$.

Since she has \$50, the first equation is $1.25r + 2g \leq 50$. It must be a $\leq$, not a $<$, because, if the total cost is \$50, she can afford it. She cannot spend more than \$50, but she can spend exactly \$50 if that's what it costs.

Since she wants to go on a minimum of 10 rides, the second equation is $r \geq 10$.

Since she wants to play a minimum of 12 games, the third equation is $g \geq 12$.

The correct choice is **(2)**.

11. The y-intercept of $f(x)$ is the y-coordinate of the point where the curve crosses the y-axis. Since the curve crosses the y-axis at the point $(0, 1)$, the y-intercept is 1.

The y-intercept of the graph of $g(x)$ is the value of $g(0)$.

For this function $g(0) = 3^0 + 2 = 1 + 2 = 3$, so the y-intercept for $g(x)$ is 3.

The y-intercept of the graph of $h(x)$ is the value of $h(0)$.

According to the chart, $h(0) = -1$, so the y-intercept for $h(x)$ is -1.

Since $3 > 1 > -1$, the y-intercept of $g(x)$ is greater than the y-intercept for both $f(x)$ and $h(x)$.

The correct choice is **(4)**.

12. When doing unit conversions, make the denominator of each fraction have the same kind of units as the numerator in the previous fraction.

    The units of the second denominator are quarts, which match the units of the first numerator.

    The units of the third denominator are pints, which match the units of the second numerator.

    The units of the fourth denominator are tablespoons, which do not match the units of the third numerator.

    This fourth fraction would need to be "flipped" to do the unit conversion properly.

    Note: The conversions of tablespoons to cups (16 tablespoons = 1 cup) and teaspoons to tablespoons (3 teaspoons = 1 tablespoon) are not on the reference sheet. Still, this question can be answered by using the above strategy while assuming that the conversion rates are accurate.

    The correct choice is **(3)**.

13. Replace y and z with the equivalent expressions in terms of x.

    $$2(y + z) = 2(3x^3 + x^2 - 5 + x^2 - 12) = 2(3x^3 + 2x^2 - 17)$$
    $$= 6x^3 + 4x^2 - 34$$

    The correct choice is **(1)**.

14. The data can first be organized into a two-way frequency table:

	Skiing	Snowboarding
Male	15	45
Female	22	38

    Out of the 60 males, 15 of them prefer to ski, so the relative frequency that a male prefers to ski is $\frac{15}{60} = 0.25$.

    The correct choice is **(2)**.

15. The graph of this piecewise function looks like this:

Since the first "piece" has the $\leq$ sign, that piece will include the endpoint with the closed circle at (3, 15).

Since the second "piece" has the $>$ sign, that piece will not contain the endpoint of (3, 13), so there will be an open circle there.

The correct choice is **(4)**.

16. The zeros of a function, $f(x)$, are the values for x that make $f(x) = 0$.

Solving the equation:

$$f(x) = 0$$
$$2x^2 + x - 3 = 0$$

The next step is to factor the quadratic equation:

$$(2x + 3)(x - 1) = 0$$

This question does not ask you to solve for x, but just to find the equation that can be used to solve for x.

The correct choice is **(2)**.

17. One way to analyze the components of this equation is to calculate the values of $E(1)$ and $E(2)$.

Plugging into the function you get:

$$E(1) = 30 + 2.40 \cdot 1 = 30 + 2.40 = 32.40$$
$$E(2) = 30 + 2.40 \cdot 2 = 30 + 4.80 = 34.80$$

Each time x increases by 1, $E(x)$ increases by 2.40, so x must represent the number of dogs at the shelter per day; therefore, Roman numeral I is one of the correct statements.

This eliminates choices (3) and (4).

Roman numeral III cannot be correct because the total expenses change depending on the value of x, so 30 must represent the shelter's nonfood expenses per day, which is the statement in Roman numeral IV.

The correct choice is **(2)**.

18. Any ordered pair in the solution set of an equation will make the left-hand side equal the right-hand side.

- Testing choice (1):

$$3 \cdot 10 + 2 \stackrel{?}{=} (-2)^2 - 5(-2) + 17$$
$$30 + 2 \stackrel{?}{=} 4 + 10 + 17$$
$$32 \neq 31$$

- Testing choice (2):

$$3 \cdot 7 + 2 \stackrel{?}{=} (-1)^2 - 5(-1) + 17$$
$$21 + 2 \stackrel{?}{=} 1 + 5 + 17$$
$$23 \stackrel{\checkmark}{=} 23$$

- Testing choice (3):

$$3 \cdot 3 + 2 \stackrel{?}{=} 2^2 - 5 \cdot 2 + 17$$
$$9 + 2 \stackrel{?}{=} 4 - 10 + 17$$
$$11 \stackrel{\checkmark}{=} 11$$

- Testing choice (4):

$$3 \cdot 5 + 2 \stackrel{?}{=} 5^2 - 5 \cdot 5 + 17$$
$$15 + 2 \stackrel{?}{=} 25 - 25 + 17$$
$$17 \stackrel{\checkmark}{=} 17$$

Of the four choices, only choice (1) is not in the solution set.

The correct choice is **(1)**.

19. A solution to this equation is an x-value that is put into the functions that will evaluate to the same output value for both functions. The graphs of these two functions intersect at approximately $(-1.5, 1.5)$ and $(8.5, -1.5)$.

This means that $f(-1.5) = g(-1.5)$ and $f(8.5) = g(8.5)$. These solutions are the x-coordinates of the points of intersection.

The correct choice is **(3)**.

20. An arithmetic sequence is one where each term is the same amount more (or less) than the previous term. For this sequence, the first term of this sequence is -27, and each term is 15 more than the previous term.

There is a formula for the nth term of an arithmetic sequence in the reference sheet provided on the test.

$a_n = a_1 + (n-1)d$, where a_1 is the first term, n is the term number, and d is the "common difference" (what you need to add to a term to get the next term).

For this sequence, $a_1 = -27$ and $d = +15$, so the formula is

$$a_n = -27 + (n-1)15 = -27 + 15(n-1)$$

The correct choice is **(4)**.

21. Since the running speed decreases when the foot size decreases, you can say, for sure, that there is a correlation between the two things. It may be true, also, that the smaller foot size actually causes the slower speed, but you cannot be sure about this. The smaller foot size and the slower speed could, for example, both be caused by some third thing, like the runner being a child. All you can say with certainty is that there is a correlation between the two things.

The correct choice is **(3)**.

22. When you have a system of equations, the solution will not change if you multiply both sides of either equation by some number. Test each of the choices to see if that system can be obtained from the original system by multiplying both sides of either one or both equations by some number.

- Testing choice (1): If you multiply both sides of the top equation by -2, you would get $-2x + 2y = -6$, not $-2x - 2y = -6$.
- Testing choice (2): If you multiply both sides of the top equation by -2, you would get $-2x + 2y = -6$, not $-2x + 2y = 3$.
- Testing choice (3): This is the system you would get if you multiply both sides of the top equation by 2.

- Testing choice (4): If you multiply both sides of the top equation by 3, you would get $3x - 3y = 9$, not $3x + 3y = 9$.

The correct choice is **(3)**.

23. An exponential function has the form $f(x) = a \cdot b^x$. The functions for each of the Roman numerals are:
    - Scenario I: $f(x) = a \cdot 2^x$
    - Scenario II: $f(x) = \dfrac{1}{4}x$
    - Scenario III: $f(x) = a(1 - 0.05)^{x/3}$

Scenarios I and III have the form of an exponential function.

Even without creating the equations, you can make a rough sketch of the graph of each scenario to see which has the shape of an exponential curve (which looks a little like a playground slide).

Graph of scenario 1:

Graph of scenario 2:

Graph of scenario 3:

The correct choice is **(4)**.

24. The volume of a rectangular box can be calculated with the formula $V = lwh$, where l is the length, w is the width, and h is the height.

    For this question, this becomes $V = (2x)(3x + 1)(5x - 6)$.

    Multiply the factors of the right-hand side to turn it into a polynomial.

    $$(2x)(3x + 1)(5x - 6)$$
    $$(6x^2 + 2x)(5x - 6)$$
    $$30x^3 - 36x^2 + 10x^2 - 12x$$
    $$30x^3 - 26x^2 - 12x$$

    For a polynomial to be in standard form, the first term must be the one with the largest exponent, and the size of the exponent must decrease as you go from left to right. Since the exponent of 3 is before the exponent of 2, which is before the exponent of 1, this is a polynomial in standard form.

    The coefficient of the second term is −26.

    The correct choice is **(3)**.

PART II

25. The zeros of a function, $f(x)$, are the values of x that make the function evaluate to 0.

To determine the zeros, solve the equation $f(x) = 0$.

$$0 = (x + 3)(x - 1)(x - 8)$$

The only way the product of three different things can be zero is if one of those things is zero.

If $(x + 3) = 0$, then $x = -3$.

If $(x - 1) = 0$, then $x = 1$.

If $(x - 8) = 0$, then $x = 8$.

So, the zeros of the function are $-3, 1,$ and 8.

26. For Roman numeral I, the graph of a function will not have more than one point with the same x-coordinate. Since there is a point at $(6, 4)$ and also a point at $(6, 6)$, this cannot be a function.

A graph where this happens is said to fail the "vertical line test," since it is possible to draw a vertical line that would pass through more than one point on the graph.

For Roman numeral II, check the list to see if there are any ordered pairs that have the same x-coordinate. Since there are repeated x-values in the ordered pairs $(1, 2)$ and $(1, -2)$ and also repeated x-values in the ordered pairs $(2, 5)$ and $(2, -5)$, this is not a function.

For Roman numeral III, check to see if there are two rows in the chart that have the same x-value. Since all the x-values are different, this is a function.

For Roman numeral IV, check to see if the graph of the equation passes the vertical line test. Since there is no vertical line that would pass through more than one point on the graph of this equation, this is a function.

Relations III and IV are functions, while relations I and II are not functions.

27. The average rate of change of a function $P(t)$ on the interval from a to b can be calculated by the formula:

$$\text{Average Rate of Change} = \frac{P(b) - P(a)}{b - a}$$

For this example,

$$\text{Average Rate of Change} = \frac{P(9) - P(3)}{9 - 3} = \frac{3.41 - 6.26}{9 - 3} = \frac{-2.85}{6}$$
$$= -0.475$$

28. The solutions to a quadratic equation $ax^2 + bx + c = 0$ are rational if the "discriminant" $D = b^2 - 4ac$ is a perfect square.

For this example, $D = 3^2 - 4 \cdot 2 \cdot (-10) = 9 + 80 = 89$. Since 89 is not a perfect square, the solutions to this quadratic equation are irrational.

Another way to answer this question is to solve the equation using the quadratic formula to show that both answers are irrational.

$$x = \frac{-b \pm \sqrt{b^2 - 4ac}}{2a} = \frac{-3 \pm \sqrt{3^2 - 4 \cdot 2 \cdot (-10)}}{2 \cdot 2} = \frac{-3 \pm \sqrt{9 + 80}}{4}$$
$$= \frac{-3 \pm \sqrt{89}}{4}$$

Since 89 is not a perfect square, $\sqrt{89}$ is irrational, and so are both $\frac{-3 + \sqrt{89}}{4}$ and $\frac{-3 - \sqrt{89}}{4}$.

ANSWERS August 2018

29. The simplest way to isolate the F is to first multiply both sides of the equation by the reciprocal of $\frac{5}{9}$, which is $\frac{9}{5}$.

$$K = \frac{5}{9}(F + 459.67)$$

$$\frac{9}{5}K = \frac{9}{5} \cdot \frac{5}{9}(F + 459.67)$$

$$\frac{9}{5}K = 1(F + 459.67)$$

$$\frac{9}{5}K = F + 459.67$$

$$-459.67 = -459.67$$

$$\frac{9}{5}K - 459.67 = F$$

30. To solve a quadratic equation in the form $x^2 + bx = c$ by the "completing the square" process, the first step is to add $\left(\frac{b}{2}\right)^2$ to both sides of the equation.

$$x^2 + 4x = 2$$

$$+\left(\frac{4}{2}\right)^2 = +\left(\frac{4}{2}\right)^2$$

$$x^2 + 4x + 4 = 2 + 4$$

$$x^2 + 4x + 4 = 6$$

The left side of this equation is now a perfect square and can be factored:

$$x^2 + 4x + 4 = 6$$
$$(x + 2)(x + 2) = 6$$
$$(x + 2)^2 = 6$$

Finish by taking the square root of both sides of the equation and then isolating the x.

$$\sqrt{(x+2)^2} = \pm\sqrt{6}$$
$$x + 2 = \pm\sqrt{6}$$
$$-2 = -2$$
$$x = -2 \pm \sqrt{6}$$

31. The "spread" is the difference between the maximum score and the minimum score.

   The spread for 4th period is $96 - 58 = 38$.
   The spread for 6th period is $96 - 59 = 37$.

   Since $38 > 37$, the class with the largest spread is period 4.

32. The first term of the sequence is given as $a_1 = 0$. To calculate the next four terms, follow the second rule for $n = 2, 3, 4,$ and 5.

   For $n = 2$,
   $$a_2 = 2(a_{2-1})^2 - 1 = 2(a_1)^2 - 1 = 2 \cdot 0^2 - 1 = 2 \cdot 0 - 1 = 0 - 1 = -1$$

   For $n = 3$,
   $$a_3 = 2(a_{3-1})^2 - 1 = 2(a_2)^2 - 1 = 2(-1)^2 - 1 = 2 \cdot 1 - 1 = 2 - 1 = 1$$

   For $n = 4$,
   $$a_4 = 2(a_{4-1})^2 - 1 = 2(a_3)^2 - 1 = 2 \cdot 1^2 - 1 = 2 \cdot 1 - 1 = 2 - 1 = 1$$

   For $n = 5$,
   $$a_5 = 2(a_{5-1})^2 - 1 = 2(a_4)^2 - 1 = 2 \cdot 1^2 - 1 = 2 \cdot 1 - 1 = 2 - 1 = 1$$

   The first five terms are $0, -1, 1, 1, 1$.

PART III

33. Each week she saves $\frac{3}{4} \times \$96 = \$72$. In w weeks, then, she saves $72w$ dollars. Since she already saved $135, after w weeks she will have saved a total of $72w + 135$ dollars.

    To find the minimum number of weeks she must work to save at least $580, the inequality is:
    $$72w + 135 \geq 580$$

    Solve as you would an equality.
    $$72w + 135 \geq 580$$
    $$-135 = -135$$
    $$\frac{72w}{72} \geq \frac{445}{72}$$
    $$w \geq 6.18$$

    Since she must work a full week, because 6 weeks will not be enough, the minimum number of full weeks she must work is 7.

34. This is an example of exponential decay. The formula for exponential decay is $A = P(1-r)^t$, where P is the starting value, r is the decay rate, and t is the amount of time.

    For this example, the formula is
    $$V(t) = 25{,}000(1 - 0.185)^t = 25{,}000(0.815)^t$$

    The value of the car at $t = 3$ is
    $$V(3) = 25{,}000(0.815)^3 \approx 13{,}533.58$$

    The value of the car at $t = 4$ is
    $$V(4) = 25{,}000(0.815)^4 \approx 11{,}029.87$$

    The difference between these two values is
    $$13{,}533.58 - 11{,}029.87 = 2503.71$$

35. Each of the inequalities will be graphed as a boundary line with one side of the line shaded.

To graph the boundary line for the first inequality, change the $\geq$ to an $=$ and convert to slope-intercept form ($y = mx + b$).

$$2y = 3x - 16$$
$$\frac{2y}{2} = \frac{3x - 16}{2}$$
$$y = \frac{3}{2}x - 8$$

Since the inequality symbol was a $\geq$, graph this line as a solid line. If it were a $>$, you would graph it as a dotted line.

To determine which side to shade, check to see if the ordered pair (0, 0) makes the inequality true or false.

$$2 \cdot 0 \geq 3 \cdot 0 - 16$$
$$0 \geq 0 - 16$$
$$0 \geq -16$$

Since it is true that $0 \geq -16$, shade the side of the line that contains $(0, 0)$.

For the second inequality, find the equation of the boundary line the same way.

$$y + 2x = -5$$
$$-2x = -2x$$
$$y = -2x - 5$$

Since the inequality symbol for the second inequality was a $>$, make this line a dotted line.

To determine which side to shade, check to see if the ordered pair $(0, 0)$ makes the second inequality true or false.

$$0 + 2 \cdot 0 > -5$$
$$0 + 0 > -5$$
$$0 > -5$$

Since it is true that $0 > -5$, shade the side of the line that contains $(0, 0)$.

The inequality graphs can also be made on the graphing calculator.

All the points in the double-shaded area are solutions to the system of inequalities. Since (6, 1) is on the solid boundary line, it is considered in the double-shaded area, but (−6, 7) is not in the double-shaded area because it is on the dotted line.

36. The simplest way to make the graph is to create a chart.

w	$f(w)$
0	$0(36 - 0) = 0$
1	$1(36 - 2 \cdot 1) = 34$
2	$2(36 - 2 \cdot 2) = 64$
3	$3(36 - 2 \cdot 3) = 90$
4	$4(36 - 2 \cdot 4) = 112$
5	$5(36 - 2 \cdot 5) = 130$
6	$6(36 - 2 \cdot 6) = 144$
7	$7(36 - 2 \cdot 7) = 154$
8	$8(36 - 2 \cdot 8) = 160$
9	$9(36 - 2 \cdot 9) = 162$

w	$f(w)$
10	$10(36 - 2 \cdot 10) = 160$
11	$11(36 - 2 \cdot 11) = 154$
12	$12(36 - 2 \cdot 12) = 144$
13	$13(36 - 2 \cdot 13) = 130$
14	$14(36 - 2 \cdot 14) = 112$
15	$15(36 - 2 \cdot 15) = 90$
16	$16(36 - 2 \cdot 16) = 64$
17	$17(36 - 2 \cdot 17) = 34$
18	$18(36 - 2 \cdot 18) = 0$

The graph looks like this:

The graph can also be created on a graphing calculator.

The shape of the graph is a parabola. The vertex of this parabola is the highest point, which is (9, 162). The meaning of this vertex is that if he makes the width 9 feet, the area of the garden will be 162 square feet, which is the greatest possible area with the fencing he has.

37. Mrs. Bee's age now is represented by b. Her son's age now is represented by s.

The first equation is $b = 4s + 6$.

Three years ago, Mrs. Bee's age was $b - 3$. Her son's age, three years ago, was $s - 3$.

The second equation is $b - 3 = 7(s - 3)$.

The system of equations is:
$$b = 4s + 6$$
$$b - 3 = 7(s - 3)$$

Using substitution, replace each b in the second equation with the expression $4s + 6$.

$$4s + 6 - 3 = 7(s - 3)$$
$$4s + 3 = 7(s - 3)$$
$$4s + 3 = 7s - 21$$
$$-4s = -4s$$
$$3 = 3s - 21$$
$$+21 = +21$$
$$\frac{24}{3} = \frac{3s}{3}$$
$$8 = s$$
$$b = 4 \cdot 8 + 6 = 38$$

The son is now 8 years old and Mrs. Bee is now 38 years old.

In x years, Mrs. Bee will be three times as old as her son will be then. In x years, Mrs. Bee will be $38 + x$ years old. In x years, her son will be $8 + x$ years old.

This becomes the equation $38 + x = 3(8 + x)$.

Solve for x:
$$38 + x = 3(8 + x)$$
$$38 + x = 24 + 3x$$
$$-x = -x$$
$$38 = 24 + 2x$$
$$-24 = -24$$
$$\frac{14}{2} = \frac{2x}{2}$$
$$7 = x$$

In 7 years, Mrs. Bee will be $38 + 7 = 45$ years old. Her son will be $8 + 7 = 15$ years old. 45 is three times 15.

SELF-ANALYSIS CHART August 2018

Topic	Question Numbers	Number of Points
1. Polynomials	4, 13, 25, 15, 29	$2+2+2+2+2=10$
2. Properties of Algebra		0
3. Functions	2, 5, 6, 11, 19, 26, 27	$2+2+2+2+2+2+2=14$
4. Creating and Interpreting Equations	17, 23, 24	$2+2+4=8$
5. Inequalities	10, 33	$2+4=6$
6. Sequences and Series	20, 32	$2+2=4$
7. Systems of Equations	9, 22, 37	$2+2+6=10$
8. Quadratic Equations and Factoring	3, 7, 16, 30	$2+2+2+2=8$
9. Regression		0
10. Exponential Equations	1	2
11. Graphing	8, 15, 18, 35, 36	$2+2+2+4+4=14$
12. Statistics	14, 21, 31	$2+2+2=6$
13. Number Properties	28	2
14. Unit Conversions	12	2

HOW TO CONVERT YOUR RAW SCORE TO YOUR ALGEBRA I REGENTS EXAMINATION SCORE

The accompanying conversion chart must be used to determine your final score on the August 2018 Regents Examination in Algebra I. To find your final exam score, locate in the column labeled "Raw Score" the total number of points you scored out of a possible 86 points. Since partial credit is allowed in Parts II, III, and IV of the test, you may need to approximate the credit you would receive for a solution that is not completely correct. Then locate in the adjacent column to the right the scale score that corresponds to your raw score. The scale score is your final Algebra I Regents Examination score.

Regents Examination in Algebra I—August 2018
Chart for Converting Total Test Raw Scores to Final Examination Scores (Scaled Scores)

Raw Score	Scale Score	Performance Level	Raw Score	Scale Score	Performance Level	Raw Score	Scale Score	Performance Level
86	100	5	57	81	4	28	66	3
85	98	5	56	81	4	27	65	3
84	97	5	55	81	4	26	64	2
83	96	5	54	81	4	25	63	2
82	94	5	53	80	4	24	61	2
81	93	5	52	80	4	23	60	2
80	92	5	51	80	4	22	58	2
79	92	5	50	80	4	21	57	2
78	91	5	49	79	3	20	55	2
77	90	5	48	79	3	19	53	1
76	89	5	47	79	3	18	51	1
75	88	5	46	78	3	17	49	1
74	88	5	45	78	3	16	47	1
73	87	5	44	78	3	15	45	1
72	87	5	43	77	3	14	43	1
71	86	5	42	77	3	13	40	1
70	86	5	41	76	3	12	38	1
69	86	5	40	76	3	11	36	1
68	85	5	39	75	3	10	33	1
67	84	4	38	75	3	9	30	1
66	84	4	37	74	3	8	27	1
65	84	4	36	74	3	7	24	1
64	83	4	35	73	3	6	21	1
63	83	4	34	72	3	5	18	1
62	83	4	33	71	3	4	15	1
61	82	4	32	71	3	3	11	1
60	82	4	31	70	3	2	8	1
59	82	4	30	69	3	1	4	1
58	82	4	29	68	3	0	0	1

Examination June 2019
Algebra I

HIGH SCHOOL MATH REFERENCE SHEET

Conversions

1 inch = 2.54 centimeters
1 meter = 39.37 inches
1 mile = 5280 feet
1 mile = 1760 yards
1 mile = 1.609 kilometers

1 kilometer = 0.62 mile
1 pound = 16 ounces
1 pound = 0.454 kilogram
1 kilogram = 2.2 pounds
1 ton = 2000 pounds

1 cup = 8 fluid ounces
1 pint = 2 cups
1 quart = 2 pints
1 gallon = 4 quarts
1 gallon = 3.785 liters
1 liter = 0.264 gallon
1 liter = 1000 cubic centimeters

Formulas

Triangle	$A = \frac{1}{2}bh$
Parallelogram	$A = bh$
Circle	$A = \pi r^2$
Circle	$C = \pi d$ or $C = 2\pi r$

Formulas (continued)

General Prisms	$V = Bh$
Cylinder	$V = \pi r^2 h$
Sphere	$V = \frac{4}{3}\pi r^3$
Cone	$V = \frac{1}{3}\pi r^2 h$
Pyramid	$V = \frac{1}{3}Bh$
Pythagorean Theorem	$a^2 + b^2 = c^2$
Quadratic Formula	$x = \dfrac{-b \pm \sqrt{b^2 - 4ac}}{2a}$
Arithmetic Sequence	$a_n = a_1 + (n-1)d$
Geometric Sequence	$a_n = a_1 r^{n-1}$
Geometric Series	$S_n = \dfrac{a_1 - a_1 r^n}{1 - r}$ where $r \neq 1$
Radians	1 radian = $\dfrac{180}{\pi}$ degrees
Degrees	1 degree = $\dfrac{\pi}{180}$ radians
Exponential Growth/Decay	$A = A_0 e^{k(t-t_0)} + B_0$

PART I

Answer all 24 questions in this part. Each correct answer will receive 2 credits. No partial credit will be allowed. For each statement or question, write in the space provided the numeral preceding the word or expression that best completes the statement or answers the question. [48 credits]

1 The expression $w^4 - 36$ is equivalent to
 (1) $(w^2 - 18)(w^2 - 18)$
 (2) $(w^2 + 18)(w^2 - 18)$
 (3) $(w^2 - 6)(w^2 - 6)$
 (4) $(w^2 + 6)(w^2 - 6)$

1 _____

2 If $f(x) = 4x + 5$, what is the value of $f(-3)$?
 (1) -2
 (2) -7
 (3) 17
 (4) 4

2 _____

3 Which relation is *not* a function?

x	y
-10	-2
-6	2
-2	6
1	9
5	13

(1)

(3)

$3x + 2y = 4$

(2)

(4)

3 _____

4 Given: $f(x) = (x - 2)^2 + 4$
   $g(x) = (x - 5)^2 + 4$

   When compared to the graph of $f(x)$, the graph of $g(x)$ is

   (1) shifted 3 units to the left
   (2) shifted 3 units to the right
   (3) shifted 5 units to the left
   (4) shifted 5 units to the right

5 Students were asked to write $6x^5 + 8x - 3x^3 + 7x^7$ in standard form. Shown below are four student responses.

   Anne: $7x^7 + 6x^5 - 3x^3 + 8x$
   Bob: $-3x^3 + 6x^5 + 7x^7 + 8x$
   Carrie: $8x + 7x^7 + 6x^5 - 3x^3$
   Dylan: $8x - 3x^3 + 6x^5 + 7x^7$

   Which student is correct?

   (1) Anne   (3) Carrie
   (2) Bob   (4) Dylan

6 The function f is shown in the table below.

x	f(x)
0	1
1	3
2	9
3	27

   Which type of function best models the given data?

   (1) exponential growth function
   (2) exponential decay function
   (3) linear function with positive rate of change
   (4) linear function with negative rate of change

7 Which expression results in a rational number?

   (1) $\sqrt{2} \cdot \sqrt{18}$ (3) $\sqrt{2} + \sqrt{2}$
   (2) $5 \cdot \sqrt{5}$ (4) $3\sqrt{2} + 2\sqrt{3}$

8 A polynomial function is graphed below.

Which function could represent this graph?

(1) $f(x) = (x + 1)(x^2 + 2)$
(2) $f(x) = (x - 1)(x^2 - 2)$
(3) $f(x) = (x - 1)(x^2 - 4)$
(4) $f(x) = (x + 1)(x^2 + 4)$

8 _____

9 When solving $p^2 + 5 = 8p - 7$, Kate wrote $p^2 + 12 = 8p$. The property she used is

(1) the associative property
(2) the commutative property
(3) the distributive property
(4) the addition property of equality

9 _____

10 David wanted to go on an amusement park ride. A sign posted at the entrance read "You must be greater than 42 inches tall and no more than 57 inches tall for this ride." Which inequality would model the height, x, required for this amusement park ride?

(1) $42 < x \leq 57$
(2) $42 > x \geq 57$
(3) $42 < x$ or $x \leq 57$
(4) $42 > x$ or $x \geq 57$

11 Which situation can be modeled by a linear function?
(1) The population of bacteria triples every day.
(2) The value of a cell phone depreciates at a rate of 3.5% each year.
(3) An amusement park allows 50 people to enter every 30 minutes.
(4) A baseball tournament eliminates half of the teams after each round.

12 Jenna took a survey of her senior class to see whether they preferred pizza or burgers. The results are summarized in the table below.

	Pizza	Burgers
Male	23	42
Female	31	26

Of the people who preferred burgers, approximately what percentage were female?

(1) 21.3
(2) 38.2
(3) 45.6
(4) 61.9

13 When $3a + 7b > 2a - 8b$ is solved for a, the result is
(1) $a > -b$
(2) $a < -b$
(3) $a < -15b$
(4) $a > -15b$

14 Three functions are shown below.

A: $g(x) = -\frac{3}{2}x + 4$

B: $f(x) = (x+2)(x+6)$

C:

Which statement is true?

(1) B and C have the same zeros.
(2) A and B have the same y-intercept.
(3) B has a minimum and C has a maximum.
(4) C has a maximum and A has a minimum.

14 ____

15 Nicci's sister is 7 years less than twice Nicci's age, a. The sum of Nicci's age and her sister's age is 41. Which equation represents this relationship?

(1) $a + (7 - 2a) = 41$ (3) $2a - 7 = 41$
(2) $a + (2a - 7) = 41$ (4) $a = 2a - 7$

15 ____

16 The population of a small town over four years is recorded in the chart below, where 2013 is represented by $x = 0$. [Population is rounded to the nearest person]

Year	2013	2014	2015	2016
Population	3810	3943	4081	4224

The population, $P(x)$, for these years can be modeled by the function $P(x) = ab^x$, where b is rounded to the nearest thousandth. Which statements about this function are true?

    I. $a = 3810$
    II. $a = 4224$
    III. $b = 0.035$
    IV. $b = 1.035$

(1) I and III   (3) II and III
(2) I and IV   (4) II and IV   16 _____

17 When written in factored form, $4w^2 - 11w - 3$ is equivalent to

(1) $(2w + 1)(2w - 3)$ (3) $(4w + 1)(w - 3)$
(2) $(2w - 1)(2w + 3)$ (4) $(4w - 1)(w + 3)$ 17 _____

18 Which ordered pair does *not* represent a point on the graph of $y = 3x^2 - x + 7$?

(1) $(-1.5, 15.25)$ (3) $(1.25, 10.25)$
(2) $(0.5, 7.25)$ (4) $(2.5, 23.25)$ 18 _____

19 Given the following three sequences:

    I.   2, 4, 6, 8, 10...
    II. 2, 4, 8, 16, 32...
    III. $a, a + 2, a + 4, a + 6, a + 8...$

Which ones are arithmetic sequences?

(1) I and II, only   (3) II and III, only
(2) I and III, only   (4) I, II, and III   19 _____

20 A grocery store sells packages of beef. The function $C(w)$ represents the cost, in dollars, of a package of beef weighing w pounds. The most appropriate domain for this function would be

(1) integers
(2) rational numbers
(3) positive integers
(4) positive rational numbers

21 The roots of $x^2 - 5x - 4 = 0$ are

(1) 1 and 4
(2) $\dfrac{5 \pm \sqrt{41}}{2}$
(3) -1 and -4
(4) $\dfrac{-5 \pm \sqrt{41}}{2}$

22 The following table shows the heights, in inches, of the players on the opening-night roster of the 2015–2016 New York Knicks.

| 84 | 80 | 87 | 75 | 77 | 79 | 80 | 74 | 76 | 80 | 80 | 82 | 82 |

The population standard deviation of these data is approximately

(1) 3.5
(2) 13
(3) 79.7
(4) 80

23 A population of bacteria can be modeled by the function $f(t) = 1000(0.98)^t$, where t represents the time since the population started decaying, and $f(t)$ represents the population of the remaining bacteria at time t. What is the rate of decay for this population?

(1) 98%
(2) 2%
(3) 0.98%
(4) 0.02%

24 Bamboo plants can grow 91 centimeters per day. What is the approximate growth of the plant, in inches per hour?

(1) 1.49
(2) 3.79
(3) 9.63
(4) 35.83

EXAMINATION June 2019

PART II

Answer all 8 questions in this part. Each correct answer will receive 2 credits. Clearly indicate the necessary steps, including appropriate formula substitutions, diagrams, graphs, charts, etc. For all questions in this part, a correct numerical answer with no work shown will receive only 1 credit. [16 credits]

25 Solve algebraically for x: $-\frac{2}{3}(x+12) + \frac{2}{3}x = -\frac{5}{4}x + 2$

x = 8 correct answer not it

$\left(-\frac{2}{3}x\right) - 8 + \left(\frac{2}{3}x\right) = -\frac{5}{4}x + 2$

$-8 \neq -\frac{5}{4}x + 2$
$ -2 -2$

$\dfrac{-10}{-\frac{5}{4}x} \neq \dfrac{-\frac{5}{4}x}{-\frac{5}{4}}$

$x = \frac{1}{2}$

26 If $C = G - 3F$, find the trinomial that represents C when $F = 2x^2 + 6x - 5$ and $G = 3x^2 + 4$.

$-3x^2 - 18x + 19$

27 Graph the following piecewise function on the set of axes below.

$$f(x) = \begin{cases} |x|, & -5 \leq x < 2 \\ -2x + 10, & 2 \leq x \leq 6 \end{cases}$$

28 Solve $5x^2 = 180$ algebraically.

$[-6, 6]$

29 A blizzard occurred on the East Coast during January, 2016. Snowfall totals from the storm were recorded for Washington, D.C. and are shown in the table below.

Washington, D.C.	
Time	Snow (inches)
1 a.m.	1
3 a.m.	5
6 a.m.	11
12 noon	33
3 p.m.	36

Which interval, 1 a.m. to 12 noon or 6 a.m. to 3 p.m., has the greatest rate of snowfall, in inches per hour? Justify your answer.

30 The formula for the volume of a cone is $V = \frac{1}{3}\pi r^2 h$. Solve the equation for h in terms of V, r, and π.

$h = \dfrac{3V}{\pi r^2}$

31 Given the recursive formula:

$$a_1 = 3$$
$$a_n = 2(a_{n-1} + 1)$$

State the values of a_2, a_3, and a_4 for the given recursive formula.

~~$a_2 = 6$~~

~~$a_3 = 12$~~

~~$a_4 = 24$~~

$a_2 = 8$

$a_3 = 18$

$a_4 = 38$

32 Determine and state the vertex of $f(x) = x^2 - 2x - 8$ using the method of completing the square.

$(1, -9)$

PART III

Answer all 4 questions in this part. Each correct answer will receive 4 credits. Clearly indicate the necessary steps, including appropriate formula substitutions, diagrams, graphs, charts, etc. For all questions in this part, a correct numerical answer with no work shown will receive only 1 credit. [16 credits]

33 A school plans to have a fundraiser before basketball games selling shirts with their school logo. The school contacted two companies to find out how much it would cost to have the shirts made. Company A charges a $50 set-up fee and $5 per shirt. Company B charges a $25 set-up fee and $6 per shirt.

Write an equation for Company A that could be used to determine the total cost, A, when x shirts are ordered. Write a second equation for Company B that could be used to determine the total cost, B, when x shirts are ordered.

A = 5x + 50

B = 6x + 25

Determine algebraically and state the *minimum* number of shirts that must be ordered for it to be cheaper to use Company A.

x = 26

EXAMINATION June 2019

34 Graph $y = f(x)$ and $y = g(x)$ on the set of axes below.

$$f(x) = 2x^2 - 8x + 3$$
$$g(x) = -2x + 3$$

Determine and state all values of x for which $f(x) = g(x)$.

$x = 0$

$x = 3$

35 The table below shows the number of hours ten students spent studying for a test and their scores.

Hours Spent Studying (x)	0	1	2	4	4	4	6	6	7	8
Test Scores (y)	35	40	46	65	67	70	82	88	82	95

Write the linear regression equation for this data set. Round all values to the *nearest hundredth*.

$$y = 7.79x + 34.27$$

State the correlation coefficient of this line, to the *nearest hundredth*.

$$r = .98$$

Explain what the correlation coefficient suggests in the context of the problem.

The equation is positive

36 A system of inequalities is graphed on the set of axes below.

State the system of inequalities represented by the graph.

$y \le 2x - 2$; $y \le -3x + 3$

State what region A represents.

A represents the solution set of the system

State what the entire gray region represents.

The gray region represents $y \le 2x - 2$

PART IV

Answer the question in this part. A correct answer will receive 6 credits. Clearly indicate the necessary steps, including appropriate formula substitutions, diagrams, graphs, charts, etc. A correct numerical answer with no work shown will receive only 1 credit. [6 credits]

37 When visiting friends in a state that has no sales tax, two families went to a fast-food restaurant for lunch. The Browns bought 4 cheeseburgers and 3 medium fries for $16.53. The Greens bought 5 cheeseburgers and 4 medium fries for $21.11.

Using *c* for the cost of a cheeseburger and *f* for the cost of medium fries, write a system of equations that models this situation.

$$4c + 3f = 16.53$$
$$5c + 4f = 21.11$$

The Greens said that since their bill was $21.11, each cheeseburger must cost $2.49 and each order of medium fries must cost $2.87 each. Are they correct? Justify your answer.

They are not correct because it would be $23.93 not $21.11

Using your equations, algebraically determine both the cost of one cheeseburger and the cost of one order of medium fries.

one cheseburger is $2.79 and one order of fries would be $1.79

Answers June 2019
Algebra I

Answer Key

PART I

1. (4)
2. (2)
3. (4)
4. (2)
5. (1)
6. (1)
7. (1)
8. (3)
9. (4)
10. (1)
11. (3)
12. (2)
13. (4)
14. (3)
15. (2)
16. (2)
17. (3)
18. (3)
19. (2)
20. (4)
21. (2)
22. (1)
23. (2)
24. (1)

PART II
25. $x = 8$
26. $-3x^2 - 18x + 19$
27.

28. $\{-6, 6\}$
29. 1 a.m. to 12 noon because $2\frac{10}{11} > 2\frac{7}{9}$
30. $h = \dfrac{3V}{\pi r^2}$
31. 8, 18, 38
32. $(1, -9)$

PART III
33. $A = 5x + 50$, $B = 6x + 25$; $x = 26$
34.

$x = 0, x = 3$

35. $y = 7.79x + 34.27$, $r = .98$
36. $y \leq 2x - 2$, $y < -3x + 3$; A represents the solution set of the system of inequalities. The gray region is the solution set to $y \leq 2x - 2$.

PART IV
37. $4c + 3f = 16.53$, $5c + 4f = 21.11$. They are not correct because it would be $23.93, not $21.11; one cheeseburger is $2.79, and one order of fries is $1.79.

In **Parts II–IV**, you are required to show how you arrived at your answers. For sample methods of solutions, see the *Answers Explained* section.

Answers Explained

PART I

1. $w^4 - 36$ can be rewritten as $(w^2)^2 - 6^2$. This can be factored with the factoring pattern $a^2 - b^2 = (a-b)(a+b)$.
$$w^4 - 36 = (w^2)^2 - 6^2 = (w^2 - 6)(w^2 + 6)$$
The correct choice is (**4**).

2. $$f(-3) = 4(-3) + 5 = -12 + 5 = -7$$
The correct choice is (**2**).

3. Testing choice (1): When a relation is described in a table, it is only a function if there are no repeats in the x column. Since there are no repeats in this x column, this relation is a function.

   Testing choice (2): When the relation is given as an equation, the simplest thing to do is to make a graph of that equation and see if it passes the vertical line test. The graph of a function will pass the "vertical line test," which means that there are no two points on the graph that have the same x-coordinate. If it is possible to draw a vertical line anywhere on the graph of a relation that will pass through more than one point, the relation will not be a function. The graph of the relation $3x + 2y = 4$ looks like this:

   Since it passes the vertical line test, it is the graph of a function.

   Testing choice (3): On this graph, there is no place where a vertical line will pass through more than one point, so this is the graph of a function.

Testing choice (4): When the relation is described with a mapping diagram, it is only a function if there is no place where the same element in the domain (the left oval) is being mapped to two different elements in the range (the right oval). Since in this mapping diagram the element 2 in the domain is mapped to both A and B, this relation is not a function.

The correct choice is (**4**).

4. Graph both of the functions on the same set of axes. The graph for $g(x)$ is 3 units to the right of the graph of $f(x)$.

For the TI-84:

For the TI-Nspire:

This question can also be done without a graphing calculator. An equation of the form $y = (x - h)^2 + k$ is called the *vertex form* of a parabola. The vertex will be (h, k).

The graph of function f will be a parabola with the vertex (2, 4). The graph of the function g will be a parabola with the vertex (5, 4). So, the graph of function g will be 3 units to the right of the graph of function f.

The correct choice is (**2**).

5. When a polynomial is written in standard form, the term with the largest exponent will be first and each term after that will have a smaller exponent than the previous term. The coefficients before the variables are not relevant to this problem.

For Anne, the sequence of exponents is 7, 5, 3, 1, so the exponents are in decreasing order.

For the other people, the exponents are not in decreasing order.
Bob: 3, 5, 7, 1
Carrie: 1, 7, 5, 3
Dylan: 1, 3, 5, 7

The correct choice is (**1**).

6. Plotting these four points on a graph looks like this:

Here is what a graph looks like for each of the four choices:

Choice (1) exponential growth	Choice (2) exponential decay	Choice (3) linear with positive rate of change	Choice (4) linear with negative rate of change

Of the four choices, the one that the four plotted points most resembles is choice (1).

The correct choice is (**1**).

7. A rule of radicals is that $\sqrt{a} \cdot \sqrt{b} = \sqrt{ab}$. When this rule is applied to choice (1), it becomes $\sqrt{2} \cdot \sqrt{18} = \sqrt{2 \cdot 18} = \sqrt{36} = 6$, which is a rational number since it is a terminating decimal number. When an irrational number is expressed as a decimal, it will be nonterminating and nonrepeating.

The correct choice is (**1**).

8. The x-intercepts of this graph are $(-2, 0)$, $(1, 0)$, and $(2, 0)$. If $(a, 0)$ is an x-intercept of the graph of a polynomial function, one of its factors will be $(x - a)$. So, for this function, $f(x) = (x - (-2))(x - 1)(x - 2) = (x + 2)(x - 1)(x - 2)$. Though this is not one of the choices, if you multiply the first and third factors together, $(x + 2)(x - 2) = x^2 - 4$. So, the function can also be written as $f(x) = (x - 1)(x^2 - 4)$.

Another way you could approach this question would be to graph the answer choices to see which one matches the given graph. When choice (3) is graphed on the graphing calculator, it looks like this:

For the TI-84:

For the TI-Nspire:

[Graphing calculator screen showing $f1(x)=(x-1)\cdot(x^2-4)$]

The correct choice is (**3**).

9. She got from $p^2 + 5 = 8p - 7$ to $p^2 + 12 = 8p$ by adding 7 to both sides of the equation.

$$\begin{array}{r} p^2 + 5 = 8p - 7 \\ +7 = +7 \\ \hline p^2 + 12 = 8p \end{array}$$

The addition property of equality states that if you have an equality and you add the same number to both sides of the equal sign, the new equation is also true. This is what was done here.

The correct choice is (**4**).

10. The solution could be written as $x > 42$ and $x \leq 57$ since a person's height must meet both of those conditions. A shorthand for this "and" statement is $42 < x \leq 57$. This is not the same thing as choice (3), $42 < x$ or $x \leq 57$. The "or" statement is true if either of the conditions is met; so, for choice (3), someone under 42 inches would be allowed on the ride because his or her height is less than or equal to 57 inches.

The correct choice is (**1**).

11. Make a sketch of the scenario in each of the answer choices and see which one looks most like a line.

Choice (1) — exponential growth (Population vs Day)

Choice (3) — linear growth (People vs Minutes)

Choice (2) — exponential decay (Value vs Year)

Choice (4) — exponential decay with points (Teams vs Rounds)

The correct choice is (**3**).

12. A total of $42 + 26 = 68$ people preferred burgers. Of those 68 people, 26 of them were female. So, the percentage of people who preferred burgers who were female is $\frac{26}{68} \approx .382 = 38.2\%$.

The correct choice is (**2**).

13. When the same number is added to or subtracted from both sides of an inequality, the inequality remains true. This inequality can be made to look like the answer choices by subtracting both $2a$ and $7b$ from both sides of the inequality.

$$\begin{array}{r} 3a + 7b > 2a - 8b \\ -2a - 7b = -2a - 7b \\ \hline a > -15b \end{array}$$

The correct choice is (**4**).

14. Finding the zeros, y-intercept, maximum, and minimum of a function is easiest to do with a graph. The zeros are the x-coordinates of the x-intercepts and the y-intercept is where the graph crosses the y-axis. Functions whose graphs are lines don't have a maximum or a minimum, but functions whose graphs are parabolas will either have a maximum or a minimum depending on whether the parabola looks more like an "n" or a "u."

The graph of the function for A looks like this:

It has one zero at the x-intercept at $x = \frac{8}{3}$. The y-intercept is at (0, 4). Because it is a line, it does not have a maximum or a minimum point.

The graph of the function for B looks like this:

It has two zeros at the x-intercepts at $x = -2$ and $x = -6$. The y-intercept is at (0, 12). Because it is a parabola shaped like a "u," it has a minimum point.

The graph of the function for C was given.

It has two zeros at the x-intercepts at $x = -2$ and $x = -4$. The y-intercept is at (0, −8). Because it is a parabola shaped like an "n," it has a maximum point.

B has a minimum and C has a maximum.

The correct choice is (**3**).

15. If Nicci's age is a, her sister's age is $2a - 7$ (<u>not</u> $7 - 2a$). The sum of their ages is $a + (2a - 7)$. Since the sum of their ages is 41, the equation is $a + (2a - 7) = 41$.

    The correct choice is (**2**).

16. Since the population is increasing each year, this is an example of exponential growth. For an equation of the form $y = ab^x$, exponential growth occurs when $b > 1$. This means that Roman numeral IV must be part of the solution.

    Since in 2013 the population is 3810, $P(0) = 1310$. This fact can be used to solve for a.

$P(0) = ab^0 = 3810$
$P(0) = a \cdot 1 = 3810$
$a = 3810$

This means that Roman numeral I must be part of the solution.

In general, in an equation of the form $y = ab^x$, the y-intercept will be a value of a.

The correct choice is (**2**).

17. The simplest way to solve this is to multiply all the answer choices to see which becomes $4w^2 - 11w - 3$.

    Testing choice (1): $(2w + 1)(2w - 3) = 4w^2 - 6w + 2w - 3 = 4w^2 - 5w - 3$
    Testing choice (2): $(2w - 1)(2w + 3) = 4w^2 + 6w - 2w - 3 = 4w^2 + 5w - 3$
    Testing choice (3): $(4w + 1)(w - 3) = 4w^2 - 12w + 1w - 3 = 4w^2 - 11w - 3$
    Testing choice (4): $(4w - 1)(w + 3) = 4w^2 + 12w - 1w - 3 = 4w^2 + 11w - 3$

    The correct choice is (**3**).

18. For each answer choice, replace the x in the equation with the x-coordinate of the ordered pair and y with the y-coordinate of the ordered pair, and see which one does not result in the left side of the equation equaling the right side of the equation.

    Testing choice (1):
    $15.25 \stackrel{?}{=} 3(-1.5)^2 - (-1.5) + 7$
    $15.25 \stackrel{?}{=} 3 \cdot 2.25 + 1.5 + 7$
    $15.25 \stackrel{\checkmark}{=} 15.25$

    Testing choice (2):
    $7.25 \stackrel{?}{=} 3 \cdot 5^2 - 0.5 + 7$
    $7.25 \stackrel{?}{=} 3 \cdot 25 - 0.5 + 7$
    $7.25 \stackrel{\checkmark}{=} 7.25$

    Testing choice (3):
    $10.25 \stackrel{?}{=} 3 \cdot 1.25^2 - 1.25 + 7$
    $10.25 \stackrel{?}{=} 3 \cdot 1.5625 - 1.25 + 7$
    $10.25 \neq 10.4375$

    Testing choice (4):
    $23.25 \stackrel{?}{=} 3 \cdot 2.5^2 - 2.5 + 7$
    $23.25 \stackrel{?}{=} 3 \cdot 6.25 - 2.5 + 7$
    $23.25 \stackrel{\checkmark}{=} 23.25$

    The correct choice is (**3**).

19. An arithmetic sequence is one in which each term is equal to the previous term plus or minus some constant.

Testing Roman numeral I: Since $4 = 2 + 2$, $6 = 4 + 2$, $8 = 6 + 2$, etc., this is an arithmetic series with a first term of 2 and a common difference of 2.

Testing Roman numeral II: Since $4 = 2 + 2$, $8 = 4 \cdot 2$, $16 = 8 \cdot 2$, etc., this is <u>not</u> an arithmetic series. It is a geometric series.

Testing Roman numeral III: Since $a + 2 = a + 2$, $a + 4 = a + 2 + 2$, $a + 6 = a + 4 + 2$, etc., this is an arithmetic series with a first term of a and a common difference of 2.

The correct choice is (**2**).

20. Choices (1) and (2) include negative numbers. Since you cannot purchase a negative amount of beef, you can eliminate these choices. Since beef does not have to be purchased in one-pound units (you can buy, for example, 1.7 pounds of beef), choice (3) can be eliminated also.

The correct choice is (**4**).

21. Using the quadratic formula with $a = 1$, $b = -5$, and $c = -4$,

$$x = \frac{-b \pm \sqrt{b^2 - 4ac}}{2a} = \frac{-(-5) \pm \sqrt{(-5)^2 - 4(1)(-4)}}{2(1)}$$

$$= \frac{5 \pm \sqrt{25 + 16}}{2} = \frac{5 \pm \sqrt{41}}{2}$$

The correct choice is (**2**).

22. Population standard deviation can be calculated with a graphing calculator.

For the TI-84:

Press [STAT] [1]. Clear all lists and enter the thirteen numbers into L1.

Press [STAT], [Right Arrow], [1] and select Calculate.

```
1-Var Stats
x̄=79.69230769
Σx=1036
Σx²=82720
Sx=3.637412436
σx=3.494712828
↓n=13
```

For the TI-NSpire:

From the home screen, select the Add Lists & Spreadsheet icon. Name column A x and fill in cells A1 to A13 with the thirteen numbers.

A x	B	C	D
1 84			
2 80			
3 87			
4 75			
5 77			

A5 77

Press [menu] [4] [1] [1] for One-Variable Statistics.

Select OK twice.

The population standard deviation is the one denoted with the symbol σx. Rounded to the nearest tenth, it is 3.5.

The correct choice is (**1**).

23. In an exponential equation of the form $y = a(1-r)^x$, r is the decay rate. For this question $1 - r = 0.98$, so $r = 0.02$.

The correct choice is (**2**).

24. From the reference sheet provided on the Regents, 1 inch = 2.54 centimeters. To convert 91 centimeters into inches, divide 91 by 2.54 to get approximately 35.8 inches. Since 1 day equals 24 hours, the growth rate in inches per hour can be calculated by dividing the total growth by the total number of hours.

$$\frac{35.8}{24} \approx 1.49 \text{ inches per hour}$$

The correct choice is (**1**).

PART II

25.
$$-\frac{2}{3}(x+12) + \frac{2}{3}x = -\frac{5}{4}x + 2$$
$$-\frac{2}{3}x - 8 + \frac{2}{3}x = -\frac{5}{4}x + 2$$
$$-8 = -\frac{5}{4}x + 2$$
$$-2 = -2$$
$$-10 = -\frac{5}{4}x$$
$$-\frac{4}{5}(-10) = -\frac{4}{5}\left(-\frac{5}{4}x\right)$$
$$8 = x$$

26. $C = G - 3F = 3x^2 + 4 - 3(2x^2 + 6x - 5) = 3x^2 + 4 - 6x^2 - 18x + 15$
    $= -3x^2 - 18x + 19$

27. A piecewise function is one that uses more than one rule, depending on what the input value to the function is. Make a table of values from $x = -5$ to $x = 6$. For the first seven values, use the top part of the piecewise function. For the last five values, use the bottom part of the piecewise function.

x	$f(x)$
-5	$|-5| = 5$
-4	$|-4| = 4$
-3	$|-3| = 3$
-2	$|-2| = 2$
-1	$|-1| = 1$
0	$|0| = 0$
1	$|1| = 1$
2	$-2(2) + 10 = -4 + 10 = 6$
3	$-2(3) + 10 = -6 + 10 = 4$
4	$-2(4) + 10 = -8 + 10 = 2$
5	$-2(5) + 10 = -10 + 10 = 0$
6	$-2(6) + 10 = -12 + 10 = -2$

Plot these twelve points.

For $x = 2$, the bottom part of the piecewise function was used. For graphing, it is also useful to substitute $x = 2$ into the top part of the piecewise function because it is the boundary between the two pieces. When $x = 2$ is substituted into the top part of the piecewise function, the expression evaluates to 2. Plot the point (2, 2) as an open circle since it is not an actual point on the graph. Then, connect the dots like this:

28.
$$\frac{5x^2}{5} = \frac{180}{5}$$
$$x^2 = 36$$
$$\sqrt{x^2} = \pm\sqrt{36}$$
$$x = \pm 6$$
$$\{-6, 6\}$$

29. To calculate the rate of snowfall for an interval of time, divide the amount of snow that fell in that interval by the amount of time that elapsed in that interval.

For the 1 a.m. to 12 noon interval, the rate of snowfall is $\frac{33-1}{12-1} = \frac{32}{11} = 2\frac{10}{11}$ inches per hour.

For the 6 a.m. to 3 p.m. interval, the rate of snowfall is $\frac{36-11}{15-6} = \frac{25}{9} = 2\frac{7}{9}$ inches per hour.

Since $2\frac{10}{11} > 2\frac{7}{9}$, the 1 a.m. to 12 noon rate of snowfall is greater than the 6 a.m. to 3 p.m. rate of snowfall.

30.
$$V = \frac{1}{3}\pi r^2 h$$
$$3V = 3 \cdot \frac{1}{3}\pi r^2 h$$
$$3V = \pi r^2 h$$
$$\frac{3V}{\pi r^2} = \frac{\pi r^2 h}{\pi r^2}$$
$$\frac{3V}{\pi r^2} = h$$

31. A recursive formula describes how to get the next term of a sequence based on the previous terms. Substituting $n = 2, 3$, and 4 into the recursive part of the equation, it becomes

$$a_2 = 2(a_{2-1} + 1) = 2(a_1 + 1) = 2(3 + 1) = 2 \cdot 4 = 8$$
$$a_3 = 2(a_{3-1} + 1) = 2(a_2 + 1) = 2(8 + 1) = 2 \cdot 9 = 18$$
$$a_4 = 2(a_{4-1} + 1) = 2(a_3 + 1) = 2(18 + 1) = 2 \cdot 19 = 38$$

32. The vertex form of a quadratic equation is $y = a(x - h)^2 + k$, where the vertex is (h, k). Completing the square is when you create a quadratic trinomial where the constant term is the square of half the coefficient on the x.

In the function $f(x) = x^2 - 2x - 8$, the coefficient on the x is -2. Calculate the square of half of this coefficient $\left(\dfrac{-2}{2}\right)^2 = (-1)^2 = +1$.

To complete the square, rewrite the -8 as $+1 - 9$. The function becomes $f(x) = x^2 - 2x + 1 - 9$. The first three terms are now a perfect square trinomial and can be factored.

$$f(x) = x^2 - 2x + 1 - 9 = (x - 1)^2 - 9$$

This is now in vertex form and the vertex is $(1, -9)$.

PART III

33. For company A, x shirts cost $5x plus the $50 set-up fee. The equation is $A = 5x + 50$. For company B, x shirts cost $6x plus the $25 set-up fee. The equation is $B = 6x + 25$.

For some values of x, A is more expensive and, for others, B is more expensive. To find the minimum number of shirts for it to be cheaper to use Company A, find the smallest integer that makes the inequality $5x + 50 < 6x + 25$ true.

Solving the inequality becomes

$$5x + 50 < 6x + 25$$
$$-6x = -6x$$
$$-x + 50 < 25$$
$$-50 = -50$$
$$-x < -25$$
$$\frac{-x}{-1} > \frac{-25}{-1}$$
$$x > 25$$

The smallest integer that makes this true is 26. 25 is not the solution since it is not cheaper to buy 25 shirts from company A; it is the same price as company B.

34. The graph of function f is a parabola, and the graph of function g is a line. An important point to graph for a parabola is the vertex. In general, the x-coordinate of the vertex of a parabola is $x = -\frac{b}{2a} = -\frac{-8}{2 \cdot 2} = 2$. Choose at least three values smaller than 2 and three values greater than 2 for the x values in your chart.

x	$f(x)$	$g(x)$
-1	$2(-1)^2 - 8(-1) + 3 = 2 + 8 + 3 = 13$	$-2(-1) + 3 = 2 + 3 = 5$
0	$2(0)^2 - 8(0) + 3 = 0 + 0 + 3 = 3$	$-2(0) + 3 = 0 + 3 = 3$
1	$2(1)^2 - 8(1) + 3 = 2 - 8 + 3 = -3$	$-2(1) + 3 = -2 + 3 = 1$
2	$2(2)^2 - 8(2) + 3 = 8 - 16 + 3 = -5$	$-2(2) + 3 = -4 + 3 = -1$
3	$2(3)^2 - 8(3) + 3 = 18 - 24 + 3 = -3$	$-2(3) + 3 = -6 + 3 = -3$
4	$2(4)^2 - 8(4) + 3 = 32 - 32 + 3 = 3$	$-2(4) + 3 = -8 + 3 = -5$
5	$2(5)^2 - 8(5) + 3 = 50 - 40 + 3 = 13$	$-2(5) + 3 = -10 + 3 = -7$

The graph looks like this:

According to the chart, $f(0) = g(0) = 3$ and $f(3) = g(3) = -3$, so the two values of x for which $f(x) = g(x)$ are 0 and 3.

It can also be seen on the graph that the x-coordinates of the two points where the line intersects the parabola are 0 and 3.

These functions can also be graphed on a graphing calculator.

For the TI-84:

For the TI-Nspire:

[Calculator screen showing graph of $f1(x) = 2 \cdot x^2 - 8 \cdot x + 3$ and $f2(x) = -2 \cdot x + 3$ with intersection points (0,3) and (3,-3)]

35. A line of best fit for a data set can be calculated with a graphing calculator.

    For the TI-84:

    To turn diagnostics on, press [2nd] [0], scroll down to select DiagnosticOn, and press [ENTER].

[Calculator CATALOG screen showing: DependAuto, det(, DiagnosticOff, ▸DiagnosticOn, dim(, Disp, DispGraph]

Press [STAT] [1] and enter the x values into L1 and the y values into L2.

Press [STAT] [Right Arrow] [4] to get the LinReg function. Select Calculate.

For the TI-NSpire:

From the home screen select the Add Lists & Spreadsheet icon. In the first row, label the A column "hours" and the B column "score." Enter the hours in cells A1 to A9. Enter the scores in cells B1 to B9.

Move the cursor to cell C1. Press [menu] [4] [1] [3]. Set X List to "hours" and Y List to "score," and select OK.

The line of best fit is $y = 7.79x + 34.27$.

This is the line of best fit graphed together with the scatter plot.

For the TI-84:

For the TI-Nspire:

The correlation coefficient (r) is a measure of how close the points are to the line of best fit. If the line of best fit has a positive slope, the correlation coefficient will be positive. If the line of best fit has a negative slope, the correlation coefficient will be negative. The closer the correlation coefficient is to 1 (or to -1), the closer to the line of best fit the points will lie.

For this data set, the correlation coefficient is .98, which is very close to 1, so the data set is very nearly a line.

36. The solid line has a y-intercept of $(0, -2)$ and, since it passes through $(1, 0)$, it has a slope of 2. The equation of the solid line is $y = 2x - 2$. Since the area underneath that line is shaded, the first inequality is $y \leq 2x - 2$.

The dotted line has a y-intercept of $(0, 3)$ and, since it passes through $(1, 0)$, it has a slope of -3. The equation of the dotted line is $y = -3x + 3$. Since the area underneath that line is shaded, the second inequality is $y < -3x + 3$. There is a "<" instead of a "≤."

So, the system of inequalities is

$$y \leq 2x - 2$$
$$y < -3x - 3$$

The region A represents all the ordered pairs that make up the solution set to the system of inequalities. Ordered pairs in this region will make both of the inequalities true.

The shaded gray region represents all the ordered pairs that make up the solution set to the inequality $y \leq 2x - 2$.

PART IV

37. If one cheeseburger costs $c, then 4 cheeseburgers are $4c and 5 cheeseburgers are $5c. If one order of fries is $f, then 3 orders of fries are $3f and 4 orders of fries are $4f.

The equation for the Browns is $4c + 3f = 16.53$. The equation for the Greens is

$$4c + 3f = 16.53$$
$$5c + 4f = 21.11$$

The Greens are not correct. If each cheeseburger was $2.49 and each order of fries was $2.87, then their total bill would be $5 \times 2.49 + 4 \times 2.87 = 23.93 \neq 21.11$.

To solve the system of equations, multiply both sides of the top equation by -4 and both sides of the bottom equation by 3. This makes the coefficients of the f variables opposites; so, when the two equations are added together, the f will be eliminated.

$$-16c - 12f = -66.12$$
$$15c + 12f = 63.33$$

Adding the two equations together becomes

$$-c = -2.79$$
$$c = 2.79$$

The price of one cheeseburger is $2.79.

To find the price of an order of fries, substitute $c = 2.79$ into either of the original equations and solve for f.

$$4(2.79) + 3f = 16.53$$
$$11.16 + 3f = 16.53$$
$$-11.16 = -11.16$$
$$\frac{3f}{3} = \frac{5.37}{3}$$
$$f = 1.79$$

The price of one order of fries is $1.79.

SELF-ANALYSIS CHART June 2019

Topic	Question Numbers	Number of Points	Your Points	Your Percentage
1. Polynomials	5, 8, 26, 32	$2 + 2 + 2 + 2 = 8$		
2. Properties of Algebra	9, 25, 28, 30	$2 + 2 + 2 + 2 = 8$		
3. Functions	2, 3, 20, 29	$2 + 2 + 2 + 2 = 8$		
4. Creating and Interpreting Equations	11, 15, 23	$2 + 2 + 2 = 6$		
5. Inequalities	10, 13, 36	$2 + 2 + 4 = 8$		
6. Sequences and Series	19, 31	$2 + 2 = 4$		
7. Systems of Equations	33, 37	$4 + 6 = 10$		
8. Quadratic Equations and Factoring	1, 17, 21	$2 + 2 + 2 = 6$		
9. Regression	35	4		
10. Exponential Equations	6, 16	$2 + 2 = 4$		
11. Graphing	4, 14, 18, 27, 34	$2 + 2 + 2 + 2 + 4 = 12$		
12. Statistics	12, 22	$2 + 2 = 4$		
13. Number Properties	7	2		
14. Unit Conversions	24	2		

HOW TO CONVERT YOUR RAW SCORE TO YOUR ALGEBRA I REGENTS EXAMINATION SCORE

The accompanying conversion chart must be used to determine your final score on the June 2019 Regents Examination in Algebra I. To find your final exam score, locate in the column labeled "Raw Score" the total number of points you scored out of a possible 86 points. Since partial credit is allowed in Parts II, III, and IV of the test, you may need to approximate the credit you would receive for a solution that is not completely correct. Then locate in the adjacent column to the right the scale score that corresponds to your raw score. The scale score is your final Algebra I Regents Examination score.

SELF-ANALYSIS CHART June 2019

Regents Examination in Algebra I—June 2019
Chart for Converting Total Test Raw Scores to Final Exam Scores (Scale Scores)

Raw Score	Scale Score	Performance Level	Raw Score	Scale Score	Performance Level	Raw Score	Scale Score	Performance Level
86	100	5	57	81	4	28	66	3
85	99	5	56	81	4	27	65	3
84	97	5	55	81	4	26	64	2
83	96	5	54	80	4	25	63	2
82	95	5	53	80	4	24	61	2
81	94	5	52	80	4	23	60	2
80	93	5	51	80	4	22	58	2
79	92	5	50	79	3	21	57	2
78	91	5	49	79	3	20	55	2
77	90	5	48	79	3	19	54	1
76	90	5	47	78	3	18	52	1
75	89	5	46	78	3	17	50	1
74	88	5	45	78	3	16	48	1
73	88	5	44	77	3	15	46	1
72	87	5	43	77	3	14	44	1
71	86	5	42	76	3	13	42	1
70	86	5	41	76	3	12	39	1
69	86	5	40	75	3	11	37	1
68	85	5	39	75	3	10	34	1
67	84	4	38	74	3	9	31	1
66	84	4	37	74	3	8	28	1
65	84	4	36	73	3	7	25	1
64	83	4	35	72	3	6	22	1
63	83	4	34	72	3	5	19	1
62	83	4	33	71	3	4	15	1
61	82	4	32	70	3	3	12	1
60	82	4	31	69	3	2	8	1
59	82	4	30	68	3	1	4	1
58	82	4	29	67	3	0	0	1

Examination
August 2019
Algebra I

HIGH SCHOOL MATH REFERENCE SHEET

Conversions

1 inch = 2.54 centimeters
1 meter = 39.37 inches
1 mile = 5280 feet
1 mile = 1760 yards
1 mile = 1.609 kilometers

1 cup = 8 fluid ounces
1 pint = 2 cups
1 quart = 2 pints
1 gallon = 4 quarts
1 gallon = 3.785 liters
1 liter = 0.264 gallon
1 liter = 1000 cubic centimeters

1 kilometer = 0.62 mile
1 pound = 16 ounces
1 pound = 0.454 kilogram
1 kilogram = 2.2 pounds
1 ton = 2000 pounds

Formulas

Triangle	$A = \frac{1}{2}bh$
Parallelogram	$A = bh$
Circle	$A = \pi r^2$
Circle	$C = \pi d$ or $C = 2\pi r$

Formulas (continued)

General Prisms	$V = Bh$
Cylinder	$V = \pi r^2 h$
Sphere	$V = \frac{4}{3}\pi r^3$
Cone	$V = \frac{1}{3}\pi r^2 h$
Pyramid	$V = \frac{1}{3}Bh$
Pythagorean Theorem	$a^2 + b^2 = c^2$
Quadratic Formula	$x = \frac{-b \pm \sqrt{b^2 - 4ac}}{2a}$
Arithmetic Sequence	$a_n = a_1 + (n-1)d$
Geometric Sequence	$a_n = a_1 r^{n-1}$
Geometric Series	$S_n = \frac{a_1 - a_1 r^n}{1 - r}$ where $r \neq 1$
Radians	1 radian = $\frac{180}{\pi}$ degrees
Degrees	1 degree = $\frac{\pi}{180}$ radians
Exponential Growth/Decay	$A = A_0 e^{k(t - t_0)} + B_0$

444　EXAMINATION August 2019

PART I

Answer all 24 questions in this part. Each correct answer will receive 2 credits. No partial credit will be allowed. For each statement or question, write in the space provided the numeral preceding the word or expression that best completes the statement or answers the question.   [48 credits]

1 Bryan's hockey team is purchasing jerseys. The company charges $250 for a onetime set-up fee and $23 for each printed jersey. Which expression represents the total cost of x number of jerseys for the team?

(1) $23x$
(2) $23 + 250x$
(3) $23x + 250$
(4) $23(x + 250)$

1 _____

2 Which table represents a function?

x	y
2	−3
3	0
4	−3
2	1

(1)

x	y
−3	0
−2	1
−3	2
2	3

(3)

x	y
1	2
1	3
1	4
1	5

(2)

x	y
−2	−4
0	2
2	4
4	6

(4)

2 _____

3. Which expression is equivalent to $2(x^2 - 1) + 3x(x - 4)$?
  (1) $5x^2 - 5$
  (2) $5x^2 - 6$
  (3) $5x^2 - 12x - 1$
  (4) $5x^2 - 12x - 2$

4. The value of x that satisfies the equation $\frac{4}{3} = \frac{x+10}{15}$ is
  (1) -6
  (2) 5
  (3) 10
  (4) 30

5. Josh graphed the function $f(x) = -3(x - 1)^2 + 2$. He then graphed the function $g(x) = -3(x - 1)^2 - 5$ on the same coordinate plane. The vertex of $g(x)$ is
  (1) 7 units below the vertex of $f(x)$
  (2) 7 units above the vertex of $f(x)$
  (3) 7 units to the right of the vertex of $f(x)$
  (4) 7 units to the left of the vertex of $f(x)$

6. A survey was given to 12th-grade students of West High School to determine the location for the senior class trip. The results are shown in the table below.

	Niagara Falls	Darien Lake	New York City
Boys	56	74	103
Girls	71	92	88

To the *nearest percent*, what percent of the boys chose Niagara Falls?
  (1) 12
  (2) 24
  (3) 44
  (4) 56

7 Which type of function is shown in the graph below?

(1) linear
(2) exponential
(3) square root
(4) absolute value

8 The expression $16x^2 - 81$ is equivalent to

(1) $(8x - 9)(8x + 9)$
(2) $(8x - 9)(8x - 9)$
(3) $(4x - 9)(4x + 9)$
(4) $(4x - 9)(4x - 9)$

9 The owner of a landscaping business wants to know how much time, on average, his workers spend mowing one lawn. Which is the most appropriate rate with which to calculate an answer to his question?

(1) lawns per employee
(2) lawns per day
(3) employee per lawns
(4) hours per lawn

10 A ball is thrown into the air from the top of a building. The height, $h(t)$, of the ball above the ground t seconds after it is thrown can be modeled by $h(t) = -16t^2 + 64t + 80$. How many seconds after being thrown will the ball hit the ground?

(1) 5
(2) 2
(3) 80
(4) 144

11 Which equation is equivalent to $y = x^2 + 24x - 18$?
  (1) $y = (x + 12)^2 - 162$
  (2) $y = (x + 12)^2 + 126$
  (3) $y = (x - 12)^2 - 162$
  (4) $y = (x - 12)^2 + 126$

12 When $(x)(x - 5)(2x + 3)$ is expressed as a polynomial in standard form, which statement about the resulting polynomial is true?
  (1) The constant term is 2.
  (2) The leading coefficient is 2.
  (3) The degree is 2.
  (4) The number of terms is 2.

13 The population of a city can be modeled by $P(t) = 3810(1.0005)^{7t}$, where $P(t)$ is the population after t years. Which function is approximately equivalent to $P(t)$?
  (1) $P(t) = 3810(0.1427)^t$
  (2) $P(t) = 3810(1.0035)^t$
  (3) $P(t) = 26{,}670(0.1427)^t$
  (4) $P(t) = 26{,}670(1.0035)^t$

14 The functions $f(x)$ and $g(x)$ are graphed on the set of axes below.

For which value of x is $f(x) \neq g(x)$?
  (1) -1
  (2) 2
  (3) 3
  (4) -2

15 What is the range of the box plot shown below?

(1) 7 (3) 3
(2) 2 (4) 4

16 Which expression is *not* equivalent to $2x^2 + 10x + 12$?

(1) $(2x + 4)(x + 3)$ (3) $(2x + 3)(x + 4)$
(2) $(2x + 6)(x + 2)$ (4) $2(x + 3)(x + 2)$

17 The quadratic functions $r(x)$ and $q(x)$ are given below.

x	r(x)
−4	−12
−3	−15
−2	−16
−1	−15
0	−12
1	7

$q(x) = x^2 + 2x - 8$

The function with the *smallest* minimum value is

(1) $q(x)$, and the value is −9 (3) $r(x)$, and the value is −16
(2) $q(x)$, and the value is −1 (4) $r(x)$, and the value is −2

18 A child is playing outside. The graph below shows the child's distance, $d(t)$, in yards from home over a period of time, t, in seconds.

Which interval represents the child constantly moving closer to home?

(1) $0 \leq t \leq 2$
(2) $2 \leq t \leq 3$
(3) $3 \leq t \leq 4$
(4) $4 \leq t \leq 6$

19 If $a_1 = 6$ and $a_n = 3 + 2(a_{n-1})^2$, then a_2 equals

(1) 75
(2) 147
(3) 180
(4) 900

20 The length of a rectangular patio is 7 feet more than its width, w. The area of a patio, $A(w)$, can be represented by the function

(1) $A(w) = w + 7$
(2) $A(w) = w^2 + 7w$
(3) $A(w) = 4w + 14$
(4) $A(w) = 4w^2 + 28w$

21 A dolphin jumps out of the water and then back into the water. His jump could be graphed on a set of axes where x represents time and y represents distance above or below sea level. The domain for this graph is best represented using a set of

(1) integers
(2) positive integers
(3) real numbers
(4) positive real numbers

22 Which system of linear equations has the same solution as the one shown below?

$$x - 4y = -10$$
$$x + y = 5$$

(1) $5x = 10$
    $x + y = 5$

(2) $-5y = -5$
    $x + y = 5$

(3) $-3x = -30$
    $x + y = 5$

(4) $-5y = -5$
    $x - 4y = -10$

22 _____

23 Which interval represents the range of the function $h(x) = 2x^2 - 2x - 4$?

(1) $(0.5, \infty)$
(2) $(-4.5, \infty)$
(3) $[0.5, \infty)$
(4) $[-4.5, \infty)$

23 _____

24 What is a common ratio of the geometric sequence whose first term is 5 and third term is 245?

(1) 7
(2) 49
(3) 120
(4) 240

24 _____

PART II

Answer all 8 questions in this part. Each correct answer will receive 2 credits. Clearly indicate the necessary steps, including appropriate formula substitutions, diagrams, graphs, charts, etc. For all questions in this part, a correct numerical answer with no work shown will receive only 1 credit. [16 credits]

25 If $g(x) = -4x^2 - 3x + 2$, determine $g(-2)$.

$-4(-2)^2 - 3(-2) + 2 = -8$

26 A student is in the process of solving an equation. The original equation and the first step are shown below.

Original: $3a + 6 = 2 - 5a + 7$
Step one: $3a + 6 = 2 + 7 - 5a$

Which property did the student use for the first step? Explain why this property is correct.

Commutative property of addition

27 On the set of axes below, graph the line whose equation is $2y = -3x - 2$.

$y = -1.5x - 1$

This linear equation contains the point $(2,k)$. State the value of k.

$k = -4$

28 The formula $a = \dfrac{v_f - v_i}{t}$ is used to calculate acceleration as the change in velocity over the period of time.

Solve the formula for the final velocity, v_f, in terms of initial velocity, v_i, acceleration, a, and time, t.

$v_f = a + t v_i$

29 Solve $\frac{3}{5}x + \frac{1}{3} < \frac{4}{5}x - \frac{1}{3}$ for x.

$x > \frac{10}{3}$

30 Is the product of two irrational numbers always irrational? Justify your answer.

No

For example

$\sqrt{2} \cdot \sqrt{2} = 2$

31 Solve $6x^2 - 42 = 0$ for the exact values of x.

$$x = \pm\sqrt{7}$$

32 Graph the function: $h(x) = \begin{cases} 2x - 3, & x < 0 \\ x^2 - 4x - 5, & 0 \leq x \leq 5 \end{cases}$

EXAMINATION August 2019 459

PART III

Answer all 4 questions in this part. Each correct answer will receive 4 credits. Clearly indicate the necessary steps, including appropriate formula substitutions, diagrams, graphs, charts, etc. For all questions in this part, a correct numerical answer with no work shown will receive only 1 credit. [16 credits]

33 On the set of axes below, graph the following system of inequalities:

$$2x + y \geq 8$$
$$y - 5 < 3x$$

Determine if the point (1,8) is in the solution set. Explain your answer.

34 On the day Alexander was born, his father invested $5000 in an account with a 1.2% annual growth rate. Write a function, $A(t)$, that represents the value of this investment t years after Alexander's birth.

$$A(t) = 5000(1.012)^t$$

Determine, to the *nearest dollar*, how much more the investment will be worth when Alexander turns 32 than when he turns 17.

1206

35 Stephen collected data from a travel website. The data included a hotel's distance from Times Square in Manhattan and the cost of a room for one weekend night in August. A table containing these data appears below.

Distance From Times Square (city blocks) (x)	0	0	1	1	3	4	7	11	14	19
Cost of a Room (dollars) (y)	293	263	244	224	185	170	219	153	136	111

Write the linear regression equation for this data set. Round all values to the *nearest hundredth*.

State the correlation coefficient for this data set, to the *nearest hundredth*.

Explain what the sign of the correlation coefficient suggests in the context of the problem.

36 A snowstorm started at midnight. For the first 4 hours, it snowed at an average rate of one-half inch per hour.

The snow then started to fall at an average rate of one inch per hour for the next 6 hours.

Then it stopped snowing for 3 hours.

Then it started snowing again at an average rate of one-half inch per hour for the next 4 hours until the storm was over.

On the set of axes below, graph the amount of snow accumulated over the time interval of the storm.

Determine the average rate of snowfall over the length of the storm. State the rate, to the *nearest hundredth of an inch per hour*.

PART IV

Answer the question in this part. A correct answer will receive 6 credits. Clearly indicate the necessary steps, including appropriate formula substitutions, diagrams, graphs, charts, etc. A correct numerical answer with no work shown will receive only 1 credit. [6 credits]

37 Allysa spent $35 to purchase 12 chickens. She bought two different types of chickens. Americana chickens cost $3.75 each and Delaware chickens cost $2.50 each.

Write a system of equations that can be used to determine the number of Americana chickens, A, and the number of Delaware chickens, D, she purchased.

Determine algebraically how many of each type of chicken Allysa purchased.

Each Americana chicken lays 2 eggs per day and each Delaware chicken lays 1 egg per day. Allysa only sells eggs by the full dozen for $2.50. Determine how much money she expects to take in at the end of the first week with her 12 chickens.

Answers August 2019
Algebra I

Answer Key

PART I

1. (3)	5. (1)	9. (4)	13. (2)	17. (3)	21. (4)
2. (4)	6. (2)	10. (1)	14. (3)	18. (1)	22. (1)
3. (4)	7. (2)	11. (1)	15. (1)	19. (1)	23. (4)
4. (3)	8. (3)	12. (2)	16. (3)	20. (2)	24. (1)

PART II

25. -8
26. Commutative property of addition
27. [graph], $k = -4$
28. $v_f = at + v_i$
29. $x > \dfrac{10}{3}$
30. No. For example, $\sqrt{2} \cdot \sqrt{2} = 2$.
31. $x = \pm\sqrt{7}$
32. [graph]

PART III

33. [graph] (1, 8) is not part of the solution set.
34. $A(t) = 5000(1.012)^t$, $1200
35. $y = -7.76x + 246.34$, $r = -0.88$, which is negative because the line of best fit has a negative slope.
36. [graph] 0.59 inches per hour

PART IV

37. $A + D = 12$
    $3.75A + 2.50D = 35$
    $A = 4, D = 8$, $22.50

In **Parts II–IV**, you are required to show how you arrived at your answers. For sample methods of solutions, see the *Answers Explained* section.

Answers Explained

PART I

1. Since each printed jersey costs 23 dollars, x jerseys cost $23x$ dollars. When the cost of the jerseys is added to the one-time setup fee of 250 dollars, the total cost is then $23x + 250$ dollars.

   In general, when an equation of the form $mx + b$ models a real-world scenario, the b represents the fixed cost and the m represents the increase in the total for each time x increases by 1.

   The correct choice is **(3)**.

2. When a function is represented in a table, there will be no repeated values in the x column. In choice (1), there are two 2's. In choice (2), there are four 1's. In choice (3), there are two -3's. In choice (4), there are no repeated values in the x column, so this table represents a function.

   Even if in choice (4) there were repeated values among the y-values, it would still be a function. All that matters is that there are no repeated values among the x-values.

   The correct choice is **(4)**.

3. Use the distributive property and then combine like terms.

   $$2(x^2 - 1) + 3x(x - 4) = 2x^2 - 2 + 3x^2 - 12x$$
   $$= 5x^2 - 12x - 2$$

   The correct choice is **(4)**.

4. Cross multiply and then solve for x.

   $$\frac{4}{3} = \frac{x + 10}{15}$$
   $$4 \cdot 15 = 3(x + 10)$$
   $$60 = 3x + 30$$
   $$-30 = -30$$
   $$30 = 3x$$
   $$\frac{30}{3} = \frac{3x}{3}$$
   $$10 = x$$

Another way to solve this is to make the two fractions have a common denominator of 15 and then use the fact that the numerators will then be equal.

$$\frac{4}{3} = \frac{x+10}{15}$$

$$\frac{20}{15} = \frac{x+10}{15}$$

$$20 = x + 10$$

$$-10 = -10$$

$$10 = x$$

The correct choice is **(3)**.

5. Graph both functions on your graphing calculator.

   On the TI-84:

   [Y1=-3(X-1)²+2 graph; X=0, Y=-1]

   On the TI-Nspire:

   [Scratchpad graph showing f1(x)=-3·(x-1)²+2 with point (1, 2) and f2(x)=-3·(x-1)²-5 with point (1, -5)]

In a parabola that opens downward, the vertex is the maximum point. The vertex of the graph of function f is $(1, 2)$. The vertex of the graph of function g is $(1, -5)$. So, the vertex of the graph of $g(x)$ is 7 units below the vertex of function $f(x)$.

This can also be done without a graphing calculator by using the properties of the vertex form of an equation for a parabola.

When a function is in the form $f(x) = a(x - h)^2 + k$, the graph will be a parabola with its vertex at (h, k). Function f is in exactly this form, so its vertex is $(1, 2)$. Function g can be expressed as $g(x) = -3(x - 1)^2 + (-5)$, so its vertex is $(1, -5)$. The point $(1, -5)$ is 7 units below the point $(1, 2)$.

The correct choice is **(1)**.

6. The percent of boys who chose Niagara Falls can be calculated by dividing the number of boys who chose Niagara Falls by the total number of boys. First find the total number of boys.

$$56 + 74 + 103 = 233$$

Now find the percent of boys.

$$\frac{56}{233} \approx 0.24 = 24\%$$

The correct choice is **(2)**.

7. Compare the shape of this graph to the shape of typical graphs for each of the answer choices.

Choice (1):
The graph of a linear function is a straight line.

Choice (2):
The graph of an exponential function looks like a playground slide.

Choice (3):
The graph of a square root function looks like half of a sideways parabola.

Choice (4):
The graph of an absolute value function looks like the letter "V."

The given graph most resembles a portion of the graph of an exponential function.

The correct choice is **(2)**.

8. The difference of perfect squares factoring pattern is $a^2 - b^2 = (a-b)(a+b)$. Because both terms of the given expression are perfect squares, the expression can be written as the difference of perfect squares.

$$16x^2 - 81 = (4x)^2 - 9^2$$
$$= (4x-9)(4x+9)$$

The correct choice is **(3)**.

9. Depending on what the owner of the business is looking to measure, any of these choices could be appropriate. Since in this problem the owner wants to know the average amount of time his team spends mowing one lawn, he would calculate the answer by dividing the total number of hours spent by the total number of lawns mowed. For example, if the workers took 8 hours to mow 4 lawns, the average amount of time to mow one lawn would be

$$\frac{8 \text{ hours}}{4 \text{ lawns}} = \frac{8}{4} = 2 \text{ hours per lawn.}$$

The correct choice is **(4)**.

10. When the ball hits the ground, the height of the ball above the ground will be 0. To find when this happens, solve the equation $0 = -16t^2 + 64t + 80$.

$$0 = -16t^2 + 64t + 80$$
$$0 = -16(t^2 - 4t - 5)$$
$$0 = -16(t-5)(t+1)$$
$$t - 5 = 0 \quad t + 1 = 0$$
$$t = 5 \quad \text{or } t = -1$$

Since time must be positive, the solution is $t = 5$.

This answer can also be determined by graphing the function on a graphing calculator and finding the x-intercept.

On the TI-84:

On the TI-Nspire:

The correct choice is (1).

11. The given equation can be expressed in a form like the answer choices by using the completing the square process.

The first step is to eliminate the coefficient from the right-hand side by adding 18 to both sides of the equation.

$$y = x^2 + 24x - 18$$
$$+18 = +18$$
$$y + 18 = x^2 + 24x$$

Next, square half the coefficient of the x-term and add that to both sides of the equation.

$$\left(\frac{24}{2}\right)^2 = 12^2 = 144$$

$$y + 18 + 144 = x^2 + 24x + 144$$
$$y + 162 = x^2 + 24x + 144$$

The right-hand side of the equation is now a perfect square trinomial and can be factored.

$$y + 162 = x^2 + 24x + 144$$
$$y + 162 = (x + 12)^2$$

Isolate the y by subtracting 162 from both sides of the equation.

$$y + 162 = (x + 12)^2$$
$$-162 = -162$$
$$y = (x + 12)^2 - 162$$

This question can also be solved without completing the square by just simplifying each of the four answer choices to see which is equivalent to $y = x^2 + 24x - 18$.

Testing choice (1):
$y = (x + 12)^2 - 162$
$y = x^2 + 24x + 144 - 162$
$y \stackrel{\checkmark}{=} x^2 + 24x - 18$

Testing choice (2):
$y = (x + 12)^2 + 126$
$y = x^2 + 24x + 144 + 126$
$y = x^2 + 24x + 270$
$y \neq x^2 + 24x - 18$

Testing choice (3):
$y = (x - 12)^2 - 162$
$y = x^2 - 24x + 144 - 162$
$y = x^2 - 24x - 18$
$y \neq x^2 + 24x - 18$

Testing choice (4):
$y = (x - 12)^2 + 126$
$y = x^2 - 24x + 144 + 126$
$y = x^2 - 24x + 270$
$y \neq x^2 + 24x - 18$

The correct choice is (**1**).

12. First multiply the factors together.

$$(x)(x - 5)(2x + 3) = (x^2 - 5x)(2x + 3)$$
$$= 2x^3 + 3x^2 - 10x^2 - 15x$$
$$= 2x^3 - 7x^2 - 15x$$

This polynomial has three terms: $2x^3$, $7x^2$, and $15x$.

The constant term is 0 since there is no term that is not multiplied by x or a power of x. Because the highest exponent is a 3, the degree of this polynomial is 3.

The leading coefficient is the number that is multiplied by the highest power of x. Since the term $2x^3$ has the highest power of x, the leading coefficient is 2.

The correct choice is **(2)**.

13. One of the laws of exponents is $x^{ab} = \left(x^a\right)^b$. Apply this to the given expression.

$$P(t) = 3810(1.0005)^{7t}$$
$$= 3810\left(1.0005^7\right)^t$$
$$= 3810(1.0035)^t$$

Another way to solve this problem is to pick any value you like for t, such as $t = 10$. Then calculate $P(10)$ in the original function.

$$P(10) = 3810(1.0005)^{7 \cdot 10}$$
$$= 3810(1.0005)^{70}$$
$$= 3810(1.03561065)$$
$$\approx 3945.7$$

Then test for $t = 10$ in each of the answer choices. When you test choice (2), you find approximately the same result.

$$P(10) = 3810(1.0035)^{10}$$
$$= 3810(1.035556427)$$
$$\approx 3945.5$$

The correct choice is **(2)**.

14. The x-coordinates of the intersection points of the line and the curve are the solutions to $f(x) = g(x)$. Since the line intersects the curve at $(-2, 5)$, $(-1, 4)$, and $(2, 1)$, the three x-values that make $f(x) = g(x)$ are -2, -1, and 2. Of the four choices, each of these three values appear. However, 3 is not one of the solutions.

    The correct choice is **(3)**.

15. In a box plot, the left endpoint represents the minimum value. For this problem, the minimum value is 1. The right endpoint represents the maximum value. For this problem, the maximum value is 8. The range is the difference between the maximum value and the minimum value, which is $8 - 1 = 7$.

    The correct choice is **(1)**.

16. Factor the given expression by first factoring out the greatest common factor of 2.

$$2x^2 + 10x + 12 = 2(x^2 + 5x + 6)$$
$$= 2(x + 3)(x + 2)$$

Since this is choice (4) and the question is asking which choice is *not* equivalent, choice (4) can be eliminated.

To find the other two equivalent expressions, distribute the 2 to each of the factors.

$$2(x + 3)(x + 2) = (2x + 6)(x + 2)$$

So, choice (2) can be eliminated.

$$2(x + 3)(x + 2) = (x + 3)2(x + 2)$$
$$= (x + 3)(2x + 4)$$

So, choice (1) can be eliminated.

Another way to solve this problem is to multiply out each of the four choices to see which one is not equivalent to the given expression.

Testing choice (1):

$$(2x + 4)(x + 3) = 2x^2 + 6x + 4x + 12$$
$$= 2x^2 + 10x + 12$$

Testing choice (2):

$$(2x + 6)(x + 2) = 2x^2 + 4x + 6x + 12$$
$$= 2x^2 + 10x + 12$$

Testing choice (3):

$$(2x + 3)(x + 4) = 2x^2 + 8x + 3x + 12$$
$$= 2x^2 + 11x + 12$$
$$\neq 2x^2 + 10x + 12$$

Testing choice (4):

$$2(x + 3)(x + 2) = (2x + 6)(x + 2)$$
$$= 2x^2 + 4x + 6x + 12$$
$$= 2x^2 + 10x + 12$$

The correct choice is **(3)**.

17. Based on the data in the table, $r(-2) = -16$ is the minimum of function r. For function q, you can graph the function on a graphing calculator and find the coordinates of the minimum point using the minimum function.

On the TI-84:

```
Minimum
X=-1.000002  Y=-9
```

On the TI-Nspire:

[Graphing calculator screenshot showing $f1(x) = x^2 + 2 \cdot x - 8$ with minimum point labeled $(-1, -9)$]

The minimum point is located at $(-1, -9)$, so the minimum value for function q is -9.

Note that the x-coordinate is not the minimum; the y-coordinate is the minimum.

Even without using a graphing calculator, you can calculate the x-coordinate of the minimum of function q with the equation $x = -\dfrac{b}{2a}$.

$$x = -\dfrac{b}{2a}$$
$$= -\dfrac{2}{2 \cdot 1}$$
$$= -1$$

Then calculate the value of q when $x = 1$.

$$q(-1) = (-1)^2 + 2(-1) - 8$$
$$= 1 - 2 - 8$$
$$= -9$$

Since $-16 < -9$, function r has the smallest minimum value.

The correct choice is **(3)**.

18. At $t = 0$, the child is 10 yards away from home. As the child gets closer to home, the distance from the child to home gets smaller. So, an interval in which the child is constantly moving closer to home will be always decreasing. This happens in the interval between $t = 0$ and $t = 2$. The other intervals have some parts where the distance is either increasing or staying the same.

    The correct choice is (**1**).

19. Substitute $n = 2$ into the second part of the recursive definition.

    $$\begin{aligned} a_2 &= 3 + 2(a_{2-1})^2 \\ &= 3 + 2(a_1)^2 \\ &= 3 + 2 \cdot 6^2 \\ &= 3 + 2 \cdot 36 \\ &= 3 + 72 \\ &= 75 \end{aligned}$$

    The correct choice is (**1**).

20. If the width is w feet and the length is 7 feet more than the width of a rectangular patio, the length can be expressed as $w + 7$ feet. The area of a rectangle is length times width.

    $$\begin{aligned} A(w) &= lw \\ &= (w + 7)w \\ &= w^2 + 7w \end{aligned}$$

    The correct choice is (**2**).

21. A sketch of the graph could look like the following.

[Graph showing a parabola-like curve with a point marked at (1.5, 10), y-axis labeled at 5, 10, 15, and x-axis labeled at 10]

The domain is the set of x-coordinates of all the points on the graph. This domain is not limited to integers since there are points that have x-coordinates that are not integers, such as (1.5, 10). Numbers between integers qualify as real numbers. Since the graph starts at $x = 0$, there are also no negative x-values. Negative x-values would be used for moments before the dolphin begins jumping. So, positive real numbers would be most appropriate for the domain of this graph.

If the question instead asked for the range of the graph, then real numbers, including negative values, would be appropriate since there are times where the dolphin is under water. The y-coordinates of the points representing those underwater moments would be negative.

The correct choice is **(4)**.

22. Solve the given system of equations with the elimination method. First multiply both sides of the second equation by -1 and then add the first equation to that result.

$$x - 4y = -10$$
$$-1(x + y) = -1(5)$$

$$x - 4y = -10$$
$$+ \quad -x - y = -5$$
$$\frac{-5y}{-5} = \frac{-15}{-5}$$
$$y = 3$$

Substitute $y = 3$ into either of the two original equations to solve for x.

$$x + 3 = 5$$
$$-3 = -3$$
$$x = 2$$

The solution to the given system is $(2, 3)$.

The answer choices are much simpler systems since in each of them, one of the equations has just one variable. Also notice that the other equation in each choice matches one of the equations in the given system. So, all that needs to be done is to check each choice to see if the solution to the one-variable equation matches either $x = 2$ or $y = 3$.

Checking choice (1):

$$\frac{5x}{5} = \frac{10}{5}$$
$$x = 2$$

Checking choice (2):

$$\frac{-5y}{-5} = \frac{-5}{-5}$$
$$y = 1$$

Checking choice (3):

$$\frac{-3x}{-3} = \frac{-30}{-3}$$

$$x = 10$$

Checking choice (4):

$$\frac{-5y}{-5} = \frac{-5}{-5}$$

$$y = 1$$

The correct choice is **(1)**.

23. The range is the set of y-coordinates of all the points on the graph of a function. Graph this function on a graphing calculator.

    On the TI-84:

    On the TI-Nspire:

The lowest y-coordinate of the graph is -4.5. There is no highest y-coordinate since both branches of the parabola go up forever.

One way to express this range is $y \geq -4.5$. The answer choices, however, use interval notation.

The only difference between choices (2) and (4) is that choice (2) has a parenthesis "(" on the left side of the interval and choice (4) has a square bracket "[."

The square bracket "[" is used when the interval does contain the first number. In this case, since the minimum point of the parabola is -4.5, the square bracket "[" is the appropriate notation.

The correct choice is **(4)**.

24. A geometric sequence is one in which each term is equal to the previous term multiplied by some constant called the common ratio. For example, 3, 6, 12, 24, ... is a geometric series with a common ratio of 2.

If the common ratio is called r and the first term of the sequence is 5, then the second term will equal $5r$ and the third term will equal $5r \cdot r = 5r^2$.

One way to solve this problem is to find the value of r that makes $5r^2$ equal to 245.

$$\frac{5r^2}{5} = \frac{245}{5}$$
$$r^2 = 49$$
$$r = \pm 7$$

There are two valid answers, but only one of them is among the four answer choices.

The correct choice is **(1)**.

PART II

25. To determine $g(-2)$, substitute -2 for x in the given equation.

$$g(-2) = -4(-2)^2 - 3(-2) + 2$$
$$= -4 \cdot 4 + 6 + 2$$
$$= -16 + 6 + 2$$
$$= -8$$

26. The commutative property of arithmetic says that $a + b = b + a$. This can be applied to an expression involving negatives, like $-5a + 7 = 7 + (-5a) = 7 - 5a$.

The first step the student used was to rearrange the second and third terms of the right-hand side of the equation. So, the student used the commutative property of addition.

This is correct because using the commutative property of addition does not change the value of the right-hand side of the equation.

Most students would skip this as their first step and instead combine like terms to get $3a + 6 = 9 - 5a$. However, this commutative property step actually allows students to combine like terms.

27. Since there are no exponents greater than or equal to 2, this is a linear equation and the graph will be a line. Only two points are needed to graph a line. To be sure, though, make a chart with more values.

Pick any value you want for x, and substitute into the equation to solve for the corresponding y value. For example, choose $x = 2$.

$$2y = -3(2) - 2$$
$$2y = -6 - 2$$
$$\frac{2y}{2} = \frac{-8}{2}$$
$$y = -4$$

x	y
0	−1
1	−2.5
2	−4
3	−5.5
4	−7

Plot the points on the graph, and connect them to draw the entire line.

This could also have been done by first converting to slope-intercept form by dividing both sides of the equation by 2.

$$\frac{2y}{2} = \frac{-3x - 2}{2}$$

$$y = -\frac{3}{2}x - 1$$

The −1 is the y-intercept and the $-\frac{3}{2}$ is the slope. Graph the point (0, −1). Then, using the slope of $-\frac{3}{2}$, move 2 units right and 3 units down to get the point (2, −4). Join these two points to form the line.

Since (2, −4) is a point on the line, −4 is the value of k.

28. First multiply both sides of the equation by t.

$$at = v_f - v_i$$

Next, add v_i to both sides of the equation to isolate the v_f term.

$$at = v_f - v_i$$
$$+v_i = +v_i$$
$$at + v_i = v_f$$

29. Solve the inequality the same way you would solve it if there were an "="
sign instead of a "<." However, be sure to reverse the direction of the
inequality sign when you multiply or divide both sides of the inequality by a
negative number.

$$\frac{3}{5}x + \frac{1}{3} < \frac{4}{5}x - \frac{1}{3}$$
$$-\frac{1}{3} = -\frac{1}{3}$$
$$\frac{3}{5}x < \frac{4}{5}x - \frac{2}{3}$$
$$-\frac{4}{5}x = -\frac{4}{5}x$$
$$-\frac{1}{5}x < -\frac{2}{3}$$
$$-5\left(-\frac{1}{5}x\right) > -5\left(-\frac{2}{3}\right)$$
$$x > \frac{10}{3}$$

30. No. The product of two irrational numbers can be rational. For example, $\sqrt{2} \cdot \sqrt{2} = 2$. The number $\sqrt{2}$ is irrational, and the number 2 is rational.

31. Rearrange the terms to solve for x.

$$6x^2 - 42 = 0$$
$$+42 = +42$$
$$\frac{6x^2}{6} = \frac{42}{6}$$
$$x^2 = 7$$
$$\sqrt{x^2} = \pm\sqrt{7}$$
$$x = \pm\sqrt{7}$$

32. Make a table of values for the piecewise functions. For negative values, use the top equation. For values between 0 and 5, use the bottom equation.

x	$h(x)$
-2	$2(-2) - 3 = -7$
-1	$2(-1) - 3 = -5$
0	$0^2 - 4(0) - 5 = -5$
1	$1^2 - 4(1) - 5 = -8$
2	$2^2 - 4(2) - 5 = -9$
3	$3^2 - 4(3) - 5 = -8$
4	$4^2 - 4(4) - 5 = -5$
5	$5^2 - 4(5) - 5 = 0$

Plot these eight points with closed circles.

Before drawing the curves, evaluate the top equation at 0 to get $2(0) - 3 = -3$. Even though the point $(0, -3)$ is not part of the graph since 0 is part of the other "piece," this point is the boundary between the two pieces. Plot $(0, -3)$ with an open circle to represent this.

Now draw the curves.

The TI-Nspire has the built-in capability for graphing piecewise functions, though it does not clearly show the open circle boundary point.

$$f1(x) = \begin{cases} 2 \cdot x - 3, & x < 0 \\ x^2 - 4 \cdot x - 5, & 0 \leq x \leq 5 \end{cases}$$

PART III

33. Rewrite both inequalities in slope-intercept form.

$$y \geq -2x + 8$$
$$y < 3x + 5$$

Graph the line $y = -2x + 8$ with a solid line because it contains the sign "$\geq$" and not "$>$." Test the ordered pair $(0, 0)$ to determine if it makes the inequality $y \geq -2x + 8$ true.

$$0 \geq -2(0) + 8$$
$$0 \geq 0 + 8$$
$$0 \geq 8$$

Since $(0, 0)$ does not make the first inequality true, shade the side of the line $y = -2x + 8$ that does not contain $(0, 0)$.

Graph the line $y = 3x + 5$ with a dotted line because it contains the sign "$<$" and not "$\leq$." Test the ordered pair $(0, 0)$ to determine if it makes the inequality $y = 3x + 5$ true.

$$0 < 3(0) + 5$$
$$0 < 0 + 5$$
$$0 < 5$$

Since (0, 0) does make the second inequality true, shade (with a different style of shading) the side of the line $y = 3x + 5$ that does contain (0, 0).

The entire double-shaded region is the solution set to the system of inequalities.

Systems of inequalities can also be graphed on a graphing calculator.

On the TI-84:

On the TI-Nspire:

Since the point (1, 8) is on the dotted line, it is not part of the solution set.

34. The formula for exponential growth for a given growth rate is $A(t) = P(1 + r)^t$, where P is the original investment and r is the growth rate. For this problem, the function can be written as follows:

$$A(t) = 5000(1 + 0.012)^t = 5000(1.012)^t$$

Subtract the values when $t = 32$ and $t = 17$ to find the solution.

$$A(32) - A(17) = 5000(1.012)^{32} - 5000(1.012)^{17}$$
$$\approx 7324 - 6124$$
$$= 1200$$

To the *nearest dollar*, the investment will be worth $1200 more when Alexander turns 32 than when he turns 17.

35. A linear regression equation, or line of best fit, for a data set can be calculated with a graphing calculator.

For the TI-84:

To turn diagnostics on, press [2nd] [0], scroll down to select DiagnosticOn, and press [ENTER].

```
CATALOG
 DependAuto
 det(
 DiagnosticOff
▶DiagnosticOn
 dim(
 Disp
 DispGraph
```

Press [STAT] [1]. Enter the *x*-values into L1 and the *y*-values into L2.

L1	L2	L3
0	293	
0	263	
1	244	
1	224	
3	185	
4	170	
7	219	

L3(1)=

Press [STAT] [right arrow] [4] to get the LinReg function. Select Calculate.

```
     LinReg
y=ax+b
a=-7.756345178
b=246.3380711
r²=.7685525674
r=-.8766712995
```

For the TI-Nspire:
From the home screen, select the Add Lists & Spreadsheet icon. In the first row, label the A column "hours" and the B column "score." Enter the hours in cells A1 to A10. Enter the scores in cells B1 to B10.

Move the cursor to cell C1. Press [menu] [4] [1] [3]. Set X List to "hours" and Y List to "score." Select OK.

The line of best fit is $y = -7.76x + 246.34$.

The correlation coefficient (r) is a measure of how close the points are to the line of best fit. If the line of best fit has a positive slope, the correlation coefficient will be positive. If the line of best fit has a negative slope, the correlation coefficient will be negative. The closer the correlation coefficient is to 1 (or to -1), the closer to the line of best fit the points will lie.

For this data set, the correlation coefficient is -0.88. Because the correlation coefficient is negative, the line of best fit must have a negative slope.

36. Make a table of values for times 0 to 17, following the increases described in the problem.

Elapsed Time	Accumulated Snowfall
0	0
1	0.5
2	1
3	1.5
4	2
5	3
6	4
7	5
8	6
9	7
10	8
11	8
12	8
13	8
14	8.5
15	9
16	9.5
17	10

Plot those seventeen points, and connect the points to form the graph.

The average rate of snowfall is the total amount of snowfall divided by the total number of hours. For this problem, that is $\frac{10}{17} \approx 0.59$ inches per hour.

PART IV

37. There are 12 chickens, so the first equation is $A + D = 12$. The chickens cost 35 dollars, so the other equation is $3.75A + 2.50D = 35$.

Write the system of equations.
$$A + D = 12$$
$$3.75A + 2.50D = 35$$

This system can be solved with either the elimination method or with the substitution method.

Elimination method:
Multiply both sides of the top equation by -2.50, and add the equations together.

$$-2.50A - 2.50D = -30$$
$$+ \quad 3.75A + 2.50D = 35$$
$$\frac{1.25A}{1.25} = \frac{5}{1.25}$$
$$A = 4$$

Solve for D by plugging $A = 4$ into one of the original equations.
$$4 + D = 12$$
$$D = 8$$

Substitution method:
To use this method, one of the equations must be changed so that one of the variables is isolated. This can be done most easily with the first equation. Subtract D from both sides to make the equation $A = 12 - D$. Next, substitute into the second equation and solve for D.

$$3.75(12 - D) + 2.50D = 35$$
$$45 - 3.75D + 2.50D = 35$$
$$45 - 1.25D = 35$$
$$-45 = -45$$
$$\frac{-1.25D}{-1.25} = \frac{-10}{-1.25}$$
$$D = 8$$

Then solve for A by substituting $D = 8$ into the first equation.

$$A + 8 = 12$$
$$A = 4$$

Allysa bought 4 Americana chickens and 8 Delaware chickens.

With 4 Americana chickens and 8 Delaware chickens, Allysa will get $4 \cdot 2 + 8 \cdot 1 = 16$ eggs per day. After one week, she will have $16 \cdot 7 = 112$ eggs. Divide this by 12 to see how many dozens of eggs she will have in one week.

$$\frac{112}{12} \approx 9.3$$

Since she sells eggs only by the full dozen, Allysa can sell 9 dozen for a total of $9 \cdot \$2.50 = \$22.50$.

SELF-ANALYSIS CHART August 2019

Topic	Question Numbers	Number of Points	Your Points	Your Percentage
1. Polynomials	3, 8, 12, 16	$2+2+2+2=8$		
2. Properties of Algebra	4, 26, 28	$2+2+2=6$		
3. Functions	2, 14, 21, 23, 25	$2+2+2+2+2=10$		
4. Creating and Interpreting Equations	1, 20	$2+2=4$		
5. Inequalities	29, 33	$2+4=6$		
6. Sequences and Series	19, 24	$2+2=4$		
7. Systems of Equations	22, 37	$2+6=8$		
8. Quadratic Equations and Factoring	5, 10, 11, 17, 31	$2+2+2+2+2=10$		
9. Regression	35	4		
10. Exponential Equations	7, 13, 34	$2+2+4=8$		
11. Graphing	18, 27, 32, 36	$2+2+2+4=10$		
12. Statistics	6, 15	$2+2=4$		
13. Number Properties	30	2		
14. Unit Conversions	9	2		

HOW TO CONVERT YOUR RAW SCORE TO YOUR ALGEBRA I REGENTS EXAMINATION SCORE

The accompanying conversion chart must be used to determine your final score on the August 2019 Regents Examination in Algebra I. To find your final exam score, locate in the column labeled "Raw Score" the total number of points you scored out of a possible 86 points. Since partial credit is allowed in Parts II, III, and IV of the test, you may need to approximate the credit you would receive for a solution that is not completely correct. Then locate in the adjacent column to the right the scale score that corresponds to your raw score. The scale score is your final Algebra I Regents Examination score.

SELF-ANALYSIS CHART August 2019

Regents Examination in Algebra I—August 2019
Chart for Converting Total Test Raw Scores to Final Exam Scores (Scale Scores)

Raw Score	Scale Score	Performance Level	Raw Score	Scale Score	Performance Level	Raw Score	Scale Score	Performance Level
86	100	5	57	81	4	28	66	3
85	99	5	56	81	4	27	65	3
84	98	5	55	81	4	26	64	2
83	96	5	54	81	4	25	63	2
82	95	5	53	80	4	24	61	2
81	94	5	52	80	4	23	60	2
80	93	5	51	80	4	22	58	2
79	92	5	50	80	4	21	57	2
78	91	5	49	79	3	20	55	2
77	90	5	48	79	3	19	53	1
76	90	5	47	79	3	18	51	1
75	89	5	46	78	3	17	49	1
74	88	5	45	78	3	16	47	1
73	87	5	44	78	3	15	45	1
72	87	5	43	77	3	14	43	1
71	86	5	42	77	3	13	41	1
70	86	5	41	76	3	12	38	1
69	86	5	40	76	3	11	36	1
68	85	5	39	75	3	10	33	1
67	84	4	38	75	3	9	30	1
66	84	4	37	74	3	8	27	1
65	84	4	36	74	3	7	24	1
64	83	4	35	73	3	6	21	1
63	83	4	34	72	3	5	18	1
62	83	4	33	71	3	4	15	1
61	82	4	32	70	3	3	11	1
60	82	4	31	70	3	2	8	1
59	82	4	30	69	3	1	4	1
58	82	4	29	68	3	0	0	1

NOTES

NOTES

NOTES